Membrane Technology for Osmotic Power Generation by Pressure Retarded Osmosis

Membrane Technology for Osmotic Power Generation by Pressure Retarded Osmosis

Edited by

Chun Feng Wan
Tai-Shung Chung

CRC Press is an imprint of the
Taylor & Francis Group, an **informa** business

CRC Press
Taylor & Francis Group
6000 Broken Sound Parkway NW, Suite 300
Boca Raton, FL 33487-2742

First issued in paperback 2021

CRC Press is an imprint of Taylor & Francis Group, an Informa business

ISBN 13: 978-0-367-25592-3 (hbk)
ISBN 13: 978-1-03-224063-3 (pbk)

Visit the Taylor & Francis Web site at
www.taylorandfrancis.com

and the CRC Press Web site at
www.crcpress.com

Publisher's Note
The publisher has gone to great lengths to ensure the quality of this reprint but points out that some imperfections in the original copies may be apparent.

Contents

Preface

In recent times, clean energy and global warming have been receiving worldwide attentions. Among various energy sources available, osmotic energy is regarded as green and sustainable, and it does not emit CO_2. Pressure-retarded osmosis (PRO) is a technology to harvest osmotic energy through an osmotically driven membrane process. Significant developments started only when researchers began to understand the transport phenomena within the PRO membranes. In this book, we have compiled the state-of-the-art PRO technology from concept validation, transport mechanisms, membrane and module fabrication, system development, to pilot tests.

Chapter 1 of this book starts with literature reviews and introduction of the fundamentals of PRO processes, where free osmotic energy is released from mixing of two solutions with different salinities via a semipermeable PRO membrane. Next, the current developments of flat sheet and hollow fiber PRO membranes are summarized in Chapters 2 and 3. Different materials and fabrication strategies are introduced in order to molecularly design PRO membranes with the desirable morphology, maximal power density, and withstanding pressure. Chapter 4 summarizes the mass transport within PRO membranes of different configurations. The water flux and power density obtained from real PRO processes are much lower than their ideal values using clean feed streams because of the concentration polarization, fouling, and scaling, Chapters 5, 6, and 7 aim to analyze the origins and mechanisms of flux reductions, development of antifouling PRO membranes, and exploration of various methods to pretreat the feed solutions, respectively. Once high-performance PRO membranes have been developed, R&D strategies must be directed to product development and pilot testing. Therefore, Chapters 8, 9, and 10 focus on module design, small pilot design, and their operation and maintenance in order to bring the RO–PRO technology closer to commercialization.

Chapter 11 presents the techno-economic analyses of RO–PRO systems to examine different scenarios of RO–PRO integrations and to provide useful information on setting up the hybrid systems for seawater desalination. In addition to the RO–PRO integration, there are many other hybrid configurations that take advantage of PRO technology and improve the overall water recovery and separation efficiency. Therefore, Chapter 12 investigates the design of other PRO-integrated systems, such as FO-PRO and MD-PRO, including their process fundamentals and technical advantages. We believe that these 12 chapters would provide valuable guidance for the commercialization of PRO technology in the years to come.

Finally, we would like to express our heartfelt thanks to the contributors of each chapter for their dedication and willingness to share their knowledge.

We would also like to thank the funding agents in the last two decades, which include Singapore National Research Foundation (NRF), Public Utility Board (PUB), BASF, NUS, and many others. Without their strong provisions, it would have been hard for us to accomplish this book. We would also like to thank the publishing team for putting this book together and for getting it published. In addition, Dr Wan is grateful for his family's strong support, while professor Chung is thankful for his family's continuous support. Last but not the least, professor Chung would also like to thank God for His grace and blessing during the last 23 years at NUS.

Tai-Shung Chung
Chun Feng Wan

Editors

Dr. Chun Feng Wan is a technical director in Urban Development & Civil Engineering, Meinhardt (Singapore) Pte Ltd. He obtained his bachelor (1st Honours) and PhD in chemical engineering from professor Chung's membrane research group in NUS. His research focuses on membrane synthesis, membrane module production, process design, and pilot testing for osmotic power generation by pressure retarded osmosis. He has been awarded the Chinese outstanding PhD student study abroad award, AICHE-SLS young researcher award, Forbes 30 under 30 Asia award, and MIT Technology Review innovators under 35 Asia Pacific. He has first-authored 8 and co-authored 12 research articles in the *Journal of Membrane Science, Applied Energy*, and other leading research journals. His H-index is 11 (Scopus) and 12 (Google Scholar); number of citations is 458 (Scopus) and 548 (Google Scholar) (30 June 2018).

Professor Tai-Shung Chung is the Provost's Chair Professor at the Department of Chemical and Biomolecular Engineering, National University of Singapore. His research focuses on polymeric membranes for clean water and clean energy. During 2005–2008, he has worked as a senior consultant for Hyflux, where he led and built a membrane research team. He became a Fellow in the Academy of Engineering Singapore in 2012 and received IChemE (Institute of Chemical Engineers, United Kingdom) 2014 Underwood Medal for his exceptional research in separations. He also received Singapore President's Technology Award in 2015. He was a highly cited researcher in Chemical Engineering & Materials Science and Engineering as per the Elsevier and Shanghai Global Ranking in 2016 and received Distinction Award in Water Reuse and Conservation from International Desalinatin Association (IDA) in 2016. He was also a highly cited researcher from Clarivate Analytics in 2018 & 2019. He is the author of 2 books, 30 book chapters, and more than or equal to 735 journal papers, and 350 conference papers. He has more than 70 patents (including 46 US patents, 34 regional and Singapore patents) in his name. He has the world's highest number of publications in *Journal of Membrane Science* (impact factor = 7.015). So far, he has trained and produced 73 PhD, 23 MEng, and 120 post-doctorates. His H-index is 101 (Scopus) and 118 (Google Scholar); number of citations is more than 39,674 (Scopus) and more than 51,192 (Google Scholar) (January 12, 2020).

Contributors

Tao Cai
Key Laboratory of Biomedical Polymers of Ministry of Education
College of Chemistry and Molecular Science, Wuhan University
Wuhan, Hubei, P. R. China

Wuhan University Shenzhen Research Institute
Shenzhen, Guangdong, P. R. China

Zhen Lei Cheng
Department of Chemical and Biomolecular Engineering
National University of Singapore
Singapore

Tai-Shung Chung
Department of Chemical and Biomolecular Engineering
National University of Singapore
Singapore

Wenxiao Gai
Department of Chemical and Biomolecular Engineering
National University of Singapore
Singapore

Chakravarthy S. Gudipati
Separation Technologies Applied Research and Translation (START) – NTUitive
Cleantech One, Singapore

Gang Han
Department of Chemical Engineering
Massachusetts Institute of Technology
Cambridge, Massachusetts

Wen Gang Huang
Key Laboratory of Biomedical Polymers of Ministry of Education
College of Chemistry and Molecular Science, Wuhan University
Wuhan, Hubei, P. R. China

Wuhan University Shenzhen Research Institute
Shenzhen, Guangdong, P. R. China

Esther Swin Hui Lee
Department of Chemical and Biomolecular Engineering
National University of Singapore
Singapore

Xue Li
Key Laboratory of Biomedical Polymers of Ministry of Education
College of Chemistry and Molecular Science, Wuhan University
Wuhan, Hubei, P. R. China

Wuhan University Shenzhen Research Institute
Shenzhen, Guangdong, P. R. China

Chun Feng Wan
Department of Chemical and Biomolecular Engineering
National University of Singapore
Singapore

Jun Ying Xiong
School of Life Science and Chemical Technology
Ngee Ann Polytechnic
Singapore

Tianshi Yang
Department of Chemical and Biomolecular Engineering
National University of Singapore
Singapore

1

Osmotic Energy and Pressure Retarded Osmosis

Chun Feng Wan and Tai-Shung Chung
Department of Chemical and Biomolecular Engineering
National University of Singapore
Singapore

CONTENTS

1.1 Water and Energy Crisis

Water and energy are crucial for human well-being and sustainable socio-economic development. Inadequate accesses to water and energy have become one of the most pervasive global problems due to the rapid increases in consumption and depletion of water and energy reserves. Demands for freshwater and energy will continue to increase significantly in the near future because of population and economic growth. The interdependence of water and energy will greatly amplify the existing pressures on limited natural resources and ecosystems (UN-Water 2013a, 2013b). For example, water is essential for energy generation, primarily for cooling power plants and fuel production. On the other hand, energy powers machines for water production, transportation, and distribution.

The current energy consumption to produce potable water varies from 0.37 to 0.48 kWh/m^3 to desalinate surface and groundwater and from 2.58 to 8.50 kWh/m^3 to desalinate seawater (UNWWAP 2014). There are growing opportunities for the joint development of water and energy technologies that maximize co-benefits and minimize negative trade-offs. A wide range of opportunities exists to coproduce energy and water and to harvest the benefits of synergies. This book aims to explore such opportunity where wastewater streams can be used to generate clean energy, which in turn can be used to compensate the energy consumption of desalination processes.

1.2 Osmotic Energy

Nature has the greatest mechanism for plants to take up water from soils and transport it across the cell membranes by forward osmosis (FO) (McElrone et al. 2013). In the past decade, FO has received increasing attentions in various water, energy, and food applications. In an FO process, as shown in Figure 1.1 (a), a semipermeable membrane is placed between two solutions with different salinities – a draw solution with a higher salinity and a feed solution with a lower salinity. Water spontaneously diffuses across the membrane, driven by the chemical potential difference that arises from the salinity difference. The osmotic pressure difference ($\Delta\pi$) between the feed and draw solutions is defined as the hydraulic pressure that has to be applied on the draw solution to stop the spontaneous water flow (Awad et al. 2019, Chung et al. 2012a, 2012b, Shaffer et al. 2015). The term "osmotic pressure (π)" implies the potential of a solution to generate power. Osmotic pressure can be calculated as follows (Van't Hoff 1901):

$$\pi = icRT \tag{1}$$

where c is the molar concentration, i is the van't Hoff factor, R is the universal gas constant, and T is the operating temperature. The average concentration of seawater is 35 g/L, corresponding to an osmotic pressure of 27 bar. The concentration of the concentrated brine from reverse osmosis (RO) desalination plants usually ranges from 50 to 75 g/L, corresponding to an osmotic pressure from 40 to 65 bar, respectively. Water from salt lakes and Dead Sea can generate an osmotic pressure higher than 200 bar (Sharqawy et al. 2012).

The global potential of osmotic energy released at the estuaries where rivers flow into the seas is estimated to be 2.6 TW (Thorsen and Holt 2009).

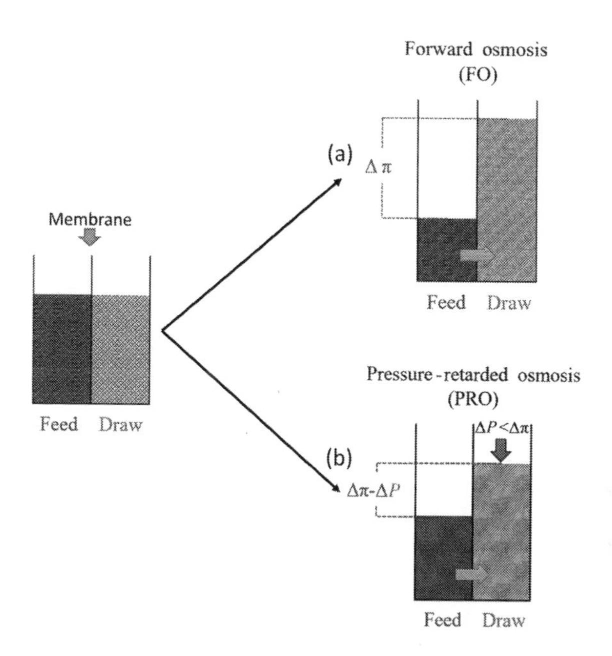

FIGURE 1.1
Steady states of (a) FO and (b) PRO processes.

More energy can be extracted if other saline solutions with higher concentrations are used, such as seawater brine (Chung et al. 2012b, Helfer et al. 2014). Various technologies have been developed to harvest the chemical potential that drives the water flow in FO. This type of energy is known as salinity gradient energy or osmotic energy (Logan and Elimelech 2012). Figure 1.2 shows the classifications of different processes to generate the osmotic energy based on whether energy is generated through transport of ions, water, or both (Micale et al. 2016). For example, reverse electro dialysis employs ionic exchange membranes to control the flow of cations and anions between solutions with different salinities to generate electricity (Tedesco et al. 2016). Accumulator mixing (AccMix) can directly generate electricity by charging the draw solution and accumulating the ions near the electrolytes, and then discharging the ions in the feed solution (Marino et al. 2015). Reverse vapor compression employs the vapor pressure difference between solutions with different salinities to drive a turbine and generate energy (Olsson et al. 1979). Energy can also be generated by cyclic operations of swelling and shrinking of hydrogels in the feed and draw solutions, respectively (Zhu et al. 2014). The hydrocratic generator pumps the feed solution to the bottom of an upwelling device filled with the draw solution, and generates net energy at the top exit of the upwelling device (Finley and Pscheidt 2001).

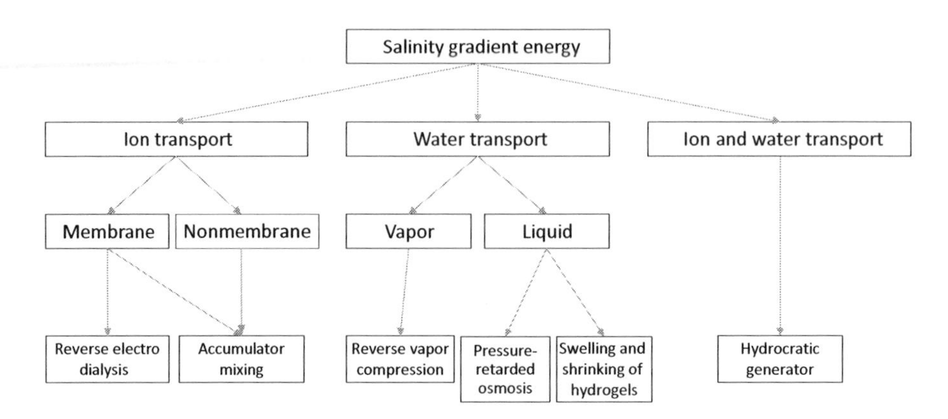

FIGURE 1.2
Classification of salinity gradient energy process based on the transport mechanisms.

Reproduced from Micale et al. 2016.

1.3 Pressure Retarded Osmosis

Pressure retarded osmosis (PRO) is the most investigated and developed process to generate osmotic energy because of its greater efficiency and higher power density. PRO was conceptualized for producing salinity gradient energy by Pattle (1954) in the 1950s and then reinvestigated in the mid-1970s due to the world's energy crisis (Loeb 1975). Loeb and his coworkers first conducted theoretical and experimental researches on the feasibility of PRO (Loeb 1976, Loeb et al. 1976). However, due to the expensive and inefficient membranes, researches were slowed down in the 1980s and 1990s. From the 1990s, membranes for desalination and wastewater treatment have advanced rapidly and have been widely commercialized. In the last decade, new perspectives have been opened up for PRO researches, thanks to the advances on the fundamental understanding of mass transports in osmotically driven processes and membrane developments for engineering osmosis (Achilli and Childress 2010, Altaee et al. 2017, Gerstandt et al. 2008, Han et al. 2015a, Kim et al. 2015, Wan et al. 2018, Zhang et al. 2014).

As depicted in Figure 1.1 (b), PRO extracts the Gibbs free energy of mixing by allowing water to spontaneously flow through a semipermeable membrane from a low-salinity feed solution to a high-salinity draw solution against an applied hydraulic pressure, ΔP (Achilli et al. 2009, Chung et al. 2012a, 2012b). The applied hydraulic pressure on the draw solution should be smaller than the osmotic pressure difference between the feed and draw solutions to maintain the spontaneous water flux. The Gibbs free energy is converted to the hydraulic pressure of the diluted draw solution, which can be further converted to mechanical energy by a pressure exchanger (Prante et al. 2014, Sarp et al. 2014) or to electrical energy by a turbine (Loeb 1976, Loeb et al. 1976, Saito et al. 2012, Thorsen and Holt 2009).

1.3.1 Thermodynamics of PRO

PRO originates from the Gibbs free energy of mixing. The Gibbs free energy of ideal mixing of strong electrolyte solutions with low salt concentrations can be calculated as follows (Feinberg et al. 2013, Yip and Elimelech 2012):

$$-\Delta G = RT\left\{\left[\sum C_i \ln(\gamma_i C_i)\right]_{\mathrm{M}} - \phi_{\mathrm{A}}\left[\sum C_i \ln(\gamma_i C_i)\right]_{\mathrm{A}} - \phi_{\mathrm{B}}\left[\sum C_i \ln(\gamma_i C_i)\right]_{\mathrm{B}}\right\} \tag{2}$$

where ΔG is the Gibbs free energy of mixing, C_i is the mole concentration, γ_i is the activity coefficient of species i in the solution, and Φ is the volumetric fraction of the solution to the total volume of the system. The subscripts A, B, and M mean solution A, solution B, and the mixture of solutions A and B, respectively.

For relative low salt concentrations, the molar salt concentration dominates over the salt activity coefficient in the logarithmic term. Equation 2 can be simplified as follows:

$$-\Delta G = iRT\left[C_D^f V_D^f \ln\left(C_D^f\right) - C_D^0 V_D^0 \ln(C_D^0) + C_F^f V_F^f \ln\left(C_F^f\right) - C_F^0 V_F^0 \ln(C_F^0)\right] / (V_D^0 + V_F^0) \tag{3}$$

where C_F^0 and C_F^f are the concentrations of the feed solution before and after mixing, respectively; C_D^0 and C_D^f are the concentrations of the draw solution before and after mixing, respectively; V_f^0 and V_F^f are the flow rates of the feed solution before and after mixing, respectively; and V_D^0 and V_D^f are the flow rates of the draw solution before and after mixing, respectively.

In a reversible PRO mixing process, the theoretical maximum amount of energy that can be harvested is equal to the Gibbs free energy of mixing. However, in the actual application of PRO, a constant hydraulic pressure, ΔP, is applied to the draw solution side. The specific energy (SE) of a constant hydraulic pressure in PRO is defined as the energy production over the total volume of the feed and draw solutions:

$$\mathrm{SE} = \frac{\Delta P \Delta V}{V_D^0 + V_F^0} \tag{4}$$

where ΔV is the total permeate volume.

The osmotic pressure difference between the feed and draw solutions diminishes as the draw solution becomes diluted and the feed solution gets concentrated. A thermodynamic equilibrium is reached and no more mixing happens when $\Delta\pi$ is reduced to ΔP. The degree of mixing is limited by the hydraulic pressure applied. Moreover, energy is lost to overcome the hydraulic resistance of the membrane. Therefore, the SE that can be harvested from a constant-pressure PRO process is less than that from a reversible PRO mixing process (Feinberg et al. 2013, Yip and Elimelech 2012). The amount of extractable work in a constant-pressure PRO process can be calculated as follows:

$$\mathrm{SE} = iRT\left[\frac{C_D^0 V_D^0}{V_D^0 + \Delta V} - \frac{C_F^0 V_F^0}{V_F^0 - \Delta V}\right]\frac{\Delta V}{V_F^0 + V_D^0} \tag{5}$$

$$\mathrm{SE}_{\mathrm{max}} = iRT\left(\frac{V_F^0}{V_F^0 + V_D^0}\right)\left(\frac{V_D^0}{V_F^0 + V_D^0}\right)\left(\sqrt{C_D^0} - \sqrt{C_F^0}\right)^2 \tag{6}$$

Figure 1.3 presents the amount of energy harvested, unutilized, and lost due to frictions. The total colored area corresponds to the Gibbs free energy of mixing defined by Eqn 3. The left upper corner represents the frictional losses of energy to overcome the hydraulic resistance of the

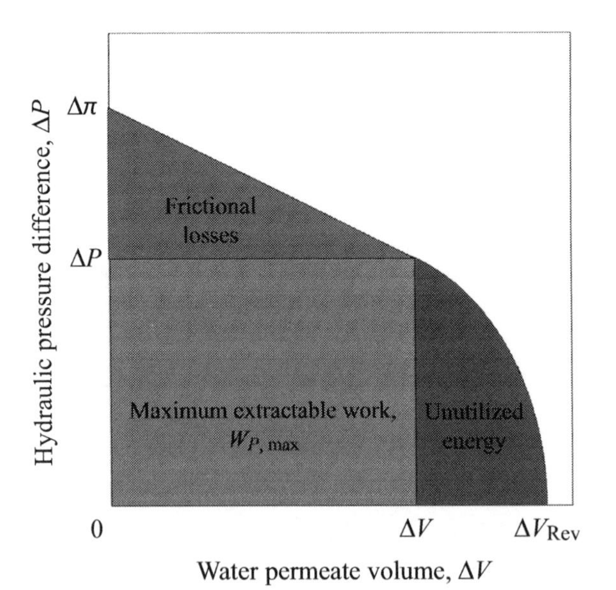

FIGURE 1.3
Maximum extractable work, unutilized energy, and frictional losses.

Reproduced from Yip et al. 2012.

membrane. The square area enclosed by ΔV and ΔP is the amount of energy that can be harvested in a constant-pressure PRO as defined by Eqn 5. PRO mixing stops when $\Delta\pi$ drops to ΔP. The incomplete mixing leaves a portion of osmotic energy unrealized, which is presented by the right lower corner (Zhang et al. 2016). Figure 1.4 shows the *SE*s of mixing in a constant-pressure PRO as functions of the feed ratio. The SE_{max} values for the pairs of (1) seawater brine–river water, (2) seawater–river water, and (3) seawater brine–seawater are 0.34, 0.16, and 0.035 kWh/m^3, respectively, at an optimal feed ratio of 0.5.

1.3.2 Mass Transfer in PRO

The PRO water flux is driven by the osmotic pressure difference ($\Delta\pi$) but slowed down by the applied pressure difference (ΔP). For an ideal and perfect semipermeable membrane, the spontaneous water permeation flux, J_w, is proportional to the driving force across the membrane and the pure water permeability (A) of the membrane.

$$J_w = A(\Delta\pi - \Delta P) \tag{7}$$

The power density (PD), which is defined as the work produced per square meter of membrane area, is the product of the spontaneous water flux and the applied pressure on the draw solution.

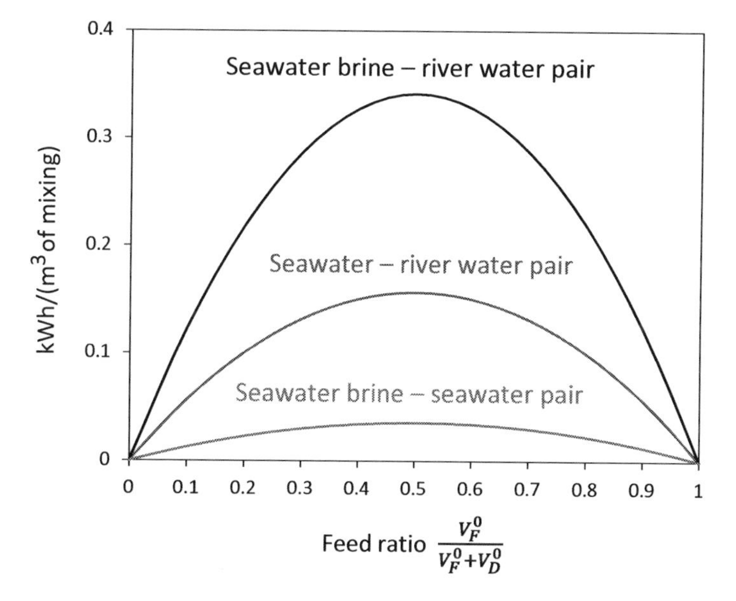

FIGURE 1.4
Specific energy of mixing in a constant-pressure PRO mixing.

$$\begin{aligned} PD &= J_w \Delta P \\ &= A(\Delta\pi - \Delta P)\Delta P \end{aligned} \tag{8}$$

Theoretically, PD is a quadratic equation of ΔP, with a maximum of $(A\Delta\pi^2)/4$ at an optimal ΔP of $\Delta\pi/2$. Figure 1.5 plots the water flux and power density of PRO as functions of ΔP.

However, actual PRO membranes usually have asymmetric structures and less than 100% salt rejections. A typical PRO membrane is composed of a thin selective layer and a porous supportive layer. Chapters 2 and 3 will discuss various types of membranes and their structures, while Chapter 8 will share how to assemble the hollow fiber membranes into an operational unit – a membrane module – for PRO tests and applications. In actual PRO processes, external concentration polarization (ECP) occurs on the external surface of the active layer. It is a function of flow configuration and operation conditions. Internal concentration polarization (ICP) occurs within the porous substrate. ICP plays a more significant role than ECP in determining power density because it can dramatically decrease the effective osmotic driving force and water flux. Unlike ECP, ICP cannot be eliminated by increasing cross-flow velocity or inducing turbulence along the surface. The ICP can be further enhanced by the reverse salt flux from the draw solution to the feed solution. As shown in Figure 1.6, due to the detrimental effects of ECP, ICP, and reverse salt flux, the effective osmotic pressure difference across the active layer is less than the bulk osmotic pressure gradient. Therefore, Eqn 7 should be rewritten as follows (Achilli et al. 2009, Lee et al. 1981, Yip et al. 2011):

$$J_w = A(\Delta\pi_{eff} - \Delta P) \tag{9}$$

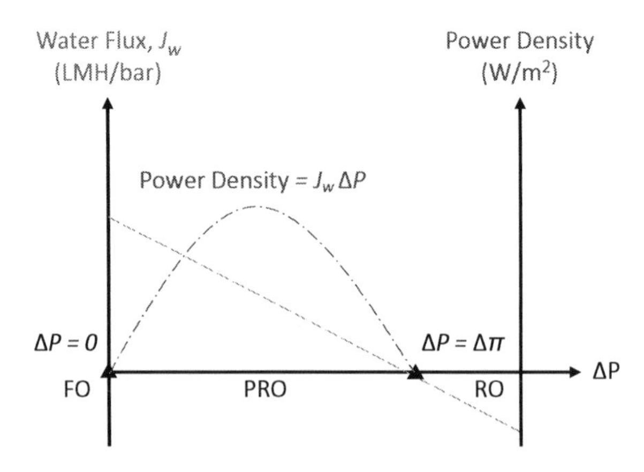

FIGURE 1.5
The PRO water flux and power density as functions of hydraulic pressure difference.

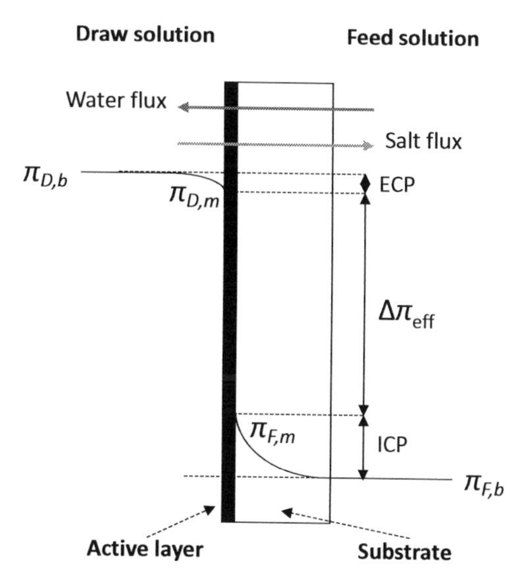

FIGURE 1.6
Concentration profile in a typical PRO membrane.

$$\Delta\pi_{\text{eff}} = \frac{\pi_{D,b}\exp\left(-\frac{J_w}{k}\right) - \pi_{F,b}\exp\left(\frac{J_w S}{D}\right)}{1 + \frac{B}{J_w}\left[\exp\left(\frac{J_w S}{D}\right) - \exp\left(-\frac{J_w}{k}\right)\right]} \tag{10}$$

where $\Delta\pi_{\text{eff}}$ is the effective osmotic pressure difference across the active layer of the membrane, B is the salt permeability, k is the mass transfer coefficient of the draw solution, and S is the structural parameter defined as follows.

$$S = \frac{\tau\lambda}{\varepsilon} \tag{11}$$

where τ, ε, and λ are the tortuosity, porosity, and wall thickness of the hollow fiber membranes, respectively. The reverse salt flux, J_s, can be calculated from the following equation.

$$J_s = \frac{B}{iRT}\left(\frac{J_w}{A} + \Delta P\right) \tag{12}$$

Figure 1.7 illustrates how the detrimental effects of ICP, ECP, and J_s reduce the PRO water flux and power density (Yip et al. 2011). It is worth mentioning that these effects indeed increase the optimal pressure where the peak power

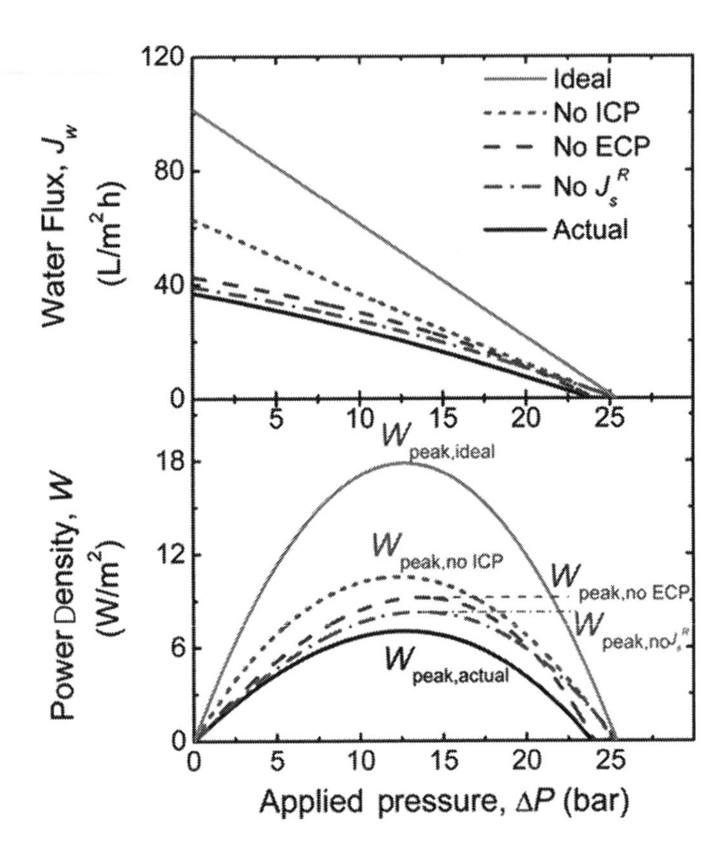

FIGURE 1.7
An illustration of the water flux and power density curves against hydraulic pressure in the PRO process. The effects of ECP, ICP, and salt reverse diffusion on membrane performance are also demonstrated.

Adapted with permission from Yip and Elimelech 2012.

density is achieved (Zhang and Chung 2013). Equation 10 is obtained based on a flat-sheet membrane configuration. In Chapter 4, we will see how this model can be extended to hollow fiber membranes with the selective layer on the inner or the outer surface of the membranes (Cheng and Chung 2017). In Chapter 5, we will take a closer look at the concentration profile within the membrane and how concentration polarization and scaling may occur within the membrane structure (Xiong et al. 2016).

1.3.3 Membrane Fouling in PRO

PRO membrane performances can be significantly compromised by membrane fouling from the feed solution – the deposition of suspended particles or colloids, organic macromolecules, sparingly soluble inorganic compounds, microorganisms, or their mixtures onto the membrane surfaces or even into the porous membrane structures (Chen et al. 2016, Chun et al. 2017, Mi and Elimelech 2010). The compositions of the feed solution have the greatest influence on membrane fouling. For instance, nature organic matters are the main foulants from river water (Thelin et al. 2013), and hydroxyapatite and silica are the main foulants from municipal wastewater or its concentrate (Chen et al. 2016, Wan et al. 2019). PRO membrane fouling is not only affected by the feed components but also by the reverse salt flux from the draw solution. For example, the Na^+ reverse salt flux may enhance alginate fouling on the feed solution side (Chen et al. 2014). General strategies to control membrane fouling are either to introduce hydrophilic groups to enhance water–surface interactions or to impart charges to induce electrostatic repulsion of the foulants (Le et al. 2017, Li et al. 2017, 2014).

The fouling mechanism is also affected by the membrane orientation. If the active layer faces the feed solution and the substrate faces the draw solution, fouling occurs by depositing foulants from the feed solution onto the surface of the active layer and forming cake layers. In contrast, if the active layer faces the draw solution and the substrate faces the feed solution orientation, the membrane fouling is more complicated because the porous substrate is facing the feed solution. While larger foulants are deposited onto the outer surface of the porous substrate, smaller foulants may be dragged into the porous substrate with the aid of water convection and then accumulated underneath the active layer and inside the pores (Han et al. 2016, She et al. 2016, Wan and Chung 2015). Chapter 6 will discuss how to develop antifouling PRO membranes, and Chapter 7 will study how to clean the fouled membrane and recover the membrane performances.

1.4 PRO Processes

1.4.1 Stand-Alone PRO Processes

To achieve continuous and stable power generation, a practical seawater and freshwater PRO system can be designed as shown in Figure 1.8 (Sivertsen et al. 2013, Skilhagen et al. 2008). First, seawater with a volumetric flow rate of V_D is pressurized by a pressure exchanger and a booster pump into the draw solution channel of a PRO module. Meanwhile, freshwater with a volumetric flow rate of V_F enters the feed channel of the PRO module. Water permeates through the membrane from the feed channel to the draw channel at a volumetric flow rate of ΔV. Seawater is diluted to brackish water with a volumetric flow rate of $V_D + \Delta V$, and its pressure is maintained. The

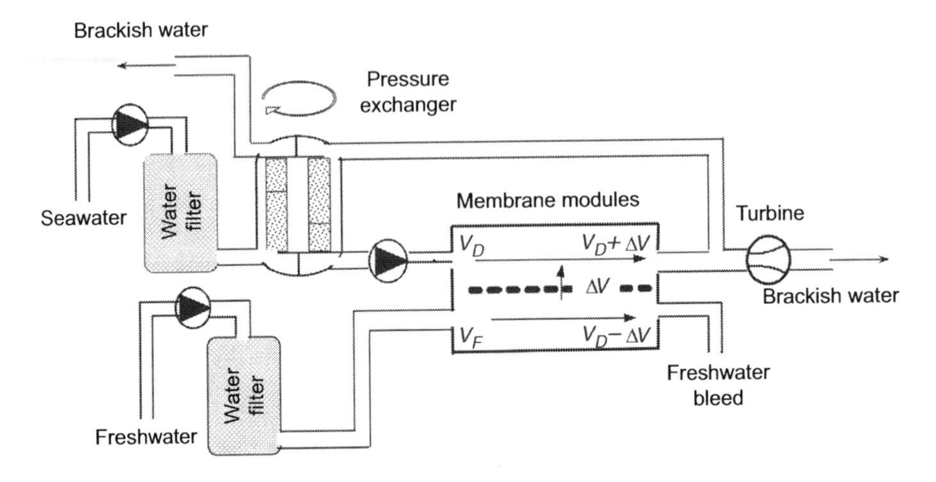

FIGURE 1.8
An illustration of the electrical power production by the PRO process.

Adapted with permission from Skilhagen et al. 2008.

cross-membrane flow (ΔV) goes into the turbine to generate electricity, while the original flow rate of V_D goes to a pressure exchanger and transfers its pressure to the incoming seawater. However, the net energy output of this PRO process is low due to the low osmotic driving force between river water and seawater and the high energy consumption to pretreat the feed solutions (Chung et al. 2015, Straub et al. 2016).

1.4.2 PRO and RO Hybrid Processes

The seawater brine from RO possesses a high pressure and a high concentration of salts, but it has fewer foulants because of the preceding seawater pretreatment. The conventional energy recovery devices (ERDs), such as pressure exchangers, only recover the energy from its high pressure. The high osmotic energy of the seawater RO brine remains unutilized. As shown in Figure 1.9 (a), RO produces a high-pressure brine. After an ERD, the pressure of the seawater brine is reduced to the operating pressure of PRO, where the osmotic energy is generated. The osmotic energy can be harvested either by another ERD to compensate the energy consumption of RO or by a turbine to generate electricity (Prante et al. 2014, Sakai et al. 2016, Wan and Chung 2016). Similarly, the warmer and more saline seawater brine from a membrane distillation (MD) process can be used as the draw solution in PRO (Han et al. 2015b, Kim et al. 2016). Potentially, using the more concentrated seawater brine will generate a higher energy output.

On the other hand, PRO can dilute the seawater feed to RO and generate the osmotic energy at the same time as shown in Figure 1.9 (b) (Chung et al. 2019,

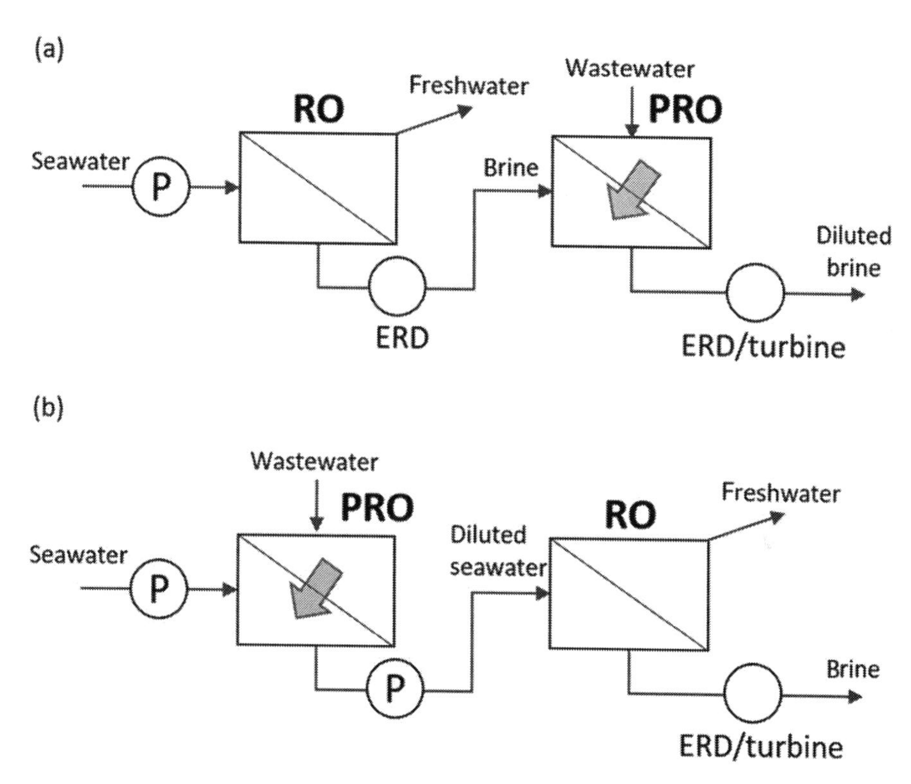

FIGURE 1.9
(a) PRO-RO and (b) RO-PRO integrated processes for energy-efficient desalination.

Adapted with permission from Chung et al. 2019.

Senthil and Senthilmurugan 2016). With PRO pre-dilution, seawater RO can operate at (1) a much lower pressure to achieve the same seawater recovery or (2) a much higher process recovery if a constant pressure is applied. More favorably, FO not only dilutes the salts but also reduces the concentration of foulants in the seawater feed. Therefore, less RO membrane fouling and scaling will occur in this hybrid process.

In Chapter 9, design, integration, and optimization of such hybrid systems will be discussed in detail. In Chapter 10, we will continue to investigate how to build, operate, and maintain such systems. In Chapter 11, a techno-economic model will be presented to break down the operating and capital expenditures and to study the economic feasibility of the integrations.

1.4.3 Closed-PRO Hybrid System

Because PRO employs a feed solution and a draw solution, closed-loop systems can be developed based on either stream. In the draw-solution-closed-loop system, the draw solution is diluted in the PRO process, then reconcentrated in a draw solution recovery unit and reused in PRO. Therefore, the closed-loop system offers greater flexibility to choose any draw solution beyond seawater or seawater brine. The draw solution recovery unit can be any desalination unit, such as nanofiltration, RO, or MD (Han et al. 2014, Wan and Chung 2016). Figure 1.10 shows one important application of the closed-loop PRO hybrid system for energy storage. When there is a surplus of power supply from the grid, the extra power can be used in RO and stored as chemical potential in the seawater brine and freshwater produced. The seawater brine and freshwater can be mixed in PRO anytime to generate electricity (Bharadwaj and Struchtrup 2017, He and Wang 2017). An osmotic heat engine utilizes the closed-loop PRO with ammonia–carbon dioxide as the draw solution (Lin et al. 2014, McGinnis et al. 2007). The ammonia–carbon dioxide draw solution can be regenerated using low-grade heat, and then generate a high-grade energy in PRO.

Figure 1.11 presents a feed-solution-closed-loop system using a 0.1 M NaCl solution as the feed solution to PRO and as the draw solution in FO at the same time. Since FO has a lower fouling propensity, the FO process can extract freshwater from wastewater with the 0.1 M NaCl solution, which is then used as the feed solution in PRO to generate energy. Experiments have verified that the FO–PRO, as shown in Figure 1.12, can generate a higher power density than a stand-alone PRO system when wastewater is utilized as the feed (Cheng et al. 2018). In Chapter 12, the models and designs of the closed-loop systems will be studied.

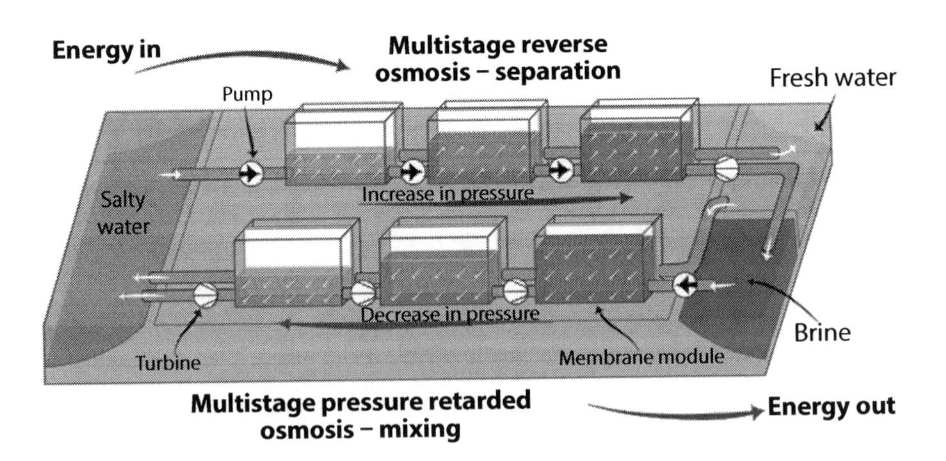

FIGURE 1.10
An osmotic energy storage system with 3-stage RO and 3-stage PRO.

Adapted with permission from Bharadwaj and Struchtrup 2017.

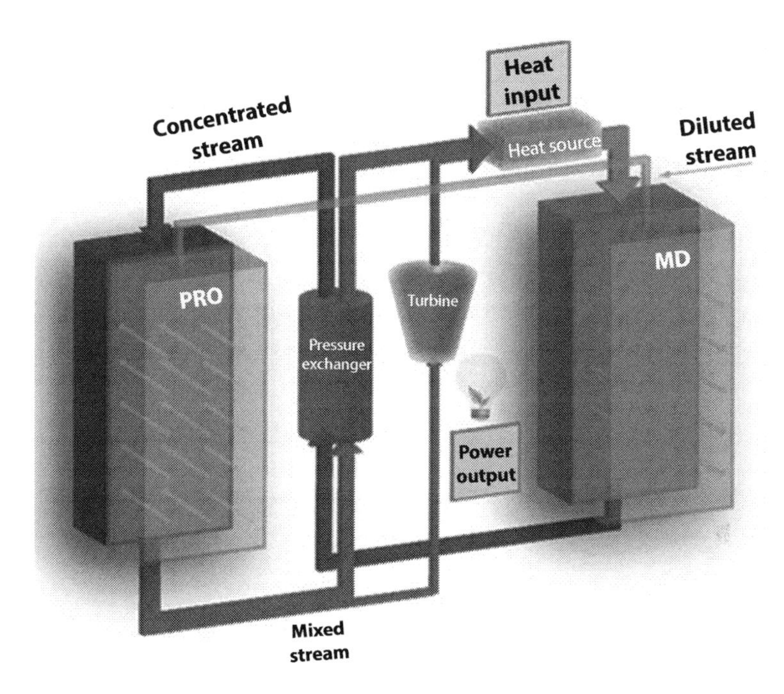

FIGURE 1.11
A PRO and MD closed-loop system to generate power from low-grade heat.

Adapted with permission from Lin et al. 2014.

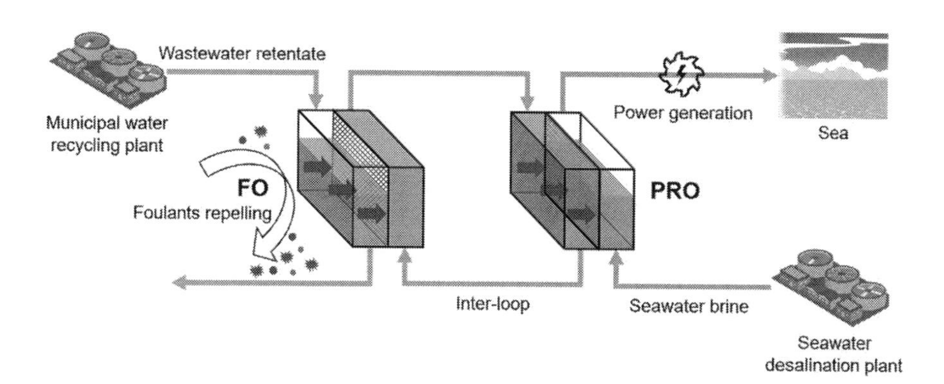

FIGURE 1.12
Illustration of an FO–PRO hybrid system utilizing wastewater retentate as the feed solution and seawater brine as the draw solution.

Adapted with permission from Cheng et al. 2018.

1.5 Summary and Outlook

In terms of material and membrane researches, efforts to develop robust, highly permeable, highly selective, cost-friendly, environment-friendly, and easy-to-scale membranes should continue. As ICP causes the most significant drop in the effective driving force and PRO performances of thin-film composite membrane, more open and interconnected membrane substrates need to be developed without compromising the mechanical properties of the substrates. Another important direction is to develop more effective antifouling and antiscaling membranes with easy-to-scale modification methods. These challenges may require researchers to develop novel methodologies or employ new materials and additives to fabricate membranes with new morphologies and structures. Although membranes with an integrally selective layer may show better antifouling performances with the selective layer facing the feed solution, their water permeability needs to be improved by two- to four-folds in order to compete with thin-film composite membranes for osmotic power generation.

In the aspect of membrane scale-up and operation, the majority of novel PRO membranes are only demonstrated in an ideal lab condition on a small scale. Fabrication of large-scale membrane modules is important to evaluate whether the membranes can be cost-effectively produced in a large area with minimal defects, high uniformity, and flexibility to roll up into a spiral wound module or assembled into a hollow fiber module. Fabrication of large-scale membrane modules, more importantly, is the foundation for pilot and long-term tests in simulated and actual conditions.

Last but not least, PRO processes should be optimized on a system level – with effective pretreatment, energy and mass integration, and energy-harvesting methods. Further pilot studies and projects will be necessary to assess how fluctuating environmental factors, such as the salinity of the feed and draw solutions, may affect the PRO operations, and how real-world PRO technologies may impact the environment and society. Life cycle analyses and techno-economic analyses should be conducted to address such concerns as more pilot data and information about PRO processes become available. To accelerate the development of PRO plants, policy makers should incorporate PRO plants in the urban master planning and colocate the necessary facilities, such as wastewater treatment plants and desalination plants, in a suitable area where PRO plants will be built.

In summary, membrane technologies for PRO have come a long way since the concept was proposed in the 1950s. There have been significant progress in fabrication of novel PRO membranes, understanding of PRO mass transport, design, optimization, operation, and maintenance of various PRO processes. With several PRO processes in pilot stages, the commercialization of PRO technologies should emerge in the near future.

References

Achilli, A., Cath, T.Y., Childress, A.E., 2009. Power generation with pressure retarded osmosis: an experimental and theoretical investigation. *J. Membr. Sci.* 343, 42–52.

Achilli, A., Childress, A.E., 2010. Pressure retarded osmosis: from the vision of Sidney Loeb to the first prototype installation – review. *Desalination* 261, 205–211.

Altaee, A., Zhou, J., Alanezi, A.A., Zaragoza, G., 2017. Pressure retarded osmosis process for power generation: feasibility, energy balance and controlling parameters. *Appl. Energy* 206, 303–311.

Awad, A.M., Jalab, R., Minier-Matar, J., Adham, S., Nasser, M.S., Judd, S.J., 2019. The status of forward osmosis technology implementation. *Desalination* 461, 10–21.

Bharadwaj, D., Struchtrup, H., 2017. Large scale energy storage using multistage osmotic processes: approaching high efficiency and energy density. *Sustain. Energ. Fuels* 1, 599–614.

Chen, S.C., Amy, G.L., Chung, T.S., 2016. Membrane fouling and anti-fouling strategies using RO retentate from a municipal water recycling plant as the feed for osmotic power generation. *Water Res.* 88, 144–155.

Chen, S.C., Fu, X.Z., Chung, T.S., 2014. Fouling behaviors of polybenzimidazole (PBI)–polyhedral oligomeric silsesquioxane (POSS)/polyacrylonitrile (PAN) hollow fiber membranes for engineering osmosis processes. *Desalination* 335, 17–26.

Cheng, Z.L., Chung, T.S., 2017. Mass transport of various membrane configurations in pressure retarded osmosis (PRO). *J. Membr. Sci.* 537, 160–176.

Cheng, Z.L., Li, X., Chung, T.S., 2018. The forward osmosis-pressure retarded osmosis (FO-PRO) hybrid system: a new process to mitigate membrane fouling for sustainable osmotic power generation. *J. Membr. Sci.* 559, 63–74.

Chun, Y., Kim, S.-J., Millar, G.J., Mulcahy, D., Kim, I.S., Zou, L., 2017. Forward osmosis as a pre-treatment for treating coal seam gas associated water: flux and fouling behaviour. *Desalination* 403, 144–152.

Chung, T.S., Li, X., Ong, R.C., Ge, Q., Wang, H., Han, G., 2012a. Emerging forward osmosis (FO) technologies and challenges ahead for clean water and clean energy applications. *Curr. Opin. Chem. Eng.* 1, 246–257.

Chung, T.S., Luo, L., Wan, C.F., Cui, Y., Amy, G., 2015. What is next for forward osmosis (FO) and pressure retarded osmosis (PRO) Sep. *Sci. Technol.* 156, 856–860.

Chung, T.S., Zhang, S., Wang, K.Y., Su, J.C., Ling, M.M., 2012b. Forward osmosis processes: yesterday, today and tomorrow. *Desalination* 287, 78–81.

Chung T.S., Zhao, D.L., Gao, J., Wan, C.F., Weber, M., Maletzko, C., 2019. Emerging R&D on membranes and systems for water reuse and desalination *Chinese. J. Chem. Eng.* in press.

Feinberg, B.J., Ramon, G.Z., Hoek, E.M., 2013. Thermodynamic analysis of osmotic energy recovery at a reverse osmosis desalination plant. *Environ. Sci. Technol.* 47, 2982–2989.

Finley, W., Pscheidt, E., 2001. *Hydrocratic generator*, US Patent 6,313,545 B1.

Gerstandt, K., Peinemann, K.V., Skilhagen, S.E., Thorsen, T., Holt, T., 2008. Membrane processes in energy supply for an osmotic power plant. *Desalination* 224, 64–70.

Han, G., Ge, Q., Chung, T.S., 2014. Conceptual demonstration of novel closed-loop pressure retarded osmosis process for sustainable osmotic energy generation. *Appl. Energy* 132, 383–393.

Han, G., Zhang, S., Li, X., Chung, T.S., 2015a. Progress in pressure retarded osmosis (PRO) membranes for osmotic power generation. *Prog. Polym. Sci.* 51, 1–27.

Han, G., Zhou, J., Wan, C., Yang, T.S., Chung, T.S., 2016. Investigations of inorganic and organic fouling behaviors, antifouling and cleaning strategies for pressure retarded osmosis (PRO) membrane using seawater desalination brine and wastewater. *Water Res.* 103, 264–275.

Han, G., Zuo, J., Wan, C.F., Chung, T.S., 2015b. Hybrid pressure retarded osmosis–membrane distillation (PRO–MD) process for osmotic power and clean water generation. *Environ. Sci. Water Res. Technol.* 1, 507–515.

He, W., Wang, J., 2017. Feasibility study of energy storage by concentrating/desalinating water: concentrated water energy storage. *Appl. Energy* 185, 872–884.

Helfer, F., Lemckert, C., Anissimov, Y.G., 2014. Osmotic power with pressure retarded osmosis: theory, performance and trends – a review. *J. Membr. Sci.* 453, 337–358.

Kim, J., Jeong, K., Park, M., Shon, H., Kim, J., 2015. Recent advances in osmotic energy generation via pressure-retarded osmosis (PRO): a review. *Energies* 8, 11821–11845.

Kim, J., Park, M., Shon, H.K., Kim, J.H., 2016. Performance analysis of reverse osmosis, membrane distillation, and pressure-retarded osmosis hybrid processes. *Desalination* 380, 85–92.

Le, N.L., Quilitzsch, M., Cheng, H., Hong, P.Y., Ulbricht, M., Nunes, S.P., Chung, T. S., 2017. Hollow fiber membrane lumen modified by polyzwitterionic grafting. *J. Membr. Sci.* 522, 1–11.

Lee, K.L., Baker, R.W., Lonsdale, H.K., 1981. Membranes for power generation by pressure-retarded osmosis. *J. Membr. Sci.* 8, 141–171.

Li, X., Cai, T., Amy, G.L., Chung, T.S., 2017. Cleaning strategies and membrane flux recovery on anti-fouling membranes for pressure retarded osmosis. *J. Membr. Sci.* 522, 116–123.

Li, X., Cai, T., Chung, T.S., 2014. Anti-fouling behavior of hyperbranched polyglycerol-grafted poly (ether sulfone) hollow fiber membranes for osmotic power generation. *Environ. Sci. Technol.* 48, 9898–9907.

Lin, S., Yip, N.Y., Cath, T.Y., Osuji, C.O., Elimelech, M., 2014. Hybrid pressure retarded osmosis-membrane distillation system for power generation from low-grade heat: thermodynamic analysis and energy efficiency. *Environ. Sci. Technol.* 48, 5306–5313.

Loeb, S., 1975. Osmotic power plants. *Science* 189, 654–655.

Loeb, S., 1976. Production of energy from concentrated brines by pressure-retarded osmosis: I. preliminary technical and economic correlations. *J. Membr. Sci.* 1, 49–63.

Loeb, S., Hessen, F.V., Shahaf, D., 1976. Production of energy from concentrated brines by pressure-retarded osmosis: II. Experimental results and projected energy costs. *J. Membr. Sci.* 1, 249–269.

Logan, B.E., Elimelech, M., 2012. Membrane-based processes for sustainable power generation using water. *Nature* 488, 313–319.

Marino, M., Misuri, L., Ruffo, R., Brogioli, D., 2015. Electrode kinetics in the "capacitive mixing" and "battery mixing" techniques for energy production from salinity differences, *Electrochim. Acta* 176, 1065–1073.

McElrone, A.J., Choat, B., Gambetta, G.A., Brodersen, C.R., 2013. Water uptake and transport in vascular plants. *Nat. Educ. Knowl.* 4. https://www.nature.com/scitable/knowledge/library/water-uptake-and-transport-in-vascular-plants-103016037/.

McGinnis, R.L., McCutcheon, J.R., Elimelech, M., 2007. A novel ammonia–carbon dioxide osmotic heat engine for power generation. *J. Membr. Sci.* 305, 13–19.

Mi, B.X., Elimelech, M., 2010. Gypsum scaling and cleaning in forward osmosis: measurements and mechanisms. *Environ. Sci. Technol.* 44, 2022–2028.

Micale, G., Cipollina, A., Tamburini, A., 2016. Salinity gradient energy. *Sustain. Energy Salinity Gradients* 1, 1–17.

Olsson, M., Wick, G.L., Isaacs, J.D., 1979. Salinity gradient power-utilizing vapor-pressure differences. *Science* 206, 452–454.

Pattle, R.E., 1954. Production of electric power by mixing fresh and salt water in the hydroelectric pile. *Nature* 174, 660.

Prante, J.L., Ruskowitz, J.A., Childress, A.E., Achilli, A., 2014. RO-PRO desalination: an integrated low-energy approach to seawater desalination. *Appl. Energy* 120, 104–114.

Saito, K., Irie, M., Zaitsu, S., Sakai, H., Hayashi, H., Tanioka, A., 2012. Power generation with salinity gradient by pressure retarded osmosis using concentrated brine from SWRO system and treated sewage as pure water. *Desalin. Water Treat.* 41, 114–121.

Sakai, H., Ueyama, T., Irie, M., Matsuyama, K., Tanioka, A., Saito, K., Kumano, A., 2016. Energy recovery by PRO in seawater desalination plant. *Desalination* 389, 52–57.

Sarp, S., Yeo, I.H., Park, Y.G., 2014. Membrane based desalination apparatus with osmotic energy recovery and membrane based desalination method with osmotic energy recovery. US Patent 20140238938 A1.

Senthil, S., Senthilmurugan, S., 2016. Reverse osmosis–pressure retarded osmosis hybrid system: modelling, simulation and optimization. *Desalination* 389, 78–97.

Shaffer, D.L., Werber, J.R., Jaramillo, H., Lin, S., Elimelech, M., 2015. Forward osmosis: where are we now? *Desalination* 356, 271–284.

Sharqawy, M.H., Lienhard, J.H., Zubair, S.M., 2012. Thermophysical properties of seawater: a review of existing correlations and data. *Desalin. Water Treat.* 16, 354–380.

She, Q., Wang, R., Fane, A.G., Tang, C.Y., 2016. Membrane fouling in osmotically driven membrane processes: a review. *J. Membr. Sci.* 499, 201–233.

Sivertsen, E., Holt, T., Thelin, W., Brekke, G., 2013. Pressure retarded osmosis efficiency for different hollow fibre membrane module flow configurations. *Desalination* 312, 107–123.

Skilhagen, S.E., Dugstad, J.E., Aaberg, R.J., 2008. Osmotic power – power production based on the osmotic pressure difference between waters with varying salt gradients. *Desalination* 220, 476–482.

Straub, A.P., Deshmukh, A., Elimelech, M., 2016. Pressure-retarded osmosis for power generation from salinity gradients: is it viable? *Energy Environ. Sci.* 9, 31–48.

Tedesco, M., Scalici, C., Vaccari, D., Cipollina, A., Tamburini, A., Micale, G., 2016. Performance of the first reverse electrodialysis pilot plant for power production from saline waters and concentrated brines. *J. Membr. Sci.* 500, 33–45.

Thelin, W.R., Sivertsen, E., Holt, T., Brekke, G., 2013. Natural organic matter fouling in pressure retarded osmosis. *J. Membr. Sci.* 438, 46–56.

Thorsen, T., Holt, T., 2009. The potential for power production from salinity gradients by pressure retarded osmosis. *J. Membr. Sci.* 335, 103–110.

UN-Water, 2013a. *UN-Water Strategy 2014–2020*. UN-Water Technical Advisory Unit, Switzerland.

UN-Water, 2013b. *Water security and the global water agenda: a UN-Water analytical brief.* United Nations University, Canada.

UNWWAP., 2014. *The United Nations world water development report 2014: water and energy*. UNESCO, France.

Van't Hoff, J.A., 1901. *Osmotic pressure and chemical equilibrium*. Nobel Prize Lecture. https://www.nobelprize.org/uploads/2018/06/hoff-lecture.pdf.

Wan, C.F., Chung, T.S., 2015. Osmotic power generation by pressure retarded osmosis using seawater brine as the draw solution and wastewater retentate as the feed. *J. Membr. Sci.* 479, 148–158.

Wan, C.F., Chung, T.S., 2016. Energy recovery by pressure retarded osmosis (PRO) in SWRO–PRO integrated processes. *Appl. Energy* 162, 687–698.

Wan, C.F., Jin, S., Chung, T.S., 2019. Mitigation of inorganic fouling on pressure retarded osmosis (PRO) membranes by coagulation pretreatment of the wastewater concentrate feed. *J. Membr. Sci.* 572, 658–667.

Wan, C.F., Yang, T.S., Gai, W.X., Lee, Y.D., Chung, T.S., 2018. Thin-film composite hollow fiber membrane with inorganic salt additives for high mechanical strength and high power density for pressure-retarded osmosis. *J. Membr. Sci.* 555, 388–397.

Xiong, J.Y., Cheng, Z.L., Wan, C.F., Chen, S.C., Chung, T.S., 2016. Analysis of flux reduction behaviors of PRO hollow fiber membranes: experiments, mechanisms, and implications. *J. Membr. Sci.* 505, 1–14.

Yip, N.Y., Elimelech, M., 2012. Thermodynamic and energy efficiency analysis of power generation from natural salinity gradients by pressure retarded osmosis. *Environ. Sci. Technol.* 46, 5230–5239.

Yip, N.Y., Tiraferri, A., Phillip, W.A., Schiffman, J.D., Hoover, L.A., Kim, Y.C., Elimelech, M., 2011. Thin-film composite pressure retarded osmosis membranes for sustainable power generation from salinity gradients. *Environ. Sci. Technol.* 45, 4360–4369.

Zhang, S., Chung, T.S., 2013. Minimizing the instant and accumulative effects of salt permeability to sustain ultrahigh osmotic power density. *Environ. Sci. Technol.* 47, 10085–10092.

Zhang, S., Han, G., Li, X., Wan, C.F., Chung, T.S., 2016. Pressure retarded osmosis: fundamentals. In A. Cipollina, G. Micale (Eds.), *Sustainable energy from salinity gradients*. Woodhead Publishing, Sawston, Cambridge, 19–53.

Zhang, S., Sukitpaneenit, P., Chung, T.S., 2014. Design of robust hollow fiber membranes with high power density for osmotic energy production. *Chem. Eng. J.* 241, 457–465.

Zhu, X., Yang, W., Hatzell, M.C., Logan, B.E., 2014. Energy recovery from solutions with different salinities based on swelling an shrinking of hydrogels. *Environ. Sci. Technol.* 48, 7157–7163.

2

Recent Development of Flat-Sheet PRO Membranes

Wenxiao Gai
Department of Chemical and Biomolecular Engineering
National University of Singapore
Singapore

Gang Han
Department of Chemical Engineering
Massachusetts Institute of Technology
Cambridge, Massachusetts

CONTENTS

2.1 Introduction

The feasibility of the pressure retarded osmosis (PRO) technology for osmotic power generation was first recognized by Loeb in the 1970s (Loeb 1976). Then the PRO technology experienced a slow development due to the shortage of effective membranes until 1990s (Han et al. 2015). The semipermeable membrane, which determines the overall power generation and profitability, is the key component of the PRO technology. It has been proved by the Norwegian company Statkraft, who built the first PRO prototype plant worldwide in 2009, that the employment of high-performance PRO membranes is vital for the commercialization of the PRO technology (Chung et al. 2015, Helfer et al. 2014). The output of the PRO process must be sufficiently higher than the energy consumption for the whole system in order to have a positive economic value. Therefore, intensive efforts have been focused on the development of high-performance PRO membranes. An ideal PRO membrane should possess the characteristics such as high water permeability, low salt permeability, good mechanical strength, and low structure parameter to achieve a high power density during the PRO process. In other words, the selective layer of PRO membranes must be thin and with excellent ion selectivity, while the support layer of PRO membranes must be strong and yet highly porous.

The support layer of PRO membranes is usually fabricated by nonsolvent-induced phase inversion or electrospinning. Currently, there are two approaches to fabricate the selective layer, which are direct phase inversion and interfacial polymerization. The direct phase inversion is the most-known route, in which the support layer and the integrally skinned membrane are formed simultaneously. The interfacial polymerization route has been widely used to prepare thin-film composite (TFC) membranes for seawater desalination, in which a cross-linked aromatic polyamide skin is formed at the interface of an aqueous solution containing one monomer and an organic solution containing a second monomer. Generally, there are two configurations for PRO membranes, which are flat sheet and hollow fiber. This chapter will mainly focus on the recent development of flat-sheet PRO membranes. The flat-sheet membranes are usually constructed by a paper-like backing material with a membrane cast on its top surface.

2.2 History of Membranes Employed in PRO Processes

2.2.1 Early RO Membranes Employed in PRO Processes

The first trial to utilize the PRO process for energy production was conducted by Loeb in 1976 (Loeb 1976). Since then, various commercial reverse osmosis (RO) and nano-filtration (NF) membranes, which were originally designed for hydraulic-pressure-driven separation processes, were utilized in PRO processes for power generation (Jellinek and Masuda 1981, Lee et al. 1981, Loeb et al. 1976, Loeb and Mehta 1979, Mehta 1982, Mehta and Loeb 1978, 1979). The commercial Permasep RO membranes employed by Loeb and Mehta for PRO tests were outer-selective hollow fibers made by DuPont for seawater desalination (Loeb 1976, Loeb et al. 1976, Mehta and Loeb 1979). The Permasep RO membranes were able to withstand a high hydraulic pressure of up to 91.2 bar but only showed a low peak power density of less than 1.74 W/m^2 at around 31 bar using a NaCl draw solution with an osmotic pressure of 25 atm (i.e., 25.33 bar) (Loeb 1976, Loeb et al. 1976). When using a NaCl draw solution with an osmotic pressure of 90 atm (i.e., 90.12 bar), a power density of 4.89 W/m^2 at 50 bar was obtained by the same Permasep RO membranes (Mehta and Loeb 1978, 1979). Besides, composite hollow fiber membranes from the Fabric Research Lab, which had a polysulfone support layer and a furan outer skin layer, were also tested by Loeb and Mehta (Loeb and Mehta 1979). The power density of the Fabric Research Lab's membranes was 1.57 W/m^2 at a hydraulic pressure gradient of 19 bar. Later, Jellinek and Masuda employed flat-sheet cellulose triacetate (CTA) RO membranes for PRO (Jellinek and Masuda 1981) and achieved a power density of 1.62 W/m^2 at about 17.2 bar. In addition, Mehta studied the PRO performance of several RO mini-modules with both spiral-wound and hollow fiber configurations (Mehta 1982). A UOP CA/SW-3 spiral-wound mini-module had the highest power density of 2.34 W/m^2 at about 20 bar.

Figure 2.1 summarizes the obtained power densities of commercially available RO and NF membranes. When seawater and fresh water were used as the draw and feed solutions ($\Delta\pi$ = 20–25 bar), respectively, a power density of 1.22 W/m^2 was obtained. When a more concentrated brine ($\Delta\pi$ > 75 bar) was used as the draw solution, a higher power density

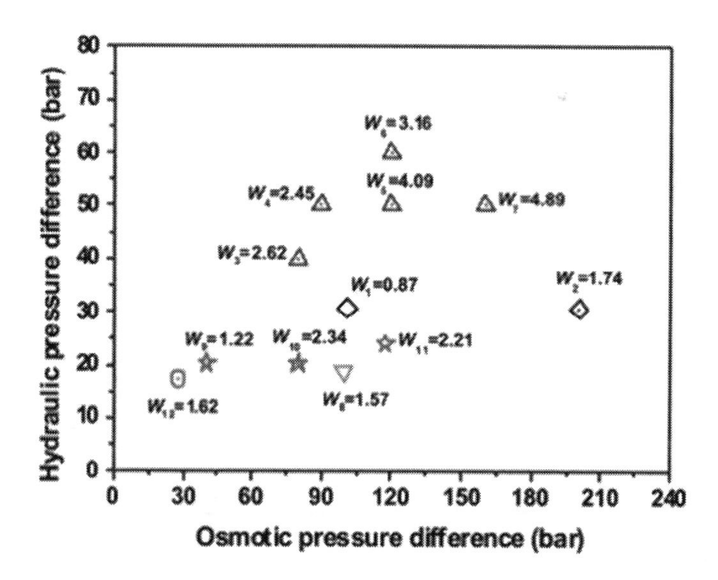

FIGURE 2.1
Power densities W (W/m^2) of commercially available RO/NF membranes. The power densities were derived from the reported osmotic pressure, hydraulic pressure, and water flux (W_1 – W_2: (Loeb 1976, Loeb et al. 1976); W_{12}: (Jellinek and Masuda 1981)) or from permeation coefficients (W_3 – W_7: (Mehta and Loeb 1978, 1979); W_8: (Loeb and Mehta 1979); W_9 – W_{11}: (Mehta 1982)).

of up to 4.89 W/m^2 could be achieved. However, the performances of these commercially available membranes were still far from applicability. The poor PRO performance was mainly attributed to the severe internal concentration polarization (ICP) occurring inside the thick substrate layer of these membranes. To withstand very high pressures of up to 100 bar, the conventional RO membranes usually have a very thick supporting layer of about 150–250 µm, consisting of a fabric and a hydrophobic polysulfone layer. The hydrophobic and thick supporting layer would significantly retard the water and ion transport and thus reduce the effective osmotic pressure across the membranes (Lee et al. 1981). Therefore, commercially available RO and NF membranes are not suitable for PRO because they will dramatically enhance ICP and reduce the PRO performance.

2.2.2 FO Membranes Employed in PRO Processes

In the early stage of developing membranes for osmotic power generation, forward osmosis (FO) membranes were employed because PRO and FO processes follow similar osmotic principles (Achilli et al. 2009, Kim and Elimelech 2013, Klaysom et al. 2013, Li et al. 2013, McCutcheon and Elimelech 2008, She et al. 2012, Xu et al. 2010, Zhang et al. 2013, Zhao et al. 2012). In other words, both PRO and FO membranes require similar characteristics, such as high water flux, low salt reverse flux, and minimal ICP (Chung et al. 2012, Klaysom et al. 2013, Zhao et al. 2012).

The asymmetric CTA-based flat-sheet membrane produced by Hydration Technology Innovations (HTI, Albany, OR) is the most investigated FO membrane for PRO. Figure 2.2 presents its morphology. It has a thickness of about 50 µm, thus its structural parameter S is much smaller

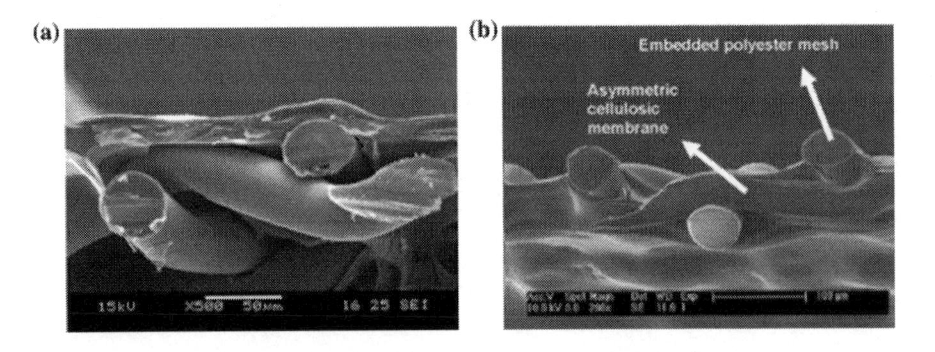

FIGURE 2.2
SEM images of the CTA-FO membranes produced by Hydration Technology Innovations (HTI): (a) FO-1, (b) FO-2 (Chung et al. 2012, McCutcheon and Elimelech 2008).

than the conventional RO membranes (i.e., 480 μm versus 37,500 μm) (Alsvik and Hägg 2013, Chung et al. 2012, McCutcheon and Elimelech 2008). Since the CTA-FO membrane is hydrophilic and thin, it is better than the conventional RO/NF membranes for PRO processes. The reported maximum power densities were 2.73 and 5.06 W/m^2 at 9.72 bar when using deionized water as the feed solution and 35 and 60 g/L NaCl solutions as the draw solutions, respectively (Achilli et al. 2009). She et al. (2012) reported that it had a peak power density of 4.5 W/m^2 at 12 bar when 1 M NaCl and 10 mM NaCl were used as the draw and feed solutions, respectively. Kim and Elimelech (2013) obtained a power density of 4.7 W/m^2 using 2 M NaCl and 0.5 M NaCl as the draw and feed solutions, respectively. Xu et al. (2010) evaluated the PRO performance of a mini Hydrowell® spiral wound FO module with an active CTA membrane area of about 0.94 m^2 produced by HTI. It had a maximum power density of about 0.5 W/m^2 when 0.5 M NaCl and deionized water were used as the draw and feed solutions, respectively. Figure 2.3 summarizes the power densities of the CTA-FO membrane in literatures using different feed pairs (Achilli et al. 2009, Gerstandt et al. 2008, Helfer et al. 2014, Klaysom et al. 2013, She et al. 2012, Skilhagen et al. 2008, Zhao et al. 2012). In general, its power densities are less than 5 W/m^2, which is the economically feasible value set by Statkraft (Gerstandt et al. 2008) for

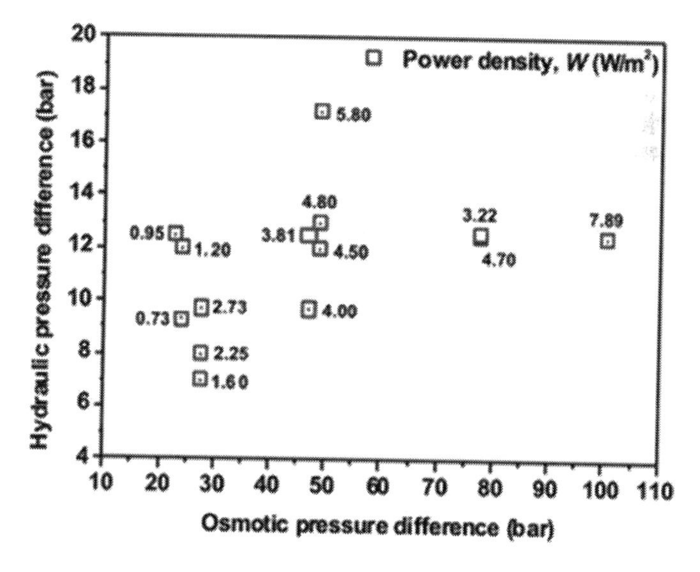

FIGURE 2.3
Power densities W (W/m^2) of CTA-FO membranes. The power densities were derived from the reported osmotic pressure, hydraulic pressure, and water flux (Achilli et al. 2009, Gerstandt et al. 2008, Helfer et al. 2014, Klaysom et al. 2013, She et al. 2012, Skilhagen et al. 2008, Zhao et al. 2012).

TABLE 2.1

Transport properties of the commercial CTA-FO membranes from HTI

Water permeability, A (10^{-3} L/m^2 h bar)	Salt permeability, B (10^{-3} L/m^2 h)	S (μm)	Reference
11.22	6.66	678	(Achilli et al. 2009)
6.12	4.62	590	(She et al. 2012)
6.00	5.34	505	(Kim and Elimelech 2013)
10.98	7.20	790	(Bui and McCutcheon 2014)
20.52	43.80	689	(Kim and Elimelech 2012)
12.48	10.56	480	(She et al. 2013)

commercializing the PRO technology. The low power densities arise from the fact that the CTA-FO membrane has a relatively low intrinsic water permeability and a high salt permeability, as shown in Table 2.1.

Besides the HTI CTA membranes, a variety of recently developed FO membranes have also been evaluated for power generation (Arena et al. 2015, Chung et al. 2012, Han et al. 2012a, 2012b, Klaysom et al. 2013, Zhao et al. 2012). However, it should be noted that the conventional FO membranes tend to be either deformed or damaged under high hydraulic pressures during PRO tests as a result of their poor mechanical strength (Li et al. 2013). Since there is no hydraulic pressure applied during common FO processes, most FO membranes are designed to be very thin and highly porous in order to reduce structural parameter and mitigate ICP. Thus, conventional FO membranes are not suitable for high-pressure PRO processes.

2.3 TFC Membranes Developed for PRO Processes

TFC membranes for seawater reverse osmosis (SWRO) desalination generally consist of an asymmetric porous support layer and a top selective layer supporting by a nonwoven fabric, as shown in Figure 2.4 (Idarraga-Mora et al. 2018). The main function of the nonwoven fabric is to provide the mechanical strength in order to withstand the applied hydraulic pressures, while the function of the top selective layer is to separate water molecules from salt molecules. Generally, the TFC membranes are prepared by (1) casting a microporous support layer on a nonwoven fabric via phase inversion of a polymer dope in a nonsolvent bath, and (2) then conducting interfacial polymerization to form a cross-linked aromatic polyamide selective layer on the substrate (Cadotte et al. 1980). The thickness of the polyamide selective layer is about a hundred nanometers as a result of the

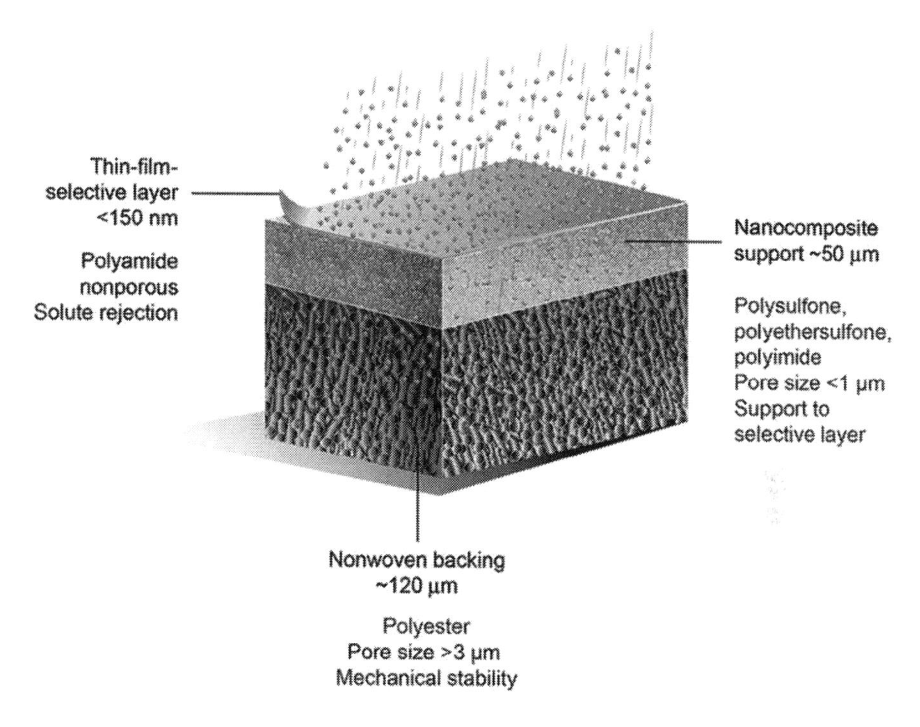

FIGURE 2.4
Schematic of a TFC membrane for SWRO and its typical characteristics in each layer (Idarraga-Mora et al. 2018).

self-terminating nature of the interfacial cross-linking reaction (Cui et al. 2017, Li et al. 2012). The advantage of fabricating TFC membranes via interfacial polymerization is that the physiochemical properties of the substrate and selective layer can be tailored and optimized independently to achieve desired properties, such as high water permeability, low salt permeability, low structural parameter, and good mechanical strength.

The morphology and mechanical strength of the substrate layer are particularly important since they directly determine the separation performance of the polyamide layer (Ghosh and Hoek 2009, Kong et al. 2010, Li et al. 2012, Singh et al. 2006), ICP effects, and pressure tolerance during PRO tests (Han et al. 2013, Lee et al. 1981, Li et al. 2013, Zhang and Chung 2013, Zhang et al. 2013). Therefore, several strategies have been employed to enhance the mechanical strength and to mitigate the ICP effects of TFC membranes, such as (1) modifying the physicochemical properties of substrates by pre-compression and polydopamine cross-linking treatments, and (2) optimizing the polymer solutions and phase inversion conditions to produce the substrates with desirable morphology (Li et al. 2013, Zhang et al. 2013). Meanwhile, efforts have been devoted to enhance the water permeability of the polyamide selective layer by modifying the interfacial polymerization reaction, such as (1) adding bulky monomers or surfactants as additives in monomer solutions to increase their intrinsic free volume (Cui et al. 2014, Li and Chung 2013), and (2) post-treating the nascent polyamide selective layer with chlorine, alcohol, and organic solvents (i.e., *N,N*-dimethylformamide) to remove unreacted monomers and low molecular components (Cui et al. 2014, Han et al. 2013, Li et al. 2013, Yip et al. 2011, Zhang et al. 2013). However, the aforementioned modifications may not only enhance water permeability but also salt permeability, which would deteriorate the overall PRO performance of the TFC membranes. One must balance these two parameters carefully to ensure a net increase in power density.

Li and Chung have fabricated several TFC flat-sheet membranes for PRO by manipulating the free volume of the polyamide selective layer (Li and Chung 2013). They added a bulky monomer of *p*-xylylenediamine into the *m*-phenylenediamine aqueous solution during interfacial polymerization, and then post-treated the fabricated TFC membranes with methanol to swell up the polyamide layer. The addition of the bulky monomer enlarged the intrinsic free volume of the polyamide layer, while the methanol immersion swollen up the polyamide chains to remove unreacted monomers and low molecular weight polymers (Shao et al. 2004, Tin et al. 2004, Zuo et al. 2012). They found that a moderate increment in free volume could enhance water permeability significantly with a slight increase in salt permeability, and thus both water flux and power density of TFC membranes could be promoted. However, an excessive increase in free volume would not only lower membrane selectivity but also deteriorate power density due to the effects of severe ICP effects. The positron annihilation lifetime spectroscopy results of the polyamide layer before and after PRO tests showed that both the thickness and free volume of the polyamide layer decreased as a result of the high pressure compression during PRO tests. The reduced thickness would increase both water and salt permeability but decrease mechanical strength, while the reduced free volume would decrease both water and salt permeability but increase mechanical strength. Therefore, the PRO performance of the TFC membranes stayed almost the same in terms of power density and resistance against high pressures (Li and Chung 2013).

Han et al. fabricated a TFC flat-sheet membrane by depositing a polyamide selective layer on a customized Matrimid® substrate, as shown in Figure 2.5 (a) and (b) (Han et al. 2013). The Matrimid® substrate was designed to have a fully sponge-like morphology with robust mechanical strength. As a result, it can withstand an applied hydraulic pressure of up to 15 bar. To promote the transporting properties of the membranes, the polyamide layer was chemically treated using hypochlorite and then with methanol. The resultant TFC membranes could achieve a higher water permeability but a lower salt rejection. After the moderate post-treatment, the modified TFC membrane showed a power density of 7–12 W/m^2 at

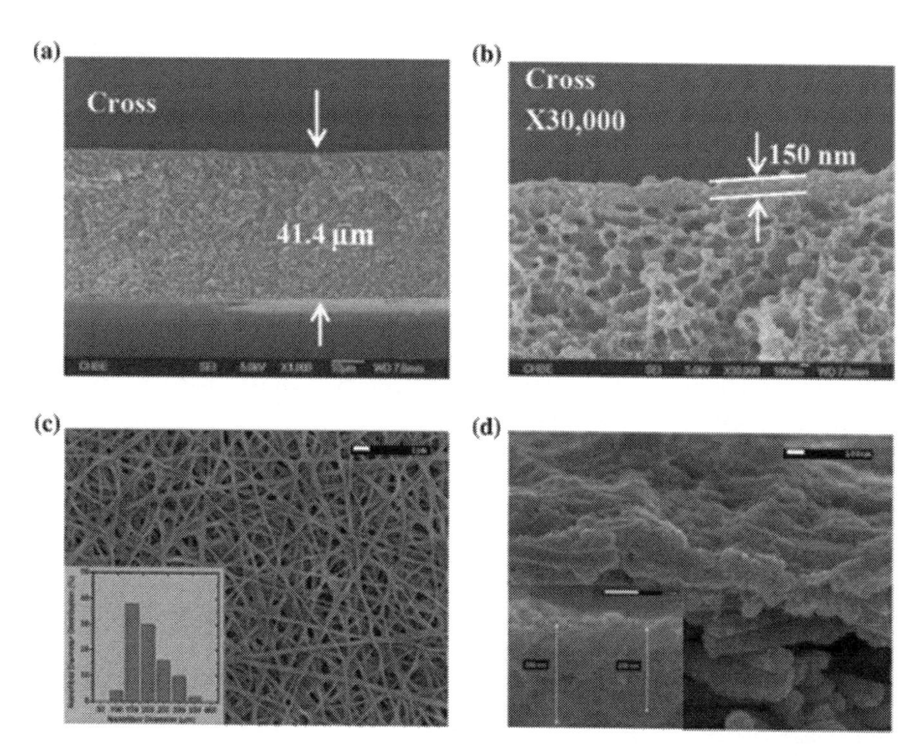

FIGURE 2.5
SEM images of (a) and (b) the cross-section of Matrimid® TFC membrane (Han et al. 2013); (c) the top surface of nanofiber membrane substrate and the distribution of its nanofiber diameter (inserted plot); and (d) the cross-section of the TFC membrane on the nanofiber substrate (Song et al. 2013).

15 bar when using various pairs of draw and feed solutions (Han et al. 2013). When synthetic seawater water (0.59 M NaCl) and deionized (DI) water were respectively employed as the draw and feed solutions, a power density of 9.0 W/m^2 could be obtained at around 13 bar (Han et al. 2013). The significantly improved membrane performance could be attributed to the combination of (1) a robust support layer with a small structural parameter and (2) a polyamide selective layer with a high water permeability (A = 5.3 L/m^2 h bar) and a moderate salt permeability (B = 2.0 L/m^2 h). When a strong post-treatment (i.e., treated by hypochlorite and then methanol) was conducted on the TFC membranes, their PRO performance decreased because the adverse effects of ICP coupling with a high reverse salt flux overwhelmed the positive effects of the enhanced water permeability on water flux and power density (Han et al. 2013, Yip and Elimelech 2011, Zhang and Chung 2013).

Cui et al. (2014) further improved the PRO performance of TFC membranes via optimizing the interfacial polymerization reaction and post-treatment conditions. Since surfactants have been reported to help form the cross-linked polyamide layer due to their amphiphilic nature (Kim et al. 2013, Mansourpanah et al. 2009), a certain amount of sodium dodecyl sulfate was added into the amine solution during interfacial polymerization. As a result, the power density of the TFC membranes was increased from 8.65 to 15.79 W/m^2. Data from positron annihilation lifetime spectroscopy showed that sodium dodecyl sulfate could really increase the free volume of the polyamide layer and thus led to a higher power density. The resultant TFC membranes were further post-treated with *N,N*-dimethylformamide to remove the loosely cross-linked parts of the polyamide selective layer. The power density was jumped to 18.09 W/m^2 at 22 bar when using 1 M NaCl and DI water as the feed pair, as illustrated in Figure 2.6. This power density was one of the highest reported in literatures for flat-sheet PRO membranes.

Song et al. and Bui et al. recently developed TFC membranes consisting of polyamide layers and customized nonwoven webs made by electrospun

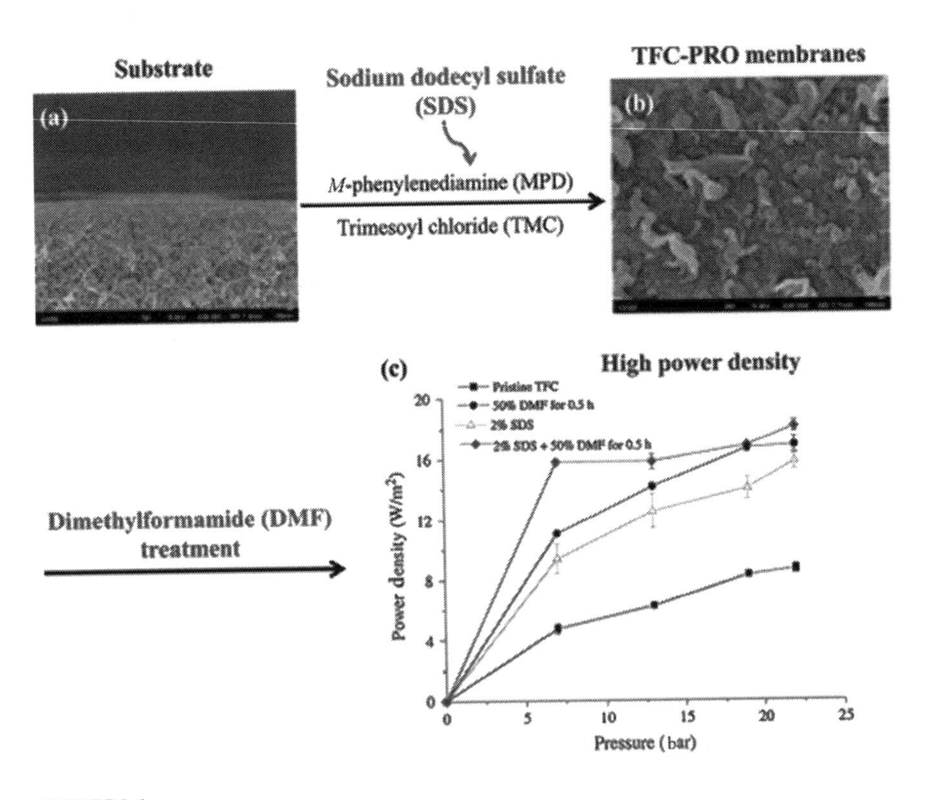

FIGURE 2.6
(a) SEM images of the cross-section of the Matrimid® membrane substrate, (b) SEM images of the top surface of the TFC membrane, and (c) the power density as a function of post-treatments using 1 M NaCl as the draw solution and DI water as the feed solution (Cui et al. 2014).

nanofibers (Bui and McCutcheon 2014, Song et al. 2013). Electrospinning is a technology using electrical potential to produce nano-polymeric fibers with diameters in the range of 40–2000 nm (Reneker and Chun 1996). Figure 2.5 (c) and (d) shows that the nanofiber membrane substrate has a highly porous morphology with interconnected pores. Both the high porosity and low tortuosity of the nanofiber substrate contribute to a very small structural parameter (Bui and McCutcheon 2014, Song et al. 2013), which could significantly mitigate the detrimental effects of ICP on the PRO performance of TFC membranes. The nanofiber-based TFC membranes developed by Song et al. had a power density of 15.2 W/m^2 at 15.2 bar using synthetic brackish water of 80 mM NaCl and seawater brine of 1.06 M NaCl as the feed and draw solutions, respectively (Song et al. 2013). A higher power density of 21.3 W/m^2 was achieved at 15.2 bar when the feed solution changed to a more dilute synthetic river water of 0.9 mM (Song et al. 2013). However, the mechanical stability of these nanofiber-based TFC membranes may be a concern under high hydraulic pressures, which could cause a high reverse salt flux during PRO tests. To overcome this concern, Hoover et al. fabricated TFC membranes composed of electrospun polyethylene terephthalate nanofibers, a microporous polysulfone layer, and a polyamide selective layer (Hoover et al. 2013). The electrospun nanofibers enmeshed with the microporous polysulfone layer showed effectiveness to minimize membrane delamination at high cross-flow velocities.

Besides its original CTA membranes, the HTI company has recently developed a TFC flat-sheet membrane that could be used for PRO applications (Straub et al. 2013). The TFC membranes had a reduced total membrane thickness of about 115 μm by incorporating an embedded woven mesh into the polysulfone porous support layer in order to minimize the ICP effects. The TFC membranes had a water flux more than double than the previous CTA membranes, while the former impressively retained a comparable salt rejection of 99.3% to the latter. Although the embedded woven mesh reinforced the mechanical strength of the TFC membranes, they might still be damaged under high hydraulic pressures. To overcome this, Straub et al. studied the PRO performance of the same HTI TFC membranes supported by tricot fabric feed spacers in a specially designed cross-flow cell for power generation (Straub et al. 2013). The TFC membranes could withstand an applied hydraulic pressure of 48 bar, at which the membranes could achieve a peak power density of 59.7 W/m^2 using 3 M NaCl as the draw solution and DI water as the feed solution. The high performance was attributed to (1) novel testing cell design, (2) special spacers, and (3) optimized membrane structure and permeation characteristics. Kim et al. reported a surface coating method to further promote the water flux of this HTI TFC-PRO membrane. After being coated with TiO_2 nanoparticles via a sol–gel-derived spray coating method on the support layer, the modified membrane showed a 25% increase in water flux and a 50% decrease in reverse salt flux (Kim et al. 2016).

2.4 Membranes Comprising Nanomaterials Developed for PRO Processes

Lim et al. reported a TFC membrane consisting of a dual-layered nanocomposite substrate cast by a dual-blade for PRO applications (Lim et al. 2018). The dual-layered nanocomposite substrate had graphene oxide (GO) and halloysite nanotubes (HNTs) incorporated in the top and bottom substrate layers, respectively, as illustrated in Figure 2.7. Since the fabricated substrate had high mechanical strength and desirable characteristics, such as high porosity, porous bottom surface, and suitable top-skin morphology for the formation of a polyamide layer formation, the resultant TFC membranes comprising a GO loading of 0.25 wt% and an HNT loading of 4 wt% showed a power density of 16.7 W/m^2 at 21 bar using 1 M NaCl and DI water as the feed pair.

Son et al. developed a thin-film nanocomposite (TFN) membrane consisting of carbon nanotubes (CNTs) embedded in the polyethersulfone (PES) supporting layer and polyamide selective layer for PRO processes (Son et al. 2016). The incorporation of CNT into the PES substrate could increase the porosity and hydrophilicity of the support layer, resulting in an increased water flux and power density, as illustrated in Figure 2.8.

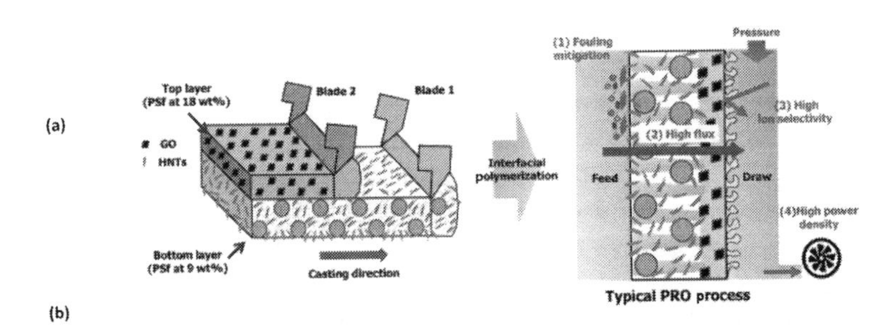

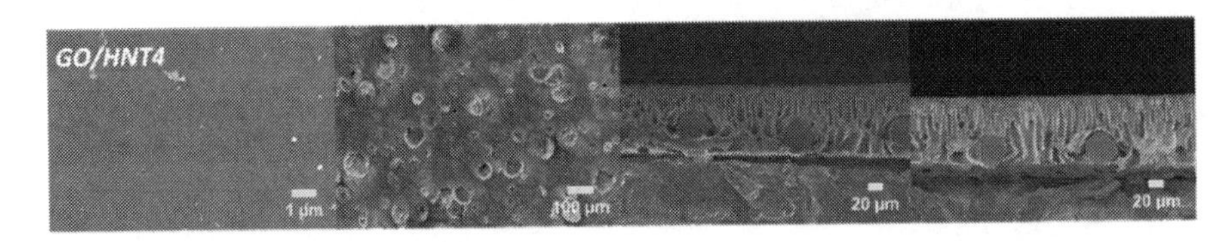

FIGURE 2.7
(a) Conceptual illustration of the dual-layered TFC PRO membranes with GO and HNTs incorporated in the top and bottom substrate layers, respectively; (b) SEM images of dual-layered flat-sheet PRO membrane substrates: top, bottom and cross-section morphologies, and cross-section morphologies for pressurized membrane substrates under the applied pressure of up to 27 bar (Lim et al. 2018).

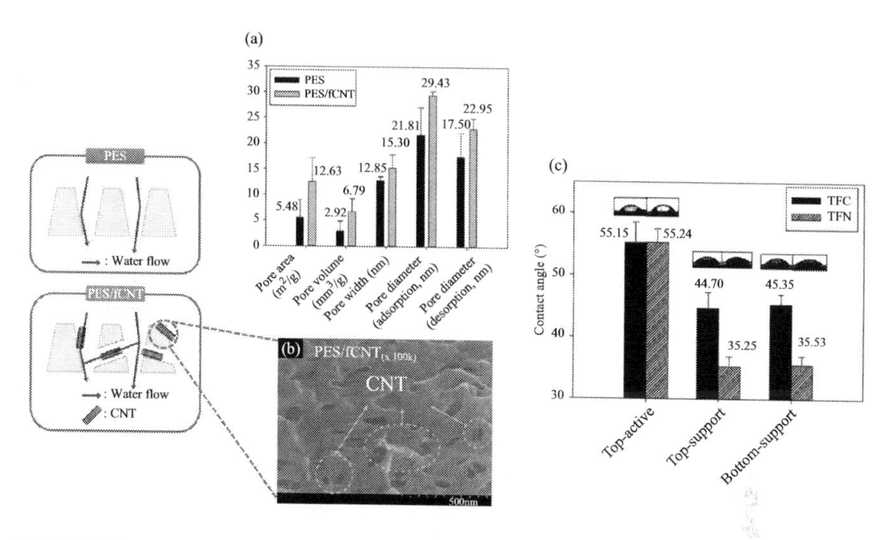

FIGURE 2.8
(a) Porosity of PES and PES/fCNT support layers, (b) SEM image of enlarged cross-section of the PES/fCNT support layer, and (c) water contact angle of top-active, top-support, and bottom-support of TFC and TFN membranes (Son et al. 2016).

Besides, the chemical etching of the active layer could further promote the water flux of the TFN membranes. The water flux and peak power density of the TFN membranes were 87% and 110% greater than those of the control TFC membranes, respectively, using artificial seawater (0.5 M NaCl) and DI water as the feed pair (Son et al. 2016). Tian et al. fabricated a TFC PRO membrane consisting of a tiered structure of polyetherimide nanofibrous support incorporated with functionalized multi-walled CNTs (f-CNTs) and an ultrathin polyamide-based selective layer (Tian et al. 2015). The well-dispersed f-CNTs reinforced the supports with greater mechanical properties and porosity, allowing the support to withstand a high hydraulic pressure in the PRO system. As a result, the optimized membrane could withstand a hydraulic pressure up to 24 bar and produce a peak power density as high as 17.3 W/m^2 at 16.9 bar using synthetic seawater brine (1.0 M NaCl) and DI water as the feed pair (Tian et al. 2015).

Tong et al. (2018) reported the application of freestanding graphene oxide membranes (GOMs) for osmotic power generation. GOM-1, which was the thinnest freestanding GOM synthesized in this work, had much better rigidity and stretch resistance than TFC membranes in terms of tensile stress (at break) and Young's modulus (Tong et al. 2018). Owing to the elimination of the porous support layer, the structure parameter of the freestanding GOMs was only 112 μm, which was much lower than that of the TFC membrane. As a result, the ICP effects of the freestanding GOMs were minimized, resulting in a high water flux and power density during the PRO tests. The freestanding GOMs achieved a power density of 24.62 W/m^2 at a hydraulic pressure of 6.90 bar using 3 M and 0.017 M of NaCl aqueous solutions as the draw and feed solutions, respectively.

Gonzales et al. (2019) incorporated a melamine-based covalent organic framework nanomaterial, Schiff base network-1 (SNW-1), into the polyamide layer and developed a novel TFN membrane for PRO. The incorporation of SNW-1 was achieved via interfacial polymerization on an open mesh fabric-reinforced polyamide-imide substrate. The surface hydrophilicity and water flux of the resultant TFN membranes were significantly enhanced as a result of the incorporation of the porous and highly hydrophilic SNW-1 nanomaterial. The membranes exhibited significantly enhanced surface hydrophilicity, water permeability, and power density. When using 1.0 M NaCl as the draw solution and DI water as the feed solution, the TFN membrane comprising 0.02 wt% SNW-1 showed the highest water flux of 42.5 L/m^2 h and power density of 12.1 W/m^2 at 20 bar.

2.5 Summary and Outlook

According to Statkraft analyses, the minimum power density of flat-sheet membranes for commercially viable PRO processes should be at least 5 W/m^2 (Achilli and Childress 2010, Thorsen and Holt 2009). Various flat-sheet membranes with power density above 5 W/m^2 have been successfully developed, as summarized in Table 2.2. The significant enhancement in PRO performance arises from the advances in membrane technologies to produce (1) polyamide layers with increased transporting properties and (2) porous support layers with greater mechanical property and smaller structural parameter.

However, the employment of flat-sheet membranes for PRO process has several inevitable shortcomings. To maintain the flow channel geometry and enhance mass transfer near the membrane surface, a feed channel spacer has to be installed for flat-sheet membrane modules. The existence of feed spacers would not only cause a hydraulic pressure loss in flow channels but also reduce water flux across the membranes due to the shadow effects of spacers (Kim and Elimelech 2012, She et al. 2013). Moreover, the current feed spacers would unavoidably deform the membranes under high hydraulic pressures during PRO tests, as shown in Figure 2.9 (a) (She et al. 2013). The degree of membrane deformation is closely related to the membrane strength, applied hydraulic pressure,

TABLE 2.2

The PRO performance of the recently developed flat-sheet membranes

Membrane	Draw solution	Feed solution	ΔP (bar)	W (W/m^2)	Reference
TFC-1	Synthetic RO brine (1 M NaCl)	Deionized water	9.0	6.0	(Li and Chung 2013)
TFC-2	Synthetic RO brine (1 M NaCl)	Deionized water	15.0	12.0	(Han et al. 2013)
TFC-3	Synthetic RO brine (1 M NaCl)	Deionized water	22.0	18.1	(Cui et al. 2014)
TFC-4	Synthetic seawater brine (1.06 M NaCl)	Synthetic river water (0.9 mM NaCl)	15.2	21.3	(Song et al. 2013)
TFC-5	Synthetic seawater (0.5 M NaCl)	Deionized water	11.5	8.0	(Bui and McCutcheon 2014)
TFC-6	Synthetic RO brine (1 M NaCl)	Deionized water	20.7	14.1	(Straub et al. 2013)
TFN-1	Synthetic RO brine (1 M NaCl)	Deionized water	21.0	16.7	(Lim et al. 2018)
TFN-2	Synthetic seawater (0.5 M NaCl)	Deionized water	6.0	1.6	(Son et al. 2016)
TFN-3	Synthetic RO brine (1 M NaCl)	Deionized water	16.9	17.3	(Tian et al. 2015)
TFN-4	Synthetic RO brine (1 M NaCl)	Deionized water	20.0	12.1	(Gonzales et al. 2019)
GOMs	Synthetic brine (3 M NaCl)	Synthetic river water (0.017 mM NaCl)	6.9	24.6	(Tong et al. 2018)

and spacer geometry (Kim and Elimelech 2013, She et al. 2013). Besides, as shown in Figure 2.10, Li et al. (2013) found that the TFC membranes were significantly compacted and deformed after being compacted at 14 bar for 250 min. Membrane deformation would not only increase the reverse salt flux of the membranes but also increase their structural parameter. Both factors result in a significant increase of ICP effects and a substantial reduction of both water flux and power density, as illustrated in Figure 2.9 (b) and (c) (She et al. 2013). Therefore, identification of spacers compatible with PRO membranes is of paramount importance for the development of effective flat-sheet PRO membrane modules. The schematic of a spiral wound module with spacers is presented in Figure 2.11 (Schwinge et al. 2004).

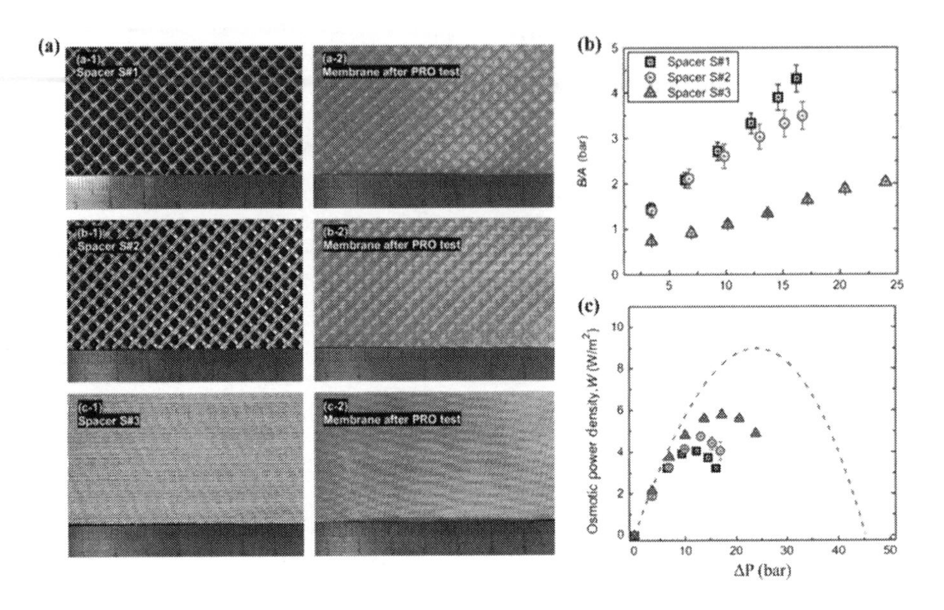

FIGURE 2.9
(a) Membrane deformation after PRO tests for different feed spacers; (b) and (c) membrane separation parameters (B/A), and power density (W) as a function of hydraulic pressure difference (ΔP) for different spacers (She et al. 2013).

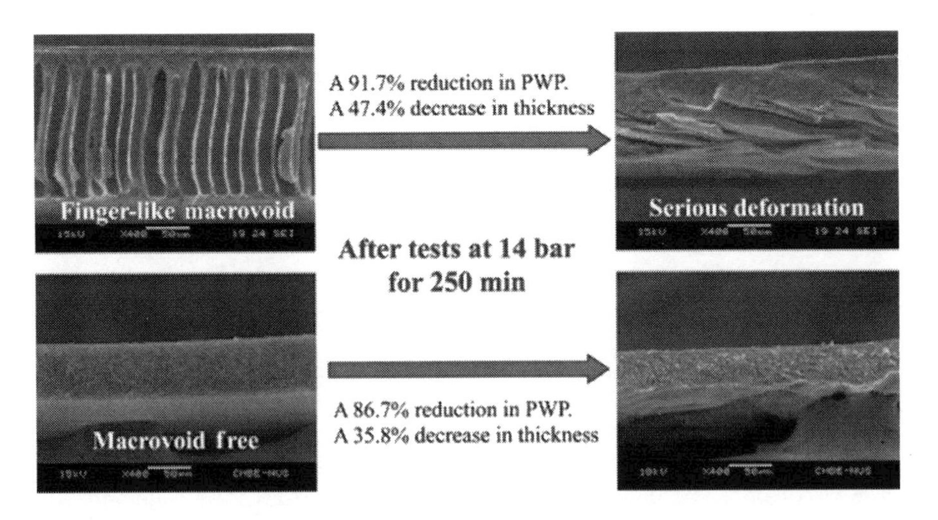

FIGURE 2.10
SEM images of the polyamide-imide membrane before and after being compacted in PRO at 14 bar for 250 min (Li et al. 2013).

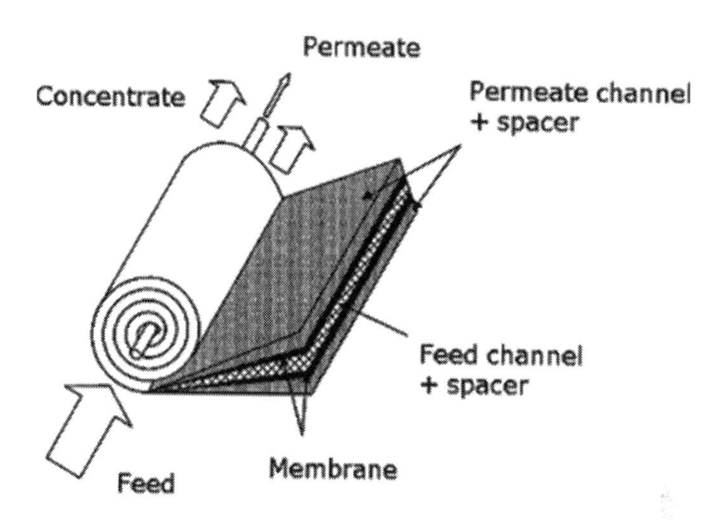

FIGURE 2.11
Schematic of a spiral wound module with spacers (Schwinge et al. 2004).

Acknowledgment

This research study was supported by the National Research Foundation, Prime Minister's Office, Republic of Singapore, under its Environmental & Water Technologies Strategic Research Programme and administered by the Environment & Water Industry Programme Office (EWI) of the PUB.

References

Achilli, A., Cath, T.Y., Childress, A.E., 2009. Power generation with pressure retarded osmosis: an experimental and theoretical investigation. *Journal of Membrane Science* 343, 42–52.

Achilli, A., Childress, A.E., 2010. Pressure retarded osmosis: from the vision of Sidney Loeb to the first prototype installation – review. *Desalination* 261, 205–211.

Alsvik, I., Hägg, M.B., 2013. Pressure retarded osmosis and forward osmosis membranes: materials and methods. *Polymers* 5, 303–327.

Arena, J.T., Manickam, S.S., Reimund, K.K., Brodskiy, P., McCutcheon, J.R., 2015. Characterization and performance relationships for a commercial thin film composite membrane in forward osmosis desalination and pressure retarded osmosis. *Industrial & Engineering Chemistry Research* 54, 11393–11403.

Bui, N.N., McCutcheon, J.R., 2014. Nanofiber supported thin-film composite membrane for pressure-retarded osmosis. *Environmental Science and Technology* 48, 4129–4136.

Cadotte, J.E., Petersen, R.J., Larson, R.E., Erickson, E.E., 1980. A new thin-film composite seawater reverse osmosis membrane. *Desalination* 32, 25–31.

Chung, T.S., Li, X., Ong, R.C., Ge, Q., Wang, H., Han, G., 2012. Emerging forward osmosis (FO) technologies and challenges ahead for clean water and clean energy applications. *Current Opinion in Chemical Engineering* 1, 246–257.

Chung, T.S., Luo, L., Wan, C.F., Cui, Y., Amy, G., 2015. What is next for forward osmosis (FO) and pressure retarded osmosis (PRO). *Separation and Purification Technology* 156, 856–860.

Cui, Y., Liu, X.Y., Chung, T.S., 2014. Enhanced osmotic energy generation from salinity gradients by modifying thin film composite membranes. *Chemical Engineering Journal* 242, 195–203.

Cui, Y., Liu, X.Y., Chung, T.S., 2017. Ultrathin polyamide membranes fabricated from free-standing interfacial polymerization: synthesis, modifications, and post-treatment. *Industrial & Engineering Chemistry Research* 56, 513–523.

Gerstandt, K., Peinemann, K.V., Skilhagen, S.E., Thorsen, T., Holtc, T., 2008. Membrane processes in energy supply for an osmotic power plant. *Desalination* 224, 64–70.

Ghosh, A.K., Hoek, E.M.V., 2009. Impacts of support membrane structure and chemistry on polyamide–polysulfone interfacial composite membranes. *Journal of Membrane Science* 336, 140–148.

Gonzales, R.R., Park, M.J., Bae, T.H., Yang, Y., Abdel-Wahab, A., Phuntsho, S., Shon, H.K., 2019. Melamine-based covalent organic framework-incorporated thin film nanocomposite membrane for enhanced osmotic power generation. *Desalination* 459, 10–19.

Han, G., Chung, T.S., Toriida, M., Tamai, S., 2012a. Thin-film composite forward osmosis membranes with novel hydrophilic supports for desalination. *Journal of Membrane Science* 423–424, 543–555.

Han, G., Zhang, S., Li, X., Chung, T.S., 2013. High performance thin film composite pressure retarded osmosis (PRO) membranes for renewable salinity-gradient energy generation. *Journal of Membrane Science* 440, 108–121.

Han, G., Zhang, S., Li, X., Chung, T.S., 2015. Progress in pressure retarded osmosis (PRO) membranes for osmotic power generation. *Progress in Polymer Science* 51, 1–27.

Han, G., Zhang, S., Li, X., Widjojo, N., Chung, T.S., 2012b. Thin film composite forward osmosis membranes based on polydopamine modified polysulfone substrates with enhancements in both water flux and salt rejection. *Chemical Engineering Science* 80, 219–231.

Helfer, F., Lemckert, C., Anissimov, Y.G., 2014. Osmotic power with pressure retarded osmosis: theory, performance and trends – a review. *Journal of Membrane Science* 453, 337–358.

Hoover, L.A., Schiffman, J.D., Elimelech, M., 2013. Nanofibers in thin-film composite membrane support layers: enabling expanded application of forward and pressure retarded osmosis. *Desalination* 308, 73–81.

Idarraga-Mora, J.A., Childress, A.S., Friedel, P.S., Ladner, D.A., Rao, A.M., Husson, S.M., 2018. Role of nanocomposite support stiffness on TFC membrane water permeance. *Membranes (Basel)* 8, 111.

Jellinek, H.H.G., Masuda, H., 1981. Osmo-power. theory and performance ofan osmo-power pilot plant. *Ocean Engineering* 8(2), 103–128.

Kim, I.C., Jeong, B.R., Kim, S.J., Lee, K.H., 2013. Preparation of high flux thin film composite polyamide membrane: the effect of alkyl phosphate additives during interfacial polymerization. *Desalination* 308, 111–114.

Kim, J., Suh, D., Kim, C., Baek, Y., Lee, B., Kim, H.J., Lee, J.C., Yoon, J., 2016. A high-performance and fouling resistant thin-film composite membrane prepared via coating TiO_2 nanoparticles by sol-gel-derived spray method for PRO applications. *Desalination* 397, 157–164.

Kim, Y.C., Elimelech, M., 2012. Adverse impact of feed channel spacers on the performance of pressure retarded osmosis. *Environmental Science and Technology* 46, 4673–4681.

Kim, Y.C., Elimelech, M., 2013. Potential of osmotic power generation by pressure retarded osmosis using seawater as feed solution: analysis and experiments. *Journal of Membrane Science* 429, 330–337.

Klaysom, C., Cath, T.Y., Depuydt, T., Vankelecom, I.F.J., 2013. Forward and pressure retarded osmosis: potential solutions for global challenges in energy and water supply. *Chemistry Society Review* 42, 6959–6989.

Kong, C., Kanezashi, M., Yamomoto, T., Shintani, T., Tsuru, T., 2010. Controlled synthesis of high performance polyamide membrane with thin dense layer for water desalination. *Journal of Membrane Science* 362, 76–80.

Lee, K.L., Baker, R.W., Lonsdale, H.K., 1981. Membranes for power generationby pressure-retarded osmosis. *Journal of Membrane Science* 8, 141–171.

Li, X., Chung, T.S., 2013. Effects of free volume in thin-film composite membranes on osmotic power generation. *AIChE Journal* 59, 4749–4761.

Li, X., Wang, K.Y., Helmer, B., Chung, T.S., 2012. Thin-film composite membranes and formation mechanism of thin-film layers on hydrophilic cellulose acetate propionate substrates for forward osmosis processes. *Industrial & Engineering Chemistry Research* 51, 10039–10050.

Li, X., Zhang, S., Fu, F., Chung, T.S., 2013. Deformation and reinforcement of thin-film composite (TFC) polyamide-imide (PAI) membranes for osmotic power generation. *Journal of Membrane Science* 434, 204–217.

Lim, S., Park, M.J., Phuntsho, S., Mai-Prochnow, A., Murphy, A.B., Seo, D., Shon, H., 2018. Dual-layered nanocomposite membrane incorporating graphene oxide and halloysite nanotube for high osmotic power density and fouling resistance. *Journal of Membrane Science* 564, 382–393.

Loeb, S., 1976. Production of energy from concentrated brines by pressure-retarded osmosis I. Preliminary technical and economic correlations. *Journal of Membrane Science* 1, 49–63.

Loeb, S., Hessen, F.V., Shahaf, D., 1976. Production of energy from concentrated brines by pressure retarded osmosis. II. Experimental results and projected energy costs. *Journal of Membrane Science* 1, 249–269.

Loeb, S., Mehta, G.D., 1979. A two-coefficient water transport equation for pressure-retarded osmosis. *Journal of Membrane Science* 4, 351–362.

Mansourpanah, Y., Madaeni, S.S., Rahimpour, A., 2009. Fabrication and development of interfacial polymerized thin-film composite nanofiltration membrane using different surfactants in organic phase; study of morphology and performance. *Journal of Membrane Science* 343, 219–228.

McCutcheon, J.R., Elimelech, M., 2008. Influence of membrane support layer hydrophobicity on water flux in osmotically driven membrane processes. *Journal of Membrane Science* 318, 458–466.
Mehta, G.D., 1982. Further results on the performance of present-day osmotic membranes in various osmotic regions. *Journal of Membrane Science* 10, 3–19.
Mehta, G.D., Loeb, S., 1978. Internal polarization in the porous substructure of a semipermeable membrane under pressure-retarded osmosis. *Journal of Membrane Science* 4, 261–265.
Mehta, G.D., Loeb, S., 1979. Performance of permasep B-9 and B-10 membranes in various osmotic regions and at high osmotic pressures. *Journal of Membrane Science* 4, 335–349.
Reneker, D.H., Chun, I., 1996. Nanometre diameter fibres of polymer, produced by electrospinning. *Nanotechnology* 7, 216–223.
Schwinge, J., Neal, P.R., Wiley, D.E., Fletcher, D.F., Fane, A.G., 2004. Spiral wound modules and spacers. *Journal of Membrane Science* 242, 129–153.
Shao, L., Chung, T.S., Goh, S.H., Pramoda, K.P., 2004. Transport properties of cross-linked polyimide membranes induced by different generations of diaminobutane (DAB) dendrimers. *Journal of Membrane Science* 238, 153–163.
She, Q., Hou, D., Liu, J., Tan, K.H., Tang, C.Y., 2013. Effect of feed spacer induced membrane deformation on the performance of pressure retarded osmosis (PRO): implications for PRO process operation. *Journal of Membrane Science* 445, 170–182.
She, Q., Jin, X., Tang, C.Y., 2012. Osmotic power production from salinity gradient resource by pressure retarded osmosis: effects of operating conditions and reverse solute diffusion. *Journal of Membrane Science* 401–402, 262–273.
Singh, P.S., Joshi, S.V., Trivedi, J.J., Devmurari, C.V., Rao, A.P., Ghosh, P.K., 2006. Probing the structural variations of thin film composite RO membranes obtained by coating polyamide over polysulfone membranes of different pore dimensions. *Journal of Membrane Science* 278, 19–25.
Skilhagen, S.E., Dugstad, J.E., Aaberg, R.J., 2008. Osmotic power – power production based on the osmotic pressure difference between waters with varying salt gradients. *Desalination* 220, 476–482.
Son, M., Park, H., Liu, L., Choi, H., Kim, J.H., Choi, H., 2016. Thin-film nanocomposite membrane with CNT positioning in support layer for energy harvesting from saline water. *Chemical Engineering Journal* 284, 68–77.
Song, X., Liu, Z., Sun, D.D., 2013. Energy recovery from concentrated seawater brine by thin-film nanofiber composite pressure retarded osmosis membranes with high power density. *Energy & Environmental Science* 6, 1199–1210.
Straub, A.P., Yip, N.Y., Elimelech, M., 2013. Raising the bar: increased hydraulic pressure allows unprecedented high power densities in pressure-retarded osmosis. *Environmental Science & Technology Letters* 1, 55–59.
Thorsen, T., Holt, T., 2009. The potential for power production from salinity gradients by pressure retarded osmosis. *Journal of Membrane Science* 335, 103–110.
Tian, M., Wang, R., Goh, K., Liao, Y., Fane, A.G., 2015. Synthesis and characterization of high-performance novel thin film nanocomposite PRO membranes with tiered nanofiber support reinforced by functionalized carbon nanotubes. *Journal of Membrane Science* 486, 151–160.
Tin, P.S., Chung, T.S., Hill, A.J., 2004. Advanced fabrication of carbon molecular sieve membranes by nonsolvent pretreatment of precursor polymers. *Industrial Engineering Chemistry Research* 43, 6476–6483.

Tong, X., Wang, X., Liu, S., Gao, H., Xu, C., Crittenden, J., Chen, Y., 2018. A freestanding graphene oxide membrane for efficiently harvesting salinity gradient power. *Carbon* 138, 410–418.

Xu, Y., Peng, X., Tang, C.Y., Fu, Q.S., Nie, S., 2010. Effect of draw solution concentration and operating conditions on forward osmosis and pressure retarded osmosis performance in a spiral wound module. *Journal of Membrane Science* 348, 298–309.

Yip, N.Y., Elimelech, M., 2011. Performance limiting effects in power generation from salinity gradients by pressure retarded osmosis. *Environmental Science & Technology* 45, 10273–10282.

Yip, N.Y., Tiraferri, A., Phillip, W.A., Schiffman, J.D., Hoover, L.A., Kim, Y.C., Elimelech, M., 2011. Thin-film composite pressure retarded osmosis membranes for sustainable power generation from salinity gradients. *Environmental Science & Technology* 45, 4360–4369.

Zhang, S., Chung, T.S., 2013. Minimizing the instant and accumulative effects of salt permeability to sustain ultrahigh osmotic power density. *Environmental Science & Technology* 47, 10085–10092.

Zhang, S., Fu, F., Chung, T.S., 2013. Substrate modifications and alcohol treatment on thin film composite membranes for osmotic power. *Chemical Engineering Science* 87, 40–50.

Zhao, S., Zou, L., Tang, C.Y., Mulcahy, D., 2012. Recent developments in forward osmosis: opportunities and challenges. *Journal of Membrane Science* 396, 1–21.

Zuo, J., Wang, Y., Sun, S.P., Chung, T.S., 2012. Molecular design of thin film composite (TFC) hollow fiber membranes for isopropanol dehydration via pervaporation. *Journal of Membrane Science* 405–406, 123–133.

3

Recent Development of Hollow Fiber PRO Membranes

Wenxiao Gai
Department of Chemical and Biomolecular Engineering
National University of Singapore
Singapore

Gang Han
Department of Chemical Engineering
Massachusetts Institute of Technology
Cambridge, Massachusetts

Tai-Shung Chung
Department of Chemical and Biomolecular Engineering
National University of Singapore
Singapore

CONTENTS

3.1 Introduction

Hollow fiber membranes were originally developed for reverse osmosis applications in the 1960s. Since then, they have become prevalent in various fields, such as water treatment, desalination, cell culture, medicine, and tissue engineering. In comparison with flat-sheet membranes, the hollow fiber membranes have several advantages, such as higher surface area per module, self-mechanical support, and ease of module fabrication (Peng et al. 2012). They could not only eliminate the membrane-spacer interactions under high hydraulic pressures in pressure retarded osmosis (PRO) operations (She et al. 2013), but also mitigate the extra energy loss in the feed flow channel of flat-sheet modules. The self-support configuration and high packing density in modules make the hollow fiber membranes potentially suitable for PRO processes (Sivertsen et al. 2012, 2013). Currently, both integrally formed phase inversion dual-layer membranes and thin-film composite (TFC) hollow fiber membranes have been explored for osmotic power generation. An ideal PRO membrane should possess the characteristics, such as (1) a thin selective layer allowing a high water flux with a reasonably low reverse salt flux; (2) a robust support layer to withstand high hydraulic pressures with small transport resistance; and (3) low affinity to foulants to maintain high power output during operation (Bui and McCutcheon 2014, Chung et al. 2012, Cui et al. 2014, She et al. 2012, Song et al. 2013, Thorsen and Holt 2009, Yip et al. 2011).

Hollow fiber membranes are usually constructed to have a microporous tubular substrate and a dense selective layer on their inside or outside surface. Therefore, there are two types of hollow fiber membranes, which are inner-selective ones and outer-selective ones. The most common method to fabricate hollow fiber membranes is by means of a dry-jet wet spinning process. As illustrated in Figure 3.1, a polymer solution, usually called the "spinning dope," is extruded via a spinneret into air first and then immersed into a coagulation bath that usually contains water (Bonyadi et al. 2007, Widjojo and Chung 2006). The spinneret consists of a needle and an annulus flow channel. A solvent

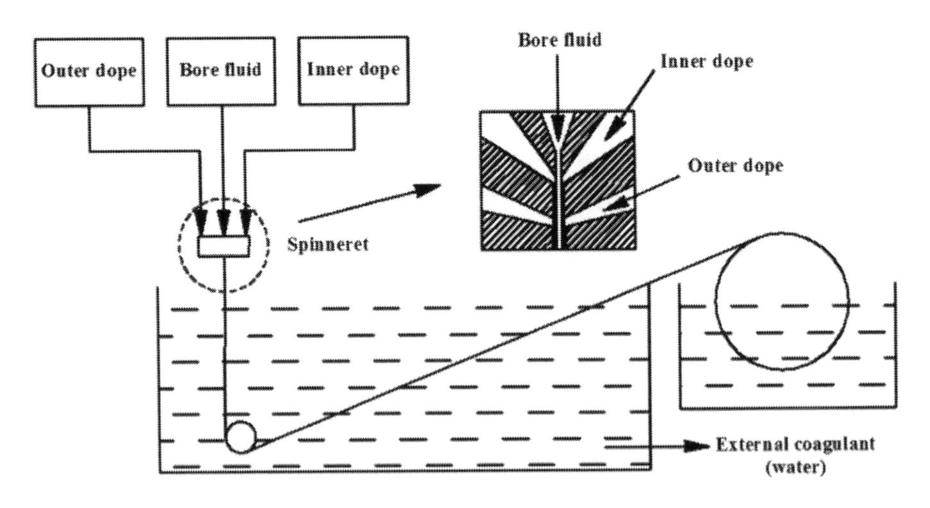

FIGURE 3.1
Schematic of a dry-jet wet spinning process.

called "bore fluid" is extruded through the lumen of the needle to induce the internal coagulation and form the "hollow" part of the hollow fiber, while the dope is extruded through the annulus channel so that the nascent membrane has a hollow cylindrical shape (Li et al. 2002, Peng and Chung 2008). As the nascent hollow fiber enters the coagulant bath, it solidifies and forms a solid hollow fiber by means of the nonsolvent-induced phase inversion process. The properties of the as-spun hollow fiber, such as morphology, average pore size, and membrane thickness, could be manipulated by changing the spinneret dimension, dope temperature, and composition, bore-fluid composition, air gap distance, coagulation temperature, and compositions, as well as take-up speed (Jiang et al. 2005, Li et al. 2008, Peng et al. 2008, Santoso et al. 2006). Extrusion of the dope and bore fluid through the spinneret can be conducted by either using gas-assistant extrusion or precision pumps.

Similar to flat-sheet PRO membranes, both phase inversion and interfacial polymerization approaches have been employed to prepare the selective layers of hollow fiber PRO membranes (Chou et al. 2013, Fu et al. 2013, 2014, Li et al. 2015, Zhang et al. 2014). The hollow fiber PRO membranes with selective layers fabricated by phase inversion usually suffer from (1) a low water flux as a result of a thick selective layer and (2) limited polymeric materials available for membrane development (Fu et al. 2014, Han et al. 2015). Since the interfacial polymerization approach has the advantage of tuning the support layer and the selective layer separately, it enables a higher degree of control on membrane morphology and performance. The resultant TFC membrane could have an ultrathin polyamide-selective layer on top of a porous and robust support layer, achieving a high water flux and a low reverse salt flux (Chou et al. 2013, Han et al. 2013, Li and Chung 2014, Li et al. 2015, Sun and Chung 2013, Zhang et al. 2014). Thus, most recent studies have been devoted to the TFC hollow fiber membranes.

3.2 Integrally Skinned Hollow Fiber Membranes

Integrally skinned asymmetric (ISA) hollow fiber membranes are directly prepared by forming a microporous substrate and a dense selective layer simultaneously via phase inversion (Fu et al. 2013, 2014). This kind of facile phase inversion process is paramount to fabricate the outer-selective hollow fiber membranes, as it is tremendously difficult to fabricate the outer-selective hollow fiber membranes via interfacial polymerization process with a defect-free selective layer (Sun and Chung 2013). In other words, since the hollow fiber substrates have to be packed into bundles and potted into tubes to form a membrane module before interfacial polymerization, contact among them is inevitable, making it difficult to form a perfect polyamide layer on the outer surface of each substrate.

Academically, Fu et al. (2013) are the pioneers to develop the integrally skinned outer-selective hollow fiber membranes utilizing the co-extrusion technique through a dual-layer spinneret. As shown in Figure 3.2, the fabricated dual-layer hollow fiber membrane consists of an outer PBI (polybenzimidazole)/POSS (polyhedral oligomeric silsesquioxane) dense selective layer and an inner PAN (polyacrylonitrile)/PVP (polyvinylpyrrolidone) support layer with a sponge-like morphology. PBI is a promising material to fabricate the dense selective layer for PRO membranes since it is hydrophilic and chemically stable. Owing to the good mechanical properties and thermal stability, PAN was used to prepare the supporting substrate to provide the mechanical strength for the PRO membranes. It was reported that the incorporation of an appropriate amount of POSS nanoparticles into the PBI outer selective layer had

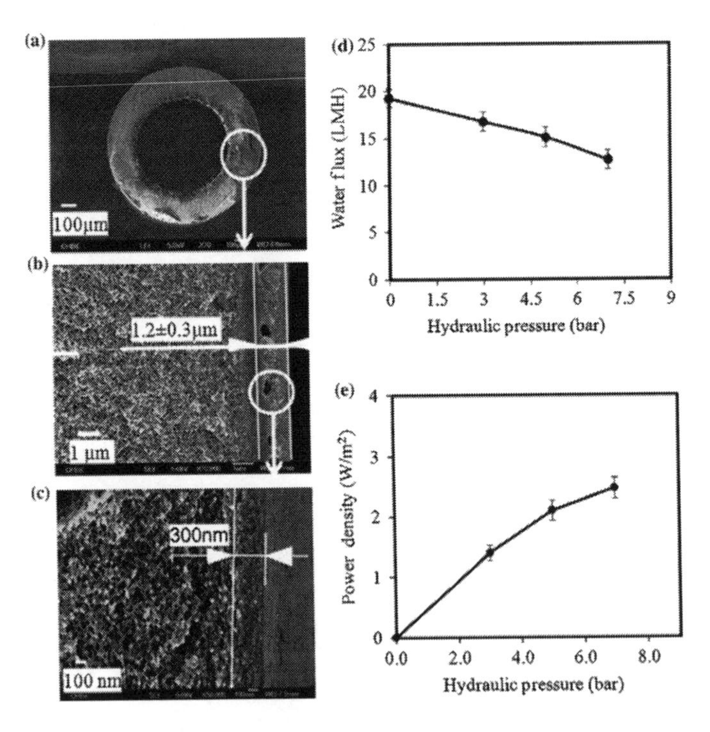

FIGURE 3.2
Morphology and PRO performances of the PBI/POSS–PAN/PVP hollow fiber membrane with a POSS loading of 0.5 wt%. (a) Cross-section of the membrane; (b) cross-section of interfaces; (c) cross-section of the outer selective layer; and (d) and (e) water flux and power density as a function of hydraulic pressures (Draw solution: 1 M NaCl; feed solution: 0.01 M NaCl) (Fu et al. 2013).

significant influences on its morphology and performance. The addition of PVP into the PAN inner support layer was reported to mitigate the delamination between the two layers because of its good compatibility with both PAN and PBI (Fu et al. 2013). Later, the PRO performance of the PBI/PAN dual-layer hollow fiber membranes was further improved by a post-treatment using an ammonium persulfate (APS) solution and water flowing counter – currently through the membranes to remove the PVP molecules entrapped in the PAN substrate while maintaining the integrity of the PBI/PAN interface (Fu et al. 2014). After the APS post-treatment, both the water flux and reverse salt flux increased; therefore, one had to balance these two parameters carefully to ensure a final net increase in power density. When the APS concentration was 5 wt%, the post-treated membrane achieved a maximum power density of 5.10 W/m^2 at 15 bar when employing 1 M NaCl and 10 mM NaCl as the draw and feed solutions, respectively (Fu et al. 2014). The fabrication of PRO membranes via nonsolvent-induced phase inversion is convenient and cost-effective, but the resultant membranes suffer from a low water flux and thus a low power density due to the low porosity of the polymeric support layer.

Li et al. developed an integral hollow fiber membrane by employing a facile cross-linking post treatment (Li et al. 2018). The results showed that the hollow fiber substrates with a sponge-like morphology and a relatively dense inner surface can tolerate a higher hydraulic pressure in PRO processes. Two steps of cross-linking endowed the hollow fiber membranes with a smaller pore size on the outer surface exhibiting high rejections against various inorganic salts. The novel integral hollow fiber membranes could achieve a stable power density of around 4.3 W/m^2 at 12–13 bar, using a real wastewater reverse osmosis (RO) retentate and 1.0 M sodium chloride as the feed and draw solutions, respectively (Li et al. 2018). Besides, their work also demonstrated that PRO operated under the active layer facing feed solution (AL–FS) mode could offer a significant advantage over the active layer facing draw solution (AL–DS) mode by alleviating membrane fouling during the osmotic power generation.

Cho et al. (2019) developed mechanically robust ISA hollow fiber membranes by combining the nonsolvent-induced phase separation (NIPS) and the thermally induced phase separation (TIPS). The new thermally assisted NIPS (T-NIPS) method aimed to combine the advantages of NIPS and TIPS methods, such as selective skin formation, high mechanical strength, and high porosity. The prepared T-NIPS membranes consisted of two distinct layers: a dense ISA outer layer and an isoporous inner layer. Comparing with conventional NIPS hollow fiber membranes, the T-NIPS ones could achieve a much higher power density and burst pressure, up to 5.5 W/m^2 and 18 bar, respectively (Cho et al. 2019). Besides, polyethylene glycols with various molecular weights and concentrations were employed as a pore-forming agent in the T-NIPS process to further promote the PRO performance. It was found that the water flux and power density of the membranes could be enhanced without sacrificing the pressure tolerance by controlling the pore structure of both the outer and inner layers. It was also noticed that the membrane dimensions played a significant role in determining the membrane performance. Optimization of membrane dimension was important to achieve high PRO performance by balancing the trade-off relationship between power density and pressure tolerance.

3.3 TFC Hollow Fiber Membranes

The TFC hollow fiber membranes, which are constructed with a highly porous polymeric substrate and an ultrathin polyamide-selective layer, could provide a high water flux with a low reverse salt flux during PRO processes. However, the fabrication of the TFC membranes is more sophisticated than that of the integrally skinned hollow fiber membranes. Recently, both inner-selective and outer-selective polyamide TFC hollow fiber membranes with good mechanical property and water permeability have been developed for PRO processes, these breakthroughs not only bring the PRO membranes with a higher power density but also make them closer to commercialization.

3.3.1 Outer-Selective TFC Hollow Fiber Membranes

Sun et al. are pioneers to develop outer-selective TFC hollow fiber membranes for PRO processes. They overcame the challenges and successfully fabricated a defect-free polyamide-selective layer on the outer surface of Matrimid® hollow fiber substrates by conducting a vacuum-assisted interfacial polymerization (Sun and Chung 2013). Figure 3.3 illustrates the scheme to produce the outer-selective TFC hollow fiber membrane. To prepare the polyamide selective layer, a vacuum pressure of 800 mbar was applied from the lumen side of a hollow fiber bundle to remove the excess amine solution. As shown in Figure 3.4 (b), the resultant polyamide layer had the typical "ridge and valley" morphology. The outer-selective TFC PRO hollow fiber membranes were further improved by optimizing the Matrimid® hollow fiber substrates in terms of pore size and mechanical strength and then modifying the polyamide selective layer with a polydopamine coating and molecular engineering of the interfacial polymerization solution. The resultant membranes could withstand a hydraulic pressure of 20 bar and achieve a peak power density of 7.63 W/m^2 when using 1 M NaCl as draw solution and deionized (DI) water as feed solution (Sun and Chung 2013). The performance was equivalent to 13.72 W/m^2, achieved by the inner-selective hollow fiber

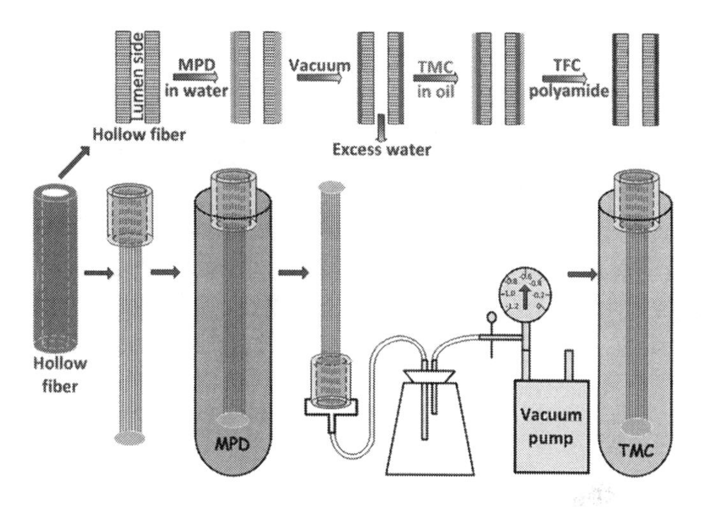

FIGURE 3.3
The schematic of a vacuum-assisted interfacial polymerization for preparing the outer-selective TFC hollow fiber membranes (Sun and Chung 2013).

counterpart with the same module size, packing density, and fiber dimensions (Sun and Chung 2013).

Le et al. (2016) manipulated the dope composition and spinning parameters to fabricate suitable hollow fiber substrates with a mixed finger- and sponge-like morphology, which could provide strong mechanical strength and low internal concentration polymerization effects for a good PRO performance. Then the hollow fiber substrates were pre-wetted with an alcohol prior to the interfacial polymerization to form a smooth and thin polyamide layer, which was favorable for a high water flux and power density. The alcohol with a lower surface tension and a closer solubility parameter to the substrate material provided more efficient pre-wetting effects, so that the *m*-phenylenediamine aqueous solution could be distributed more uniformly. The polyamide selective layer on the outer surface of the Ultem® hollow fiber substrates was fabricated via a dip-coating approach before being potting into lab-scale modules. The TFC membranes pre-wetted with *n*-propanol showed the optimal performance with a peak power density of 9.59 W/m^2, which was achieved at 17 bar using 1 M NaCl as the draw solution and DI water as the feed solution (Le et al. 2016).

Cheng et al. (2016) developed a simplified vacuum-assisted interfacial polymerization approach to fabricate the outer-selective TFC hollow fiber membranes with an extremely low reverse salt flux and robustness for harvesting salinity-gradient energy, as illustrated in Figure 3.5. Nearly defect-free polyamide layers with an impressive high salt rejection were

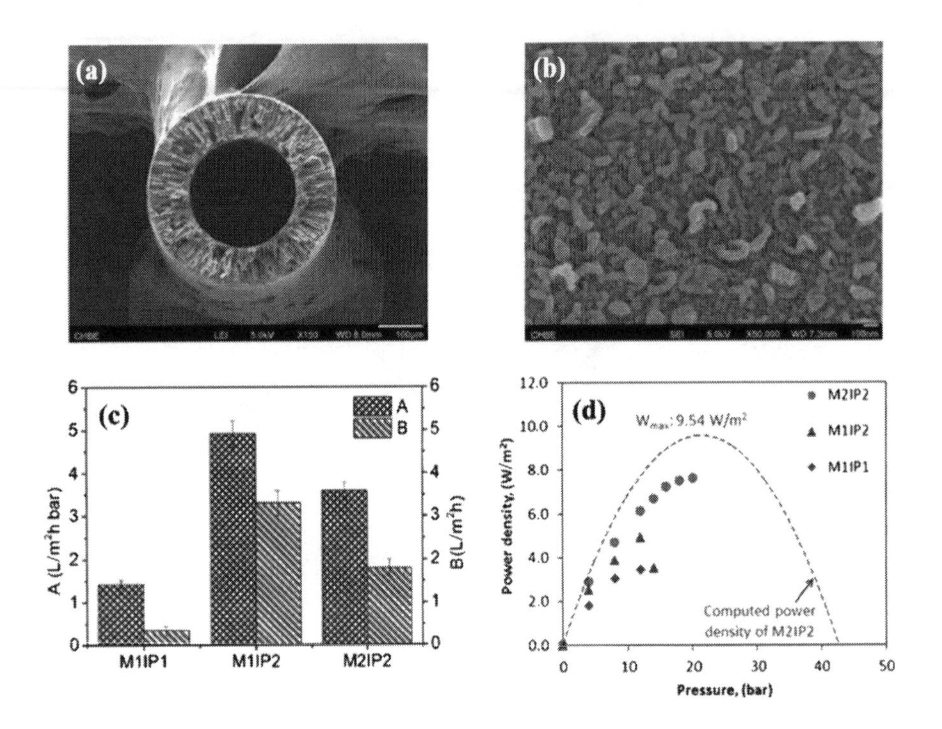

FIGURE 3.4
(a) Cross-section morphology of the Matrimid® hollow fiber substrate, (b) outer surface morphology of the outer-selective TFC hollow fiber membrane, (c) water permeability (A) and salt permeability (B) of the outer-selective TFC hollow fiber membranes tested under RO mode with a constant pressure of 1 bar using 200 ppm NaCl aqueous solution as the feed, (d) power density (W) of the outer-selective TFC hollow fiber membrane as a function of hydraulic pressures (Draw solution: 1 M NaCl; Feed solution: DI water) (Sun and Chung 2013).

prepared on top of the outer surface of the polyethersulfone (PES) porous hollow fiber substrates with the aid of a vacuum pressure of 800 mbar to remove excess *m*-phenylenediamine residuals during interfacial polymerization. The newly developed TFC-II membrane had a maximum power density of 7.81 W/m^2 at 20 bar using 1 M NaCl and DI water as the draw and feed solutions, respectively (Cheng et al. 2016). As compared with the other reported PRO membranes, this new membrane showed not only the smallest slope between water flux decline and hydraulic pressure increase but also the lowest ratio of reverse salt flux to water flux (Cheng et al. 2016). Both characteristics would help maintain an effective osmotic driving force even under high pressure operations. Later, Cheng et al. further promoted the membrane performance by tuning the water content in polymer dopes, which could simultaneously enhance the mechanical robustness and water transport properties of the membranes. With a water content of 2 wt% in the

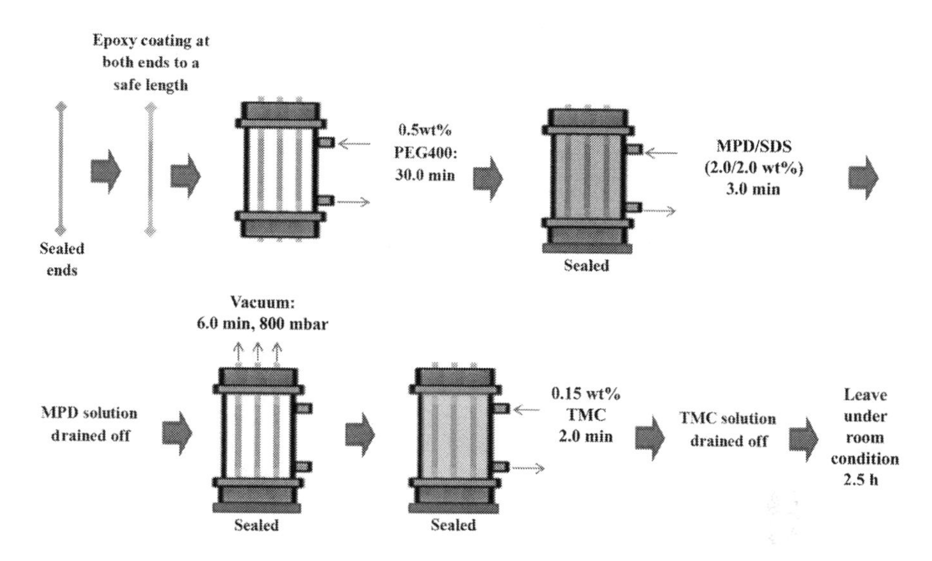

FIGURE 3.5
The schematic of a simplified vacuum-assisted interfacial polymerization approach to fabricate the outer-selective TFC hollow fiber membranes (Cheng et al. 2016).

polymer dope, the outer-selective TFC membrane not only had the smallest structural parameter, highest toughness, and largest water permeability among all membranes studied but also displayed a peak power density of 10.05 W/m^2 at 22 bar using 1 M NaCl and DI water as the draw and feed solutions, respectively (Cheng et al. 2017).

In comparison with the inner-selective hollow fiber membranes, the outer-selective ones show the advantages of (1) a lower pressure drop in the pressurized draw solution side for the actual PRO applications, (2) a more surface area of selective layers per module, (3) possibly less external concentration polarization, and (4) higher ease of fouling control on the outer surface of hollow fiber membranes (Le et al. 2016, Naguib et al. 2015, Ren and McCutcheon 2015, Sivertsen et al. 2013, Sun and Chung 2013, Zhang and Chung 2016). In addition, the outer selective configuration is more mechanically resistant to pressurize from the shell side of modules. However, the reported outer-selective TFC hollow fiber membranes still possess lower PRO performance than the inner-selective ones due to the difficulties of forming a defect-free polyamide layer on the outer surface of each hollow fiber substrate.

3.3.2 Inner-Selective TFC Hollow Fiber Membranes

Han et al. have developed a series of inner-selective TFC hollow fiber membranes with polyamide-selective layers formed on the lumen side of the well-constructed Matrimid® hollow fiber substrates for PRO applications (Han and Chung 2014, Han et al. 2013). The optimal TFC hollow fiber membranes exhibited a power density of 16.5 W/m^2 at 15 bar when using 1 M NaCl and DI water as the feed pair (Han and Chung 2014). In their work, new fabrication perspectives and design strategies were demonstrated to molecularly construct robust support layers. By manipulating the chemistry of polymer solutions and the kinetics of phase inversion processes, hollow fiber substrates with different micromorphology and mechanical strengths were fabricated, as shown in Figure 3.6. It could be observed that the strength of the microstructure of the hollow fiber substrates may dominate the overall membrane robustness instead of the apparent cross-section morphology. In addition, their work also illustrated that pre-stabilization of the inner-selective TFC hollow fiber membranes at a high hydraulic pressure could not only enhance the water permeability with a slight decrease in selectivity but also reduce the structural parameter of the substrate, as presented in Figure 3.7. A similar phenomenon was also observed by Gai et al. in which the peak power density of the inner-selective TFC hollow fiber membranes increased from 15.37 to 22.05 W/m^2 after pre-stabilization at 20 bar for 30 minutes when using 1 M NaCl and DI water as the feed pair (Gai et al. 2016). This phenomenon could be attributed to the membrane expansion induced by the high hydraulic pressure applied in the lumen side, resulting in an increased membrane surface area, stretched

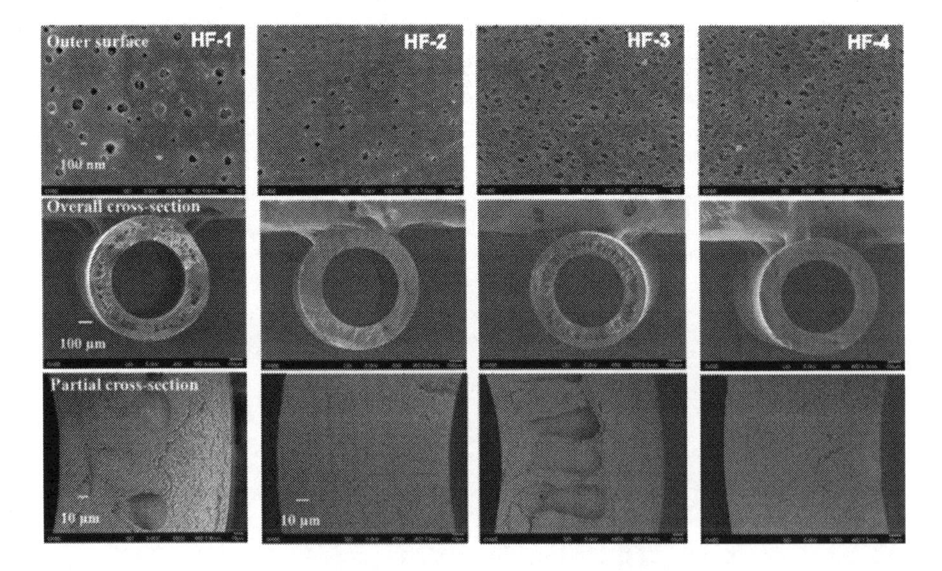

FIGURE 3.6
SEM images of the outer surface and cross-section of the Matrimid® hollow fiber membrane substrates spun under different conditions (Han and Chung 2014).

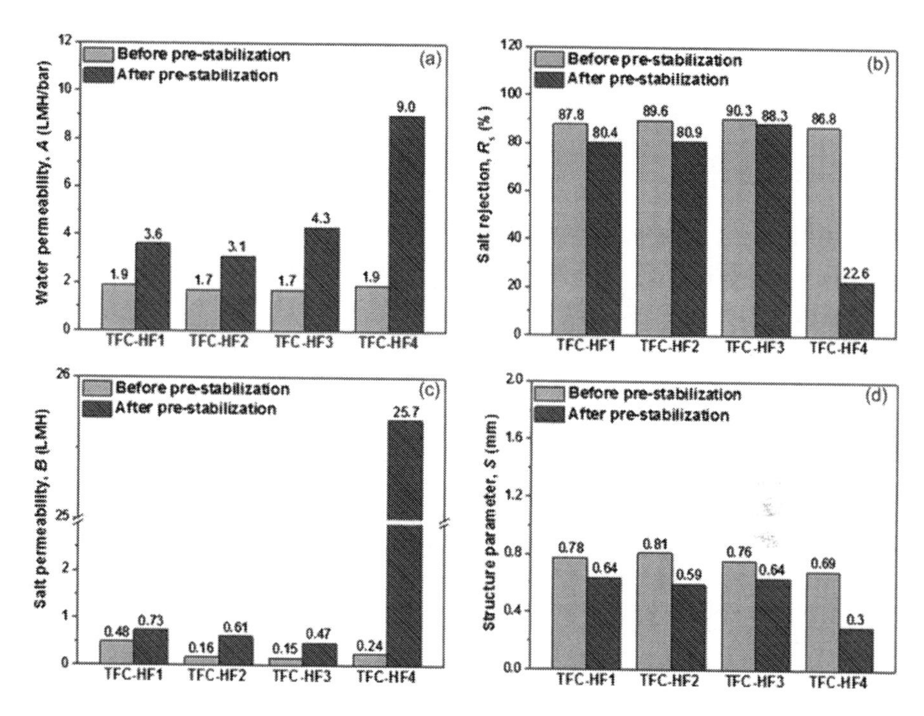

FIGURE 3.7
The effects of pre-stabilization on the transport properties of the inner-selective TFC hollow fiber membranes with different substrates are shown in Figure 3.6 (the stabilization pressures were 16, 15, 15, and 6 bar, respectively) (Han and Chung 2014).

polyamide-selective layer and lower tortuosity, as illustrated in Figure 3.8 (a) (Gai et al. 2016, Han and Chung 2014).

Chou et al. (2012) also developed an inner-selective PES TFC hollow fiber membrane with a relatively large lumen diameter of 0.98 mm, which could achieve a peak power density of 10.6 W/m^2 at 9 bar when using 1 M NaCl and 40 mM NaCl as the draw and feed solutions, respectively. To improve the power density, Chou et al. (2013) replaced PES with polyether-imide to fabricate the hollow fiber substrates. In addition, the fiber diameter was reduced and the cross-section morphology was tuned from a finger-like to fully sponge-like morphology. The resultant inner-selective TFC hollow fiber membrane had a power density of 20.9 W/m^2 at 15 bar using 1 M NaCl and 1 mM NaCl as the feed pair (Chou et al. 2013). However, the salt permeability and structural parameter of the TFC hollow fiber membrane increased noticeably with an increase in hydraulic pressure imposed upon the fiber lumen (Chou et al. 2013).

Li et al. strategically designed a series of P84 co-polyimide hollow fiber membranes with various structures, dimensions, pore characteristics, and

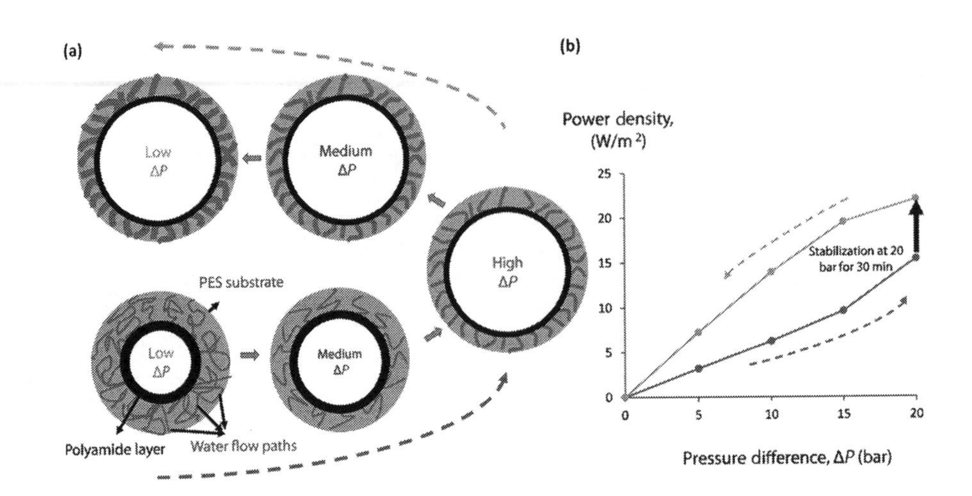

FIGURE 3.8
(a) Schematic of the effects of pre-stabilization on the inner-selective TFC hollow fiber membranes; (b) power density of the inner-selective TFC hollow fiber membranes before and after being stabilized at 20 bar for 30 min using 1 M NaCl and DI water as the feed pair (Gai et al. 2016).

mechanical properties as the substrates for inner-selective TFC membranes by manipulating the phase inversion process during spinning (Li and Chung 2014). The addition of *N*-methyl-2-pyrrolidinone and ethylene glycol into the bore fluid could promote delayed demixing, increase the interconnectivity of pores, and broaden the pore size distribution of the hollow fiber substrates. The TFC hollow fiber membranes with a small dimension, which were made by the delayed demixing, not only had better mechanical properties and a higher burst pressure, but also showed a power density of 12 W/m^2 at 21 bar using DI water and 1 M NaCl as the feed pair (Li and Chung 2014).

Zhang et al. developed highly asymmetric TFC hollow fiber membranes that showed both high water flux and high mechanical strength to effectively harvest the osmotic energy by molecularly engineering the PES hollow fiber substrates (Zhang and Chung 2013, Zhang et al. 2014). Hollow fiber substrates with various morphologies from macrovoid to sponge-like were fabricated by varying the water content in dope solutions. As illustrated in Figure 3.9, the asymmetric PES hollow fiber substrates comprised three layers: (1) a highly porous outer part, (2) a layer of macrovoids at the inner part, and (3) a relatively dense and thin sponge-like layer with small surface pores on the lumen side. The highly porous outer part and macrovoids could reduce the internal concentration polymerization effects; the relatively dense and thin sponge-like layer underneath the polyamide layer was beneficial for maintaining good mechanical stability and stress

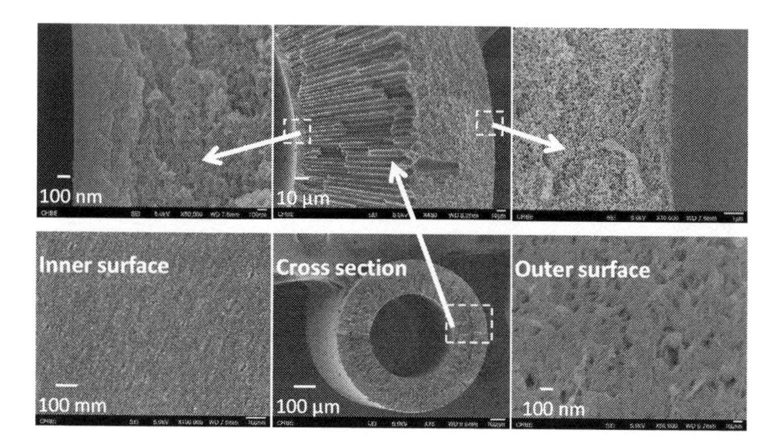

FIGURE 3.9
SEM images of the cross-section and surface morphologies of the PES hollow fiber supports (Zhang and Chung 2013).

dissipation; and the small pore size on the inner surface helped form a less-defective and mechanically stable polyamide layer. As a result, the newly developed TFC hollow fiber membranes achieved a peak power density of 24.0 W/m^2 at 20 bar using 1 M NaCl as draw solution and DI water as feed solution (Zhang et al. 2014).

Gai et al. (2018) developed a novel TFC hollow fiber membrane by incorporating carbon quantum dots (CQDs) into the TFC layer for osmotic power generation. CQDs are a new class of carbon nanomaterials with characteristics of excellent hydrophilicity, easy synthesis, and low cost. CQDs were incorporated into the polyamide layers via the interfacial polymerization reaction by dissolving them into the *m*-phenylenediamine aqueous solution. The addition of Na^+-functionalized CQDs could not only increase the existence of hydrophilic oxygen-containing groups and the effective surface area of the polyamide layer but also change the membrane morphology with a looser and thinner polyamide network. The TFC membrane comprising 1 wt% Na–CQD-9 could achieve a peak power density as high as 34.2 W/m^2 at 23 bar using 1.0 M NaCl and DI water as the draw and feed solutions, respectively (Gai et al. 2018). Compared with the control, the water flux and power density at 23 bar were increased by 20.3%, while the reverse salt flux remained unchanged.

Wan et al. modified Zhang et al.'s membranes (Zhang et al. 2014) and developed a TFC PES membrane with a peak density of 27 W/m^2 at 20 bar using 1.0 M NaCl and DI water as the feed pair (Wan and Chung 2015). Recently, they further enhanced the PRO performance by adding an inorganic salt ($CaCl_2$) into the dope solution and fine-tuning the spinning conditions (Wan et al. 2018). The addition of $CaCl_2$ at an appropriate amount could not only increase the dope viscosity but also reduce the mean pore size and narrow the pore size distribution of the hollow fiber substrates. Besides, a high flow rate ratio of dope solution to bore fluid could effectively enhance the mechanical strength of the PES substrates. The resultant TFC membrane could withstand a hydraulic pressure more than 35 bar and produce a peak power density of 38 W/m^2 at 30 bar by using 1.2 M NaCl and DI water as the draw and feed solutions, respectively (Wan et al. 2018). The TFC membranes with a higher withstanding pressure and a higher power density enable the utilization of high salinity draw solutions in the PRO process and taking full advantage of the increased osmotic gradient for greater osmotic energy production.

Park et al. developed novel TFC hollow fiber membranes by incorporating hydrophilic graphene oxide (GO) nanosheets into the hollow fiber PES substrates for osmotic power generation. The addition of GO (≤0.2 wt%) into the PES substrates could lead to noticeable improvements in pure water permeability, structural morphologies, and hydrophilicity within the support layer without deteriorating the mechanical properties. The TFC hollow fiber membranes incorporated with 0.2 wt% GO had an initial PRO flux (without any applied pressure) of 43.74 L/m^2 h and a specific reverse salt flux of 0.04 g/L (Park et al. 2019). The optimal membrane in their work could achieve a peak power density of 14.6 W/m^2 at an operating pressure of 16.5 bar using DI water and 1 M NaCl as the feed pair (Park et al. 2019).

3.4 Summary and Outlook

Recently, both inner- and outer-selective TFC hollow fiber membranes have been successfully developed for the PRO process to harvest osmotic energy, as summarized in Table 3.1. Significant breakthroughs have been made in terms of (1) approaches to fabricate defect-free polyamide-selective layers for outer-selective TFC membranes and (2) water permeability and mechanical strength for the inner-selective TFC membranes. Effective TFC hollow fiber membranes with proper membrane structure, balanced permeation properties, and high mechanical strength have been demonstrated. The best membrane can withstand a hydraulic pressure of more than 35 bar and harvest a power density of 38 W/m^2 at 30 bar using 1.2 M NaCl and DI water as the feed pair (Wan et al. 2018). The newly developed inner-selective TFC hollow fiber membranes show great potential for osmotic power generation.

TABLE 3.1

The PRO performance of the recently developed hollow fiber membranes

Membrane	Configuration	Draw Solution	Feed Solution	ΔP (bar)	W (W/m^2)	Reference
ISA-1	Outer-selective	1 M NaCl	10 mM NaCl	15.0	5.1	(Fu et al. 2014)
ISA-2	Outer-selective	1 M NaCl	Real wastewater reverse osmosis retentate	12.0–13.0	4.3	(Li et al. 2018)
ISA-3	Outer-selective	1 M NaCl	Deionized water	14.0	5.5	(Cho et al. 2019)
TFC-1	Outer-selective	1 M NaCl	Deionized water	20.0	7.63	(Sun and Chung 2013)
TFC-2	Outer-selective	1 M NaCl	Deionized water	17.0	9.59	(Le et al. 2016)
TFC-3	Outer-selective	1 M NaCl	Deionized water	22.0	10.05	(Cheng et al. 2017)
TFC-4	Inner-selective	1 M NaCl	Deionized water	15.0	16.5	(Han and Chung 2014)
TFC-5	Inner-selective	1 M NaCl	Deionized water	20.0	22.05	(Gai et al. 2016)
TFC-6	Inner-selective	1 M NaCl	Deionized water	9.0	10.6	(Chou et al. 2012)
TFC-7	Inner-selective	1 M NaCl	Deionized water	15.0	20.9	(Chou et al. 2013)
TFC-8	Inner-selective	1 M NaCl	Deionized water	21.0	12.0	(Li and Chung 2014)
TFC-9	Inner-selective	1 M NaCl	Deionized water	20.0	24.0	(Zhang et al. 2014)
TFC-10	Inner-selective	1 M NaCl	Deionized water	23.0	34.2	(Gai et al. 2018)
TFC-11	Inner-selective	1 M NaCl	Deionized water	20.0	27.0	(Wan and Chung 2015)
TFC-12	Inner-selective	1.2 M NaCl	Deionized water	30.0	38.0	(Wan et al. 2018)
TFC-13	Inner-selective	1 M NaCl	Deionized water	16.5	14.6	(Park et al. 2019)

For the recently developed high-performance inner-selective TFC hollow fiber membranes, the significant breakthroughs depend not only on the improvement of the polyamide-selective layer but also on the enhanced mechanical characteristics of hollow fiber substrates. Theoretically, the optimal operation pressure on the draw solution side should be equal to one-half of the osmotic pressure difference across the membranes for the maximum power generation (i.e., around 23 bar for the feed pair consisting of 1 M NaCl and DI water). However, the operation pressure during PRO processes is usually hindered by the limited mechanical strength of the TFC hollow fiber membranes. Figure 3.10 presents the polymeric materials that have been used for the development of effective hollow fiber substrates thus far. All of them are intrinsically robust materials comprising mechanically strong benzene rings. In addition to the careful selection of the substrates material, the microstructure across the hollow fiber substrates should be carefully designed with a balanced asymmetry. Han et al. (Han and Chung 2014) and Zhang et al. (Zhang et al. 2014) pointed out that the desirable hollow fiber substrates to achieve high performance during the PRO process should have a highly porous support layer to reduce the internal concentration polymerization effects, while a relatively dense cushion layer beneath the polyamide-selective layer is desired for mechanical stability, as demonstrated in Figure 3.11 (Zhang et al. 2014). In addition, the pore size on the inner surface of the hollow fiber substrates should be small with a narrow distribution to ensure the mechanical stability of the polyamide-selective layer under high hydraulic pressures. Moreover, the hollow fiber substrates should have interconnected morphology to have high fracture resistance that can effectively dissipate the stresses from the high-pressure water (Han and Chung 2014, Zhang et al. 2014). The dimension and wall thickness of the hollow fiber membranes also significantly affect the mechanical strength and performance of the TFC membranes. A critical wall thickness was observed to ensure sufficient mechanical stability of the TFC hollow fiber membranes and thus a low salt permeability at high operation pressures (Zhang and Chung 2013).

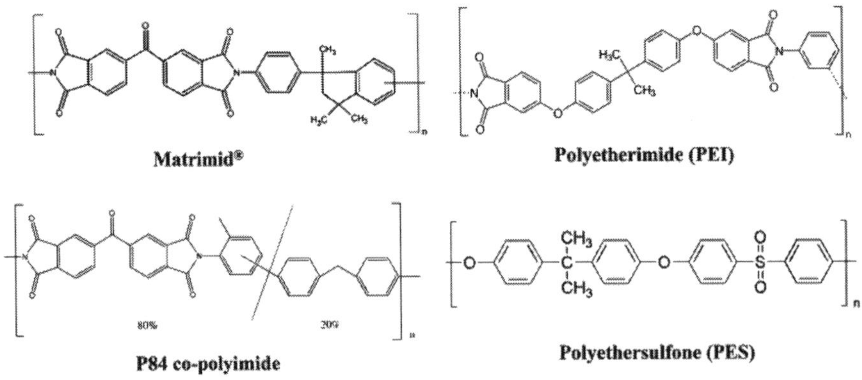

Matrimid®

Polyetherimide (PEI)

P84 co-polyimide

Polyethersulfone (PES)

FIGURE 3.10
Structures of the reported polymer materials for the fabrication of high-performance TFC hollow fiber membranes (Han et al. 2015).

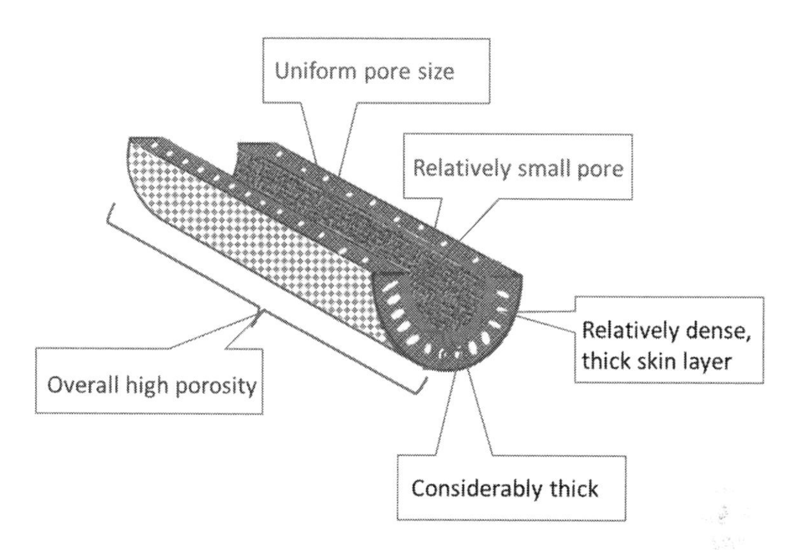

FIGURE 3.11
Schematic illustration of the desirable highly asymmetric hollow fiber substrate for inner-selective TFC hollow fiber membranes for osmotic power generation (Zhang et al. 2014).

Acknowledgment

This research study was supported by the National Research Foundation, Prime Minister's Office, Republic of Singapore, under its Environmental & Water Technologies Strategic Research Programme and administered by the Environment & Water Industry Programme Office (EWI) of the PUB.

References

Bonyadi, S., Chung, T.S., Krantz, W.B., 2007. Investigation of corrugation phenomenon in the inner contour of hollow fibers during the non-solvent induced phase-separation process. *Journal of Membrane Science* 299, 200–210.

Bui, N.N., McCutcheon, J.R., 2014. Nanofiber supported thin-film composite membrane for pressure-retarded osmosis. *Environmental Science & Technology* 48, 4129–4136.

Cheng, Z.L., Li, X., Feng, Y., Wan, C.F., Chung, T.S., 2017. Tuning water content in polymer dopes to boost the performance of outer-selective thin-film composite (TFC) hollow fiber membranes for osmotic power generation. *Journal of Membrane Science* 524, 97–107.

Cheng, Z.L., Li, X., Liu, Y.D., Chung, T.S., 2016. Robust outer-selective thin-film composite polyethersulfone hollow fiber membranes with low reverse salt flux for renewable salinity-gradient energy generation. *Journal of Membrane Science* 506, 119–129.

Cho, Y.H., Kim, S.D., Kim, J.F., Choi, H.G., Kim, Y., Nam, S.E., Park, Y.I., Park, H., 2019. Tailoring the porous structure of hollow fiber membranes for osmotic power generation applications via thermally assisted nonsolvent induced phase separation. *Journal of Membrane Science* 579, 329–341.

Chou, S., Wang, R., Fane, A.G., 2013. Robust and high performance hollow fiber membranes for energy harvesting from salinity gradients by pressure retarded osmosis. *Journal of Membrane Science* 448, 44–54.

Chou, S., Wang, R., Shi, L., She, Q., Tang, C., Fane, A.G., 2012. Thin-film composite hollow fiber membranes for pressure retarded osmosis (PRO) process with high power density. *Journal of Membrane Science* 389, 25–33.

Chung, T.S., Li, X., Ong, R.C., Ge, Q., Wang, H., Han, G., 2012. Emerging forward osmosis (FO) technologies and challenges ahead for clean water and clean energy applications. *Current Opinion in Chemical Engineering* 1, 246–257.

Cui, Y., Liu, X.Y., Chung, T.S., 2014. Enhanced osmotic energy generation from salinity gradients by modifying thin film composite membranes. *Chemical Engineering Journal* 242, 195–203.

Fu, F.J., Sun, S.P., Zhang, S., Chung, T.S., 2014. Pressure retarded osmosis dual-layer hollow fiber membranes developed by co-casting method and ammonium persulfate (APS) treatment. *Journal of Membrane Science* 469, 488–498.

Fu, F.J., Zhang, S., Sun, S.P., Wang, K.Y., Chung, T.S., 2013. POSS-containing delamination-free dual-layer hollow fiber membranes for forward osmosis and osmotic power generation. *Journal of Membrane Science* 443, 144–155.

Gai, W., Li, X., Xiong, J.Y., Wan, C.F., Chung, T.S., 2016. Evolution of micro-deformation in inner-selective thin film composite hollow fiber membranes and its implications for osmotic power generation. *Journal of Membrane Science* 516, 104–112.

Gai, W., Zhao, D.L., Chung, T.S., 2018. Novel thin film composite hollow fiber membranes incorporated with carbon quantum dots for osmotic power generation. *Journal of Membrane Science* 551, 94–102.

Han, G., Chung, T.S., 2014. Robust and high performance pressure retarded osmosis hollow fiber membranes for osmotic power generation. *AIChE Journal* 60, 1107–1119.

Han, G., Wang, P., Chung, T.S., 2013. Highly robust thin-film composite pressure retarded osmosis (PRO) hollow fiber membranes with high power densities for renewable salinity-gradient energy generation. *Environmental Science & Technology* 47, 8070–8077.

Han, G., Zhang, S., Li, X., Chung, T.S., 2015. Progress in pressure retarded osmosis (PRO) membranes for osmotic power generation. *Progress in Polymer Science* 51, 1–27.

Jiang, L.Y., Chung, T.S., Cao, C., Huang, Z., Kulprathipanja, S., 2005. Fundamental understanding of nano-sized zeolite distribution in the formation of the mixed matrix single- and dual-layer asymmetric hollow fiber membranes. *Journal of Membrane Science* 252, 89–100.

Le, N.L., Bettahalli, N.M.S., Nunes, S.P., Chung, T.S., 2016. Outer-selective thin film composite (TFC) hollow fiber membranes for osmotic power generation. *Journal of Membrane Science* 505, 157–166.

Li, D.F., Chung, T.S., Wang, R., Liu, Y., 2002. Fabrication of fluoropolyimide/polyethersulfone (PES) dual-layer asymmetric hollow fiber membranes for gas separation. *Journal of Membrane Science* 198, 211–223.

Li, X., Chung, T.S., 2014. Thin-film composite P84 co-polyimide hollow fiber membranes for osmotic power generation. *Applied Energy* 114, 600–610.

Li, Y., Chung, T.S., Xiao, Y., 2008. Superior gas separation performance of dual-layer hollow fiber membranes with an ultrathin dense-selective layer. *Journal of Membrane Science* 325, 23–27.

Li, Y., Wang, R., Qi, S., Tang, C., 2015. Structural stability and mass transfer properties of pressure retarded osmosis (PRO) membrane under high operating pressures. *Journal of Membrane Science* 488, 143–153.

Li, Y., Zhao, S., Setiawan, L., Zhang, L., Wang, R., 2018. Integral hollow fiber membrane with chemical cross-linking for pressure retarded osmosis operated in the orientation of active layer facing feed solution. *Journal of Membrane Science* 550, 163–172.

Naguib, M.F., Maisonneuve, J., Laflamme, C.B., Pillay, P., 2015. Modeling pressure-retarded osmotic power in commercial length membranes. *Renewable Energy* 76, 619–627.

Park, M.J., Lim, S., Gonzales, R.R., Phuntsho, S., Han, D.S., Abdel-Wahab, A., Adham, S., Shon, H.K., 2019. Thin-film composite hollow fiber membranes incorporated with graphene oxide in polyethersulfone support layers for enhanced osmotic power density. *Desalination* 464, 63–75.

Peng, N., Chung, T.S., 2008. The effects of spinneret dimension and hollow fiber dimension on gas separation performance of ultra-thin defect-free Torlon® hollow fiber membranes. *Journal of Membrane Science* 310, 455–465.

Peng, N., Chung, T.S., Wang, K.Y., 2008. Macrovoid evolution and critical factors to form macrovoid-free hollow fiber membranes. *Journal of Membrane Science* 318, 363–372.

Peng, N., Widjojo, N., Sukitpaneenit, P., Teoh, M.M., Lipscomb, G.G., Chung, T.S., Lai, J.Y., 2012. Evolution of polymeric hollow fibers as sustainable technologies: past, present, and future. *Progress in Polymer Science* 37, 1401–1424.

Ren, J., McCutcheon, J.R., 2015. Polyacrylonitrile supported thin film composite hollow fiber membranes for forward osmosis. *Desalination* 372, 67–74.

Santoso, Y.E., Chung, T.S., Wang, K.Y., Weber, M., 2006. The investigation of irregular inner skin morphology of hollow fiber membranes at high-speed spinning and the solutions to overcome it. *Journal of Membrane Science* 282, 383–392.

She, Q., Hou, D., Liu, J., Tan, K.H., Tang, C.Y., 2013. Effect of feed spacer induced membrane deformation on the performance of pressure retarded osmosis (PRO): implications for PRO process operation. *Journal of Membrane Science* 445, 170–182.

She, Q., Jin, X., Tang, C.Y., 2012. Osmotic power production from salinity gradient resource by pressure retarded osmosis: effects of operating conditions and reverse solute diffusion. *Journal of Membrane Science* 401–402, 262–273.

Sivertsen, E., Holt, T., Thelin, W., Brekke, G., 2012. Modelling mass transport in hollow fibre membranes used for pressure retarded osmosis. *Journal of Membrane Science* 417–418, 69–79.

Sivertsen, E., Holt, T., Thelin, W., Brekke, G., 2013. Pressure retarded osmosis efficiency for different hollow fibre membrane module flow configurations. *Desalination* 312, 107–123.

Song, X., Liu, Z., Sun, D.D., 2013. Energy recovery from concentrated seawater brine by thin-film nanofiber composite pressure retarded osmosis membranes with high power density. *Energy & Environmental Science* 6, 1199.

Sun, S.P., Chung, T.S., 2013. Outer-selective pressure-retarded osmosis hollow fiber membranes from vacuum-assisted interfacial polymerization for osmotic power generation. *Environmental Science & Technology* 47, 13167–13174.

Thorsen, T., Holt, T., 2009. The potential for power production from salinity gradients by pressure retarded osmosis. *Journal of Membrane Science* 335, 103–110.

Wan, C.F., Chung, T.S., 2015. Osmotic power generation by pressure retarded osmosis using seawater brine as the draw solution and wastewater retentate as the feed. *Journal of Membrane Science* 479, 148–158.

Wan, C.F., Yang, T., Gai, W., Lee, Y.D., Chung, T.S., 2018. Thin-film composite hollow fiber membrane with inorganic salt additives for high mechanical strength and high power density for pressure-retarded osmosis. *Journal of Membrane Science* 555, 388–397.

Widjojo, N., Chung, T.S., 2006. Thickness and air gap dependence of macrovoid evolution in phase-inversion asymmetric hollow fiber membranes. *Industrial & Engineering Chemistry Research* 45, 7618–76266.

Yip, N.Y., Tiraferri, A., Phillip, W.A., Schiffman, J.D., Hoover, L.A., Kim, Y.C., Elimelech, M., 2011. Thin-film composite pressure retarded osmosis membranes for sustainable power generation from salinity gradients. *Environmental Science & Technology* 45, 4360–4369.

Zhang, S., Chung, T.S., 2013. Minimizing the instant and accumulative effects of salt permeability to sustain ultrahigh osmotic power density. *Environmental Science & Technology* 47, 10085–10092.

Zhang, S., Chung, T.S., 2016. Osmotic power production from seawater brine by hollow fiber membrane modules: net power output and optimum operating conditions. *AIChE Journal* 62, 1216–1225.

Zhang, S., Sukitpaneenit, P., Chung, T.S., 2014. Design of robust hollow fiber membranes with high power density for osmotic energy production. *Chemical Engineering Journal* 241, 457–465.

4

Mass Transport within Pressure Retarded Osmosis Membranes of Different Configurations

Zhen Lei Cheng

Department of Chemical and Biomolecular Engineering
National University of Singapore
Singapore

CONTENTS

Nomenclature

A (overall) water permeability coefficient (L m^{-2} h^{-1} bar^{-1}, abbreviated as LMH/bar)

A_f (overall) water permeability coefficient of the fouled membrane (LMH/bar)

A_i	water permeability coefficient of the inner skin layer (LMH/bar)
A_o	water permeability coefficient of the outer skin layer (LMH/bar)
B	(overall) salt permeability coefficient (L m^{-2} h^{-1}, abbreviated as LMH)
B_D	salt permeability coefficient of the skin layer facing the draw solution (LMH)
B_F	salt permeability coefficient of the skin layer facing the feed solution (LMH)
B_i	salt permeability coefficient of the inner skin layer (LMH)
B_o	salt permeability coefficient of the outer skin layer (LMH)
$C_{D,b}$	bulk concentration of the draw solution (M)
$C_{D,m}$	surface salinity of the selective skin on the draw side (M)
$C_{D/s}$	surface salinity of the selective skin on the support side (M)
$C_{F,b}$	bulk concentration of the feed solution (M)
$C_{F,m}$	surface salinity of the protective skin on the feed side (M)
$C_{F,s}$	salinity at the support back surface (M)
$C_{F/s}$	surface salinity of the protective skin on the support side (M)
D	bulk diffusion coefficient of NaCl (m^2 s^{-1})
D_s	diffusion coefficient of the draw solute (m^2 s^{-1})
J_s	reverse salt flux (g m^{-2} h^{-1}, abbreviated as gMH)
J_w	water flux (LMH)
$J_{w,f}$	water flux of the fouled membrane (LMH)
J_w^{FO}	water flux under FO mode of FO processes (LMH)
P_s	hydraulic pressure within the support layer (bar)
R	universal gas constant (m^3 Pa K^{-1} mol^{-1})
R_f	hydraulic resistance induced by the foulants (m^{-1})
R_m	membrane hydraulic resistance (m^{-1})
S	structural parameter (μm)
T	absolute temperature of the solution (K)
i	van't Hoff factor ($i = 2$ for NaCl)
k_D	boundary layer mass transfer coefficient on the draw side (m s^{-1})
k_F	boundary layer mass transfer coefficient on the feed side (m s^{-1})
r	radial coordinate with a positive direction pointing outwards
r_i	inner radius of a HF membrane (μm)
r_o	outer radius of a HF membrane (μm)
t_c	thickness of the cake layer (μm)
t_s	thickness of the support layer (μm)
t_s^{eq}	equivalent membrane thickness (μm)

Greek symbols

ΔP	applied hydraulic pressure difference (bar)
$\Delta \pi$	osmotic pressure difference (bar)

$\Delta\pi_{eff}$ effective osmotic pressure difference (bar)
Ω resistance
δ_D boundary layer thickness on the draw side (μm)
δ_F boundary layer thickness on the feed side (μm)
ε porosity of the support layer
ε_c porosity of the cake layer
μ viscosity of the solution (Pa·s)
ξ_s linear reverse salt flux (g m^{-1} h^{-1})
ξ_w linear water flux (L m^{-1} h^{-1})
$\pi_{D,b}$ bulk osmotic pressure of the draw solution (bar)
$\pi_{F,b}$ bulk osmotic pressure of the feed solution (bar)
τ tortuosity of the support layer
τ_c tortuosity of the cake layer

Abbreviations

BC boundary condition
DI deionized
ECP external concentration polarization
cECP concentrative external concentration polarization
dECP dilutive external concentration polarization
FO forward osmosis
FS flat sheet
$F_{ECP,D}$ factor of ECP on the draw side
$F_{ECP,F}$ factor of ECP on the feed side
F_{ICP} factor of ICP
GO graphene oxide
HF hollow fiber
ICP internal concentration polarization
NPPD normalized peak power density
PD power density (W m^{-2})
PES polyethersulfone
PRO pressure retarded osmosis
RO reverse osmosis
TFC thin-film composite

Subscripts

D draw solution
F feed solution

b	bulk
c	cake layer
eff	effective
f	fouled/foulants
i	inner
m	membrane/membrane surface
o	outer
s	support layer (except as solute/salt for D_s, J_s and ξ_s)
w	water

Superscript

FO	under FO mode of FO processes
eq	equivalent

4.1 Introduction

To move the pressure retarded osmosis (PRO) technology closer to commercialization, researchers have developed high-performance PRO membranes with (1) high water permeability, (2) reasonably low salt permeability, (3) good mechanical strength, and (4) small transport resistance (Bui and McCutcheon, 2014; Cheng et al., 2017; Chou et al., 2013; Cui et al., 2014; Fu et al., 2015; Han and Chung, 2014; Han et al., 2015; Ingole et al., 2014; Kumano et al., 2016; Le et al., 2016; Li and Chung, 2014; Li et al., 2015; Song et al., 2013; Sun and Chung, 2013; Wan and Chung, 2015; Zhang et al., 2014). To date, a wide range of materials, fabrication methods, and membrane configurations [i.e., flat sheet (FS) and hollow fiber (HF)] have been explored (Table 4.1). In parallel, modeling PRO processes, at the scale of membrane coupons (Lee and Kim, 2016; Nagy et al., 2016; Sivertsen et al., 2012; Yip and Elimelech, 2011; Yip et al., 2011) as well as modules (Kumano et al., 2016; Kurihara et al., 2016; Sivertsen et al., 2013; Xiong et al., 2017; Zhang and Chung, 2016), have been conducted to examine the effects of newly developed membranes and guide the membrane development. These models often use the mass transfer equations derived from the FS geometry (Achilli et al., 2009; Lee et al., 1981; Yip et al., 2011). However, HF membranes with inner- and outer-selective (i.e., single-skinned) configurations have attracted significant attention in recent years owing to their promising prospect for PRO applications (Chung et al., 2015; Han et al., 2015). Although Sivertsen et al. (2012) pioneered the modeling work particularly for the HF

TABLE 4.1

Summary of selected state-of-the-art PRO membranes reported in the literature

Config.	Material[a]	ID (μm)	OD (μm)	Thickness, t_s (μm)	$t_s/2r_o$	A (LMH/bar)	B (LMH)	S^b (μm)	τ/ε	Reference
FS	CTA	–	–	~100	–	$1.5^{\dagger}$	$2.2^{\dagger}$	$716^{\dagger}$	7.16	She et al., 2013
	CTA	–	–	~100	–	$1.44^{\dagger}$	$2.05^{\dagger}$	$757^{\dagger}$	7.57	
	CTA	–	–	~100	–	$1^{\dagger}$	$0.8^{\dagger}$	$686^{\dagger}$	6.86	
	TFC-PEI	–	–	68.5	–	2.28	0.67	510	7.45	Li et al., 2015
	TFC-PEI	–	–	76.2	–	2.09	0.87	554	7.27	
	TFC-PEI	–	–	82.5	–	1.65	0.75	687	8.33	
	TFC-PI	–	–	$55^{\dagger}$	–	2.77	$1.19^{\ddagger}$	$503^{\|}$	9.15	Cui et al., 2014
	TFC-PAN	–	–	$55^{\dagger}$	–	2.83	0.44	273	4.96	Bui and McCutcheon, 2014
	TFC-PAN	–	–	$50^{\dagger}$	–	3.82	1.19	135	2.7	Song et al., 2013
Inner-selective HF	TFC-coPI	630	1011	190.5	0.19	0.91	0.09	685	4.6	Li and Chung, 2014
	TFC-PI	520	880	180	0.20	3.6	0.73	640	4.68	Han and Chung, 2014
	TFC-PI	540	860	160	0.19	3.1	0.61	590	4.7	
	TFC-PI	520	820	150	0.18	4.3	0.47	640	5.4	
	TFC-PEI	975	1260	142.5	0.11	1.52	0.24	610	4.88	Chou et al., 2013
	TFC-PES	575	1025	225	0.22	3.5	0.32	450	2.71	Wan and Chung, 2015
Outer-selective HF	CTA	100	200	50	0.25	–	–	–	–	Kumano et al., 2016
	PBI-PAN	520	870	175	0.2	1.63	$0.43^{\ddagger}$	$846^{\|}$	3.78	Fu et al., 2015
	PBI-PAN	450	810	180	0.22	0.57	$0.03^{\ddagger}$	$1269^{\|}$	5.33	
	TFC-PES	944	1320	188	0.14	–	–	–	–	Ingole et al., 2014

(Continued)

TABLE 4.1 (Cont.)

Config.	Material[a]	ID (μm)	OD (μm)	Thickness, t_s (μm)	$t_s/2r_o$	A (LMH/bar)	B (LMH)	S^b (μm)	τ/ε	Reference
	TFC-PI	234	404	85	0.21	1.42	0.4	1176[‖]	10.7	Sun and Chung, 2013
	TFC-PI	232	417	92.5	0.22	3.59	1.82	705[‖]	5.77	
	TFC-PEI	294	468	87	0.19	1.59[†]	0.7[†]	1374[‖]	12.6	Le et al., 2016
	TFC-PEI	321	527	103	0.2	2.81[†]	0.97[†]	928[‖]	7.1	
	TFC-PES	500	1070	285	0.27	1.48	0.15	1363	3.35	Cheng et al., 2017
	TFC-PES	500	1090	295	0.27	1.9	0.16	1116	2.63	
Double-skinned (draw-on-lumen) HF	TFC-PES	525	1125	300	0.27	1.5	0.02	996[*]	2.32	Han et al., 2017

[a] Abbreviations. CTA: cellulose triacetate; PEI: polyetherimide; PI: polyimide; PAN: polyacrylonitrile; coPI: co-polyimide; PES: polyethersulfone; PBI: polybenzimidazole.

[b] Fitting results based on the RO–FO method. No adjustment given to consider the curvature effect of HF geometry.

[†] Extracted from figures in the literature using a free software PlotDigitizer.

[‡] Calculated from reported NaCl rejection data.

[‖] Fitting results using the FS model based on the PRO performance at $\Delta P = 0$ bar.

[*] Taken from TFC-II of (Cheng et al., 2016), which was obtained based on the outer-selective HF configuration.

configuration, it was less commonly used probably due to the inherent mathematical complexity. Most researchers still applied the classic FS model to evaluate the PRO performance of HF membranes (Cheng et al., 2017; Chou et al., 2013; Wan and Chung, 2015; Xiong et al., 2016). A previous work has assessed the validity of using the FS model for HF membranes (Sivertsen et al., 2012). Generally, as the outer radius of the fiber increases, the FS model would have the same solution as the HF model. However, no in-depth study has been further carried out. For future PRO development, comprehensive analyses are urgently needed to (1) investigate the feasibility of generalizing the FS model for HF membranes and (2) build a universal platform to assess PRO membranes of various configurations.

An ideal PRO evaluation platform should also consider the effects of membrane fouling, as it substantially reduces the osmotic power production for real applications. In order to tackle this problem, one promising way is to install an antifouling barrier on the back surface of porous membrane support. As a result, foulants are blocked to enter the support layer while water molecules can still pass through it (Han et al., 2017; Hu et al., 2016; Li et al., 2014; She et al., 2016). Hu et al. demonstrated that a dense graphene oxide (GO) layer assembled on the back of an asymmetric FS membrane could effectively reduce the organic fouling during PRO processes (Hu et al., 2016). However, this GO layer would inevitably contact with the feed spacers and possibly be damaged severely because of the high pressure applied to the draw solution (She et al., 2013). In contrast, the HF configuration is preferred because of its mechanically self-supported nature (Chung et al., 2015; Fu et al., 2015; Han et al., 2015; Wan and Chung, 2015). Li et al. (2014) grafted a layer of hyperbranched polyglycerol on the outer surface of poly(ether sulfone) HF substrate, which showed antifouling behavior for osmotic power generation. Recently, Han et al. (2017) deposited a polyamide layer on the outer surface of an inner-selective thin-film composite (TFC) HF membrane as the antifouling barrier. The resultant membrane consisted of a special double-skinned (i.e., inner-selective skin and outer protective skin) design and showed superior antifouling performance than other membranes. Despite there being many merits of the double-skinned HF configuration, the mathematical understanding is still lacking on (1) how the additional antifouling barrier behaves under high-pressure PRO processes and (2) how it influences the antifouling features compared to the single-skinned one.

In this chapter, analytical PRO models with FS, single-, and double-skinned HF configurations are derived, providing a universal platform to evaluate the existing state-of-the-art PRO membranes. Subsequently, the derived mass transfer equations for single-skinned HF membranes are systematically compared to those of FS membranes. The applicability of the FS model to predict the PRO performance of HF membranes is assessed. Finally, the mass transfer equations for the double-skinned HF configuration are employed to analyze the transport properties of skin layers of a fabricated membrane reported in the literature. New expressions are made by considering the foulant build-up as an external cake layer on top of the protective skin of a double-skinned HF membrane. In such a way, its antifouling performance can be mathematically compared with the single-skinned configuration. This chapter may serve as a useful guidance on how to access and design high-performance PRO membranes.

4.2 Derivation of Mass Transfer Equations

4.2.1 The FS Model

The schematic of the salt concentration profile across an asymmetric FS membrane is illustrated in Figure 4.1(a). In a typical PRO process, the selective skin always faces the pressurized draw solution to provide a higher water flux. Ideally, the water flux, J_w, is a product of the membrane water permeability coefficient, A, and the driving force across the membrane as follows (Achilli et al., 2009; Lee et al., 1981; Thorsen and Holt, 2009; Yip et al., 2011).

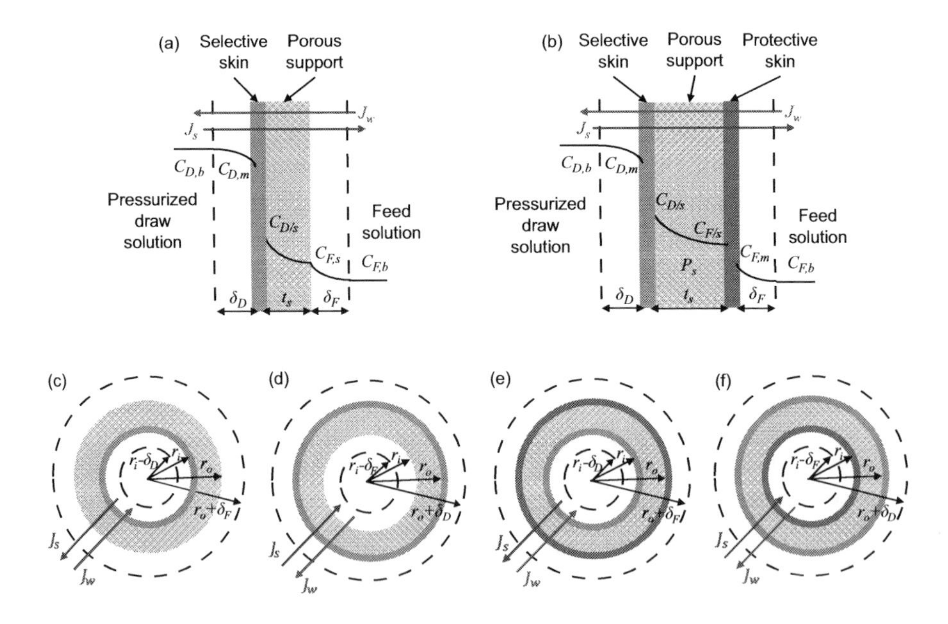

FIGURE 4.1
Schematics of the salt concentration profile across (a) single-skinned and (b) double-skinned PRO membranes and the key geometric parameters of (c) inner-selective, (d) outer-selective, (e) double-skinned (draw-on-lumen), and (f) double-skinned (draw-on-shell) HF configurations.

$$J_w = A(\Delta\pi - \Delta P) \tag{1}$$

where $\Delta\pi$ is the osmotic pressure difference between the bulk osmotic pressures of the draw solution, $\pi_{D,b}$, and feed solution, $\pi_{F,b}$. For simplicity, they can be estimated from the bulk concentrations of the draw and feed solutions, $C_{D,b}$ and $C_{F,b}$, respectively, using the van't Hoff equation. Compared to the commercially available OLI Stream Analyzer, the van't Hoff equation can give a reasonable estimation of the osmotic pressure for salinity up to 2 M (Lee and Kim, 2016). ΔP is the applied hydraulic pressure difference across the membrane.

However, in reality, the effective osmotic pressure difference, $\Delta\pi_{\text{eff}}$, is much smaller than $\Delta\pi$ because of the following factors (Achilli et al., 2009; Lee et al., 1981; Sivertsen et al., 2012; Yip et al., 2011): (1) the dilutive external concentration polarization (dECP) on the draw side reduces $C_{D,b}$ to $C_{D,m}$ (i.e., the surface salinity on the selective skin facing the draw solution); (2) the concentrative external concentration polarization (cECP) on the feed side raises $C_{F,b}$ to $C_{F,s}$ (i.e., the salinity on the back surface of the support); (3) the internal concentration polarization (ICP) within the porous support elevates the salinity from $C_{F,s}$ to $C_{D/s}$ (i.e., the surface salinity on the selective skin facing the feed solution); and (4) the reverse salt ions diffuse from the draw solution to the feed solution. The severity of dECP and cECP is dependent on the boundary layer thickness on the draw and feed sides, that is, δ_D and δ_F, respectively. Their values can be calculated using $\delta_D = D/k_D$ and $\delta_F = D/k_F$, where D is the bulk diffusion coefficient of NaCl, and k_D and k_F are the boundary layer mass transfer coefficients on the draw and feed sides, respectively (Yip et al., 2011). Experimentally, k can be determined via Sherwood number, Sh, and the hydraulic diameter of the flow channel, d_h, using the following equation (Mulder, 1996; Zhang and Chung, 2016).

$$k = \frac{ShD}{d_h} \tag{2}$$

The Sh value is related to Reynolds number, Re, Schmidt number, Sc, and the flow channel length, L, depending on which flow regime exhibits.

For laminar flow ($Re \leq 2320$):

$$Sh = 1.62\left(ReSc\frac{d_h}{L}\right)^{0.33} \tag{3}$$

For turbulent flow ($Re > 2320$):

$$Sh = 0.04Re^{0.75}Sc^{0.33} \tag{4}$$

The reverse salt flux, J_s, is quantified using the solution-diffusion model (Lee et al., 1981; Thorsen and Holt, 2009; Yip et al., 2011).

$$J_s = B(C_{D,m} - C_{D/s}) \tag{5}$$

where B is the membrane salt permeability coefficient. Meanwhile, J_s can also be expressed as the sum of the diffusive flux, which is described by the Fick's first law, and the convective flux induced by the bulk water flux as follows (Lee et al., 1981; Sivertsen et al., 2012; Xiong et al., 2016; Yip et al., 2011):

$$J_s = D_s \frac{dC(x)}{dx} - J_w C(x) \tag{6}$$

where x is the axial coordinate with a positive direction in line with J_w. D_s is the diffusion coefficient of the draw solute. It is equal to the bulk diffusion coefficient, D, within the boundary layers on the draw and feed sides, and is related to the porosity, ε, and tortuosity, τ, within the support layer in the form of $D_s = D\varepsilon/\tau$. We note that the flux terms in Equation (6) are the areal fluxes defined as the volumetric or mass flow rate through a unit area of the selective skin. Conventionally, it adopts the unit of L m^{-2} h^{-1} (abbreviated as LMH) for J_w and g m^{-2} h^{-1} (abbreviated as gMH) for J_s, respectively. At steady state, the salt flux across the selective skin (Equation (5)) equals the salt flux across the boundary layers on the draw and feed sides, as well as on the support layer (Equation (6)).

$$B(C_{D,m} - C_{D/s}) = D_s \frac{dC(x)}{dx} - J_w C(x) \tag{7}$$

The corresponding boundary conditions (BCs) are depicted in Figure 4.1(a) as follows:

$$C(x = 0) = C_{F,b} \tag{8}$$

$$C(x = \delta_F) = C_{F,s} \tag{9}$$

$$C(x = \delta_F + t_s^-) = C_{D/s} \tag{10}$$

$$C(x = \delta_F + t_s^+) = C_{D,m} \tag{11}$$

$$C(x = \delta_F + t_s + \delta_D) = C_{D,b} \tag{12}$$

where t_s is the thickness of the support layer. We note that the thickness of the selective skin is negligibly small compared to those of the support and boundary layers. Thus, the positions on two sides of the selective skin are not differentiated (i.e., $t_s^- = t_s^+ = t_s$). Integrating Equation (7) with Equations (8)–(12) yields the salinity difference across the selective skin ($C_{D,m} - C_{D/s}$). The derivation of Equation (13) can follow the similar approach employed in other studies (Xiong et al., 2016; Yip et al., 2011).

$$C_{D,m} - C_{D/s} = \frac{C_{D,b}\exp\left(-\frac{J_w}{k_D}\right) - C_{F,b}\exp\left(\frac{J_w}{k_F} + \frac{J_w S}{D}\right)}{1 + \frac{B}{J_w}\left[\exp\left(\frac{J_w}{k_F} + \frac{J_w S}{D}\right) - \exp\left(-\frac{J_w}{k_D}\right)\right]} \tag{13}$$

where $S = \tau/\varepsilon \times t_s$ is defined as the structural parameter, describing the mass transport resistance of the support layer in osmosis-driven processes (Bui et al., 2015; Gerstandt et al., 2008; Lin, 2016; Sivertsen et al., 2012; Thorsen and Holt, 2009; Yip et al., 2011). $\Delta\pi_{\text{eff}}$ can be further calculated using the van't Hoff equation (Lee and Kim, 2016).

$$\Delta\pi_{\text{eff}} = \pi_{D,m} - \pi_{D/s} = iRT\left(C_{D,m} - C_{D/s}\right) \tag{14}$$

where i is the van't Hoff factor, R is the universal gas constant, and T is the absolute temperature of the solution. By substituting $\Delta\pi$ with $\Delta\pi_{\text{eff}}$ in Equation (1) and then combining it with Equation (13), one is able to get the following equation for the actual J_w in the PRO process.

$$J_w = A\left\{\frac{\pi_{D,b}\exp\left(-\frac{J_w}{k_D}\right) - \pi_{F,b}\exp\left(\frac{J_w}{k_F} + \frac{J_w S}{D}\right)}{1 + \frac{B}{J_w}\left[\exp\left(\frac{J_w}{k_F} + \frac{J_w S}{D}\right) - \exp\left(-\frac{J_w}{k_D}\right)\right]} - \Delta P\right\} \tag{15}$$

By substituting $\Delta\pi$ with $\Delta\pi_{\text{eff}}$ in Equation (1) and combining it with Equation (5), the expression of J_s can be rewritten as follows:

$$J_s = \frac{B}{iRT}\left(\frac{J_w}{A} + \Delta P\right) \tag{16}$$

Compared to Equation (5), Equation (16) is more readily used, as all the terms on the right-hand side are experimentally measurable or literarily available.

The performance indicator of a PRO process, power density (PD), quantifying the osmotic power production is calculated as follows:

$$PD = J_w \Delta P \tag{17}$$

By combining Equations (1) and (17) and then differentiating it by ΔP, one can obtain the peak power density (PPD) occurring at $\Delta P = \Delta\pi/2$ for the ideal case. In reality, the maximal ΔP will take place depending on the concentration polarization effects and operating parameters (Achilli et al., 2009; Lee et al., 1981; Zhang and Chung, 2013).

4.2.2 The Single-Skinned HF Model

Following the derivation steps for the FS model shown above, a similar equation to express J_s as the sum of the diffusive and convective fluxes can be written for the single-skinned HF configuration.

$$\text{For inner-selective HF}: J_s(r) = -D_s\frac{dC(r)}{dr} - J_w(r)C(r) \tag{18}$$

$$\text{For outer-selective HF}: J_s(r) = D_s\frac{dC(r)}{dr} - J_w(r)C(r) \tag{19}$$

where r is the radial coordinate with a positive direction pointing outwards. Contrary to Equation (6), the areal flux terms (J_w and J_s) in Equations (18) and (19) are no longer constant but dependent on r since the available flow area changes along the radial position. In order to integrate Equations (18) and (19) easily, a concept of linear water flux, ξ_w, and linear reverse salt flux, ξ_s, defined as the volumetric and mass flow rate through a unit length of a given HF membrane, respectively, is introduced. Mathematically, $\xi_w = 2\pi r J_w(r)$ and $\xi_s = 2\pi r J_s(r)$ (Lin, 2016). Substituting them into Equations (18) and (19), respectively, obtains

$$\text{For inner-selective HF}: \xi_s = -2\pi r D_s\frac{dC(r)}{dr} - \xi_w C(r) \tag{20}$$

$$\text{For outer-selective HF}: \xi_s = 2\pi r D_s\frac{dC(r)}{dr} - \xi_w C(r) \tag{21}$$

Since both ξ_w and ξ_s are constant and independent of r, Equations (20) and (21) can be handled by the same approach that was previously applied to the FS model. The corresponding BCs (i.e., the salt concentration profile across an asymmetric HF membrane) are depicted in Figure 4.1(a), with key geometric parameters highlighted in Figure 4.1(c) and (d) for the inner- and outer-selective HF configurations, respectively. For the inner-selective HF configuration, the BCs are as follows:

$$C(r = r_i - \delta_D) = C_{D,b} \tag{22}$$

$$C(r = r_i^-) = C_{D,m} \tag{23}$$

$$C(r = r_i^+) = C_{D/s} \tag{24}$$

$$C(r = r_o) = C_{F,s} \tag{25}$$

$$C(r = r_o + \delta_F) = C_{F,b} \tag{26}$$

where r_i and r_o are the inner and outer radii of the HF membrane, respectively. For the outer-selective HF, the BCs can be formulated in the same way except that the geometric parameters need to be flipped. The other two important mass transfer equations concerning the water and salt transport across the selective skin at position r_i for the inner-selective HF configuration and r_o for the outer-selective HF configuration are as follows:

For inner-selective HF:

$$\xi_w = 2\pi r_i J_w(r_i) = 2\pi r_i A(\pi_{D,m} - \pi_{D/s} - \Delta P) \tag{27}$$

$$\xi_s = 2\pi r_i J_s(r_i) = 2\pi r_i B(C_{D,m} - C_{D/s}) \tag{28}$$

For outer-selective HF:

$$\xi_w = 2\pi r_o J_w(r_o) = 2\pi r_o A(\pi_{D,m} - \pi_{D/s} - \Delta P) \tag{29}$$

$$\xi_s = 2\pi r_o J_s(r_o) = 2\pi r_o B(C_{D,m} - C_{D/s}) \tag{30}$$

By solving Equations (20), (22)–(26), (27), and (28), one is able to obtain Equation (31) for an inner-selective HF membrane (Lin, 2016; Sivertsen et al., 2012; Xiong et al., 2017).

$$\xi_w = 2\pi r_i A \left\{ \frac{\pi_{D,b}\left(\frac{r_i}{r_i - D/k_D}\right)^{-\frac{\xi_w}{2\pi D}} - \pi_{F,b}\left(\frac{r_o + D/k_F}{r_o}\right)^{\frac{\xi_w}{2\pi D}}\left(\frac{r_o}{r_i}\right)^{\frac{\xi_w \tau}{2\pi D\varepsilon}}}{1 + \frac{B}{\xi_w}2\pi r_i\left[\left(\frac{r_o + D/k_F}{r_o}\right)^{\frac{\xi_w}{2\pi D}}\left(\frac{r_o}{r_i}\right)^{\frac{\xi_w \tau}{2\pi D\varepsilon}} - \left(\frac{r_i}{r_i - D/k_D}\right)^{-\frac{\xi_w}{2\pi D}}\right]} - \Delta P \right\} \tag{31}$$

Replacing ξ_w with J_w, Equation (31) becomes

$$J_w = A \left\{ \frac{\pi_{D,b}\left(\frac{r_i}{r_i - D/k_D}\right)^{-\frac{J_w r_i}{D}} - \pi_{F,b}\left(\frac{r_o + D/k_F}{r_o}\right)^{\frac{J_w r_i}{D}}\left(\frac{r_o}{r_i}\right)^{\frac{J_w r_i \tau}{D\varepsilon}}}{1 + \frac{B}{J_w}\left[\left(\frac{r_o + D/k_F}{r_o}\right)^{\frac{J_w r_i}{D}}\left(\frac{r_o}{r_i}\right)^{\frac{J_w r_i \tau}{D\varepsilon}} - \left(\frac{r_i}{r_i - D/k_D}\right)^{-\frac{J_w r_i}{D}}\right]} - \Delta P \right\} \tag{32}$$

For an outer-selective HF membrane, it can be solved similarly.

$$J_w = A\left\{\frac{\pi_{D,b}\left(\frac{r_o+D/k_D}{r_o}\right)^{-\frac{J_w r_o}{D}} - \pi_{F,b}\left(\frac{r_i}{r_i-D/k_F}\right)^{\frac{J_w r_o}{D}}\left(\frac{r_o}{r_i}\right)^{\frac{J_w r_o \tau}{D\varepsilon}}}{1+\frac{B}{J_w}\left[\left(\frac{r_i}{r_i-D/k_F}\right)^{\frac{J_w r_o}{D}}\left(\frac{r_o}{r_i}\right)^{\frac{J_w r_o \tau}{D\varepsilon}} - \left(\frac{r_o+D/k_D}{r_o}\right)^{-\frac{J_w r_o}{D}}\right]} - \Delta P\right\} \tag{33}$$

We note that Equation (16) is still applicable to predict J_s for both inner- and outer-selective HF configurations, as it is derived based on the mass transport across the selective skin and regardless of the geometry change of the support layer.

4.2.3 The Double-Skinned HF Model

An analytical model for double-skinned HF PRO membranes is not yet established in the literature. However, this can be developed by coupling the solution-diffusion model for the skin layers and the diffusion-convection transport across the boundary layers and porous support in a way similar to that of single-skinned HF membranes. Figure 4.1(b) shows the salt concentration profile across a double-skinned HF membrane. In this configuration, the selective skin facing the draw solution is normally tighter (i.e., smaller A and B/A) than the protective skin facing the feed solution because of their different functions. The selective skin needs to reject the small salt ions diffusing from the draw side, while the protective skin blocks the big foulant molecules brought by the feed side (Han et al., 2017; Hu et al., 2016; Tang et al., 2011). To illustrate the derivation steps, a double-skinned HF membrane with a draw-on-lumen configuration is chosen, as shown in Figure 4.1(e). The solution to another configuration of draw-on-shell (Figure 4.1(f)) can be obtained similarly.

By applying the solution-diffusion model to describe the water and salt transport across the selective skin at position r_i, one obtains

$$\xi_w = 2\pi r_i J_w(r_i) = 2\pi r_i A_i\left(\pi_{D,m} - \pi_{D/s} + P_s - \Delta P\right) \tag{34}$$

$$\xi_s = 2\pi r_i J_s(r_i) = 2\pi r_i B_i\left(C_{D,m} - C_{D/s}\right) \tag{35}$$

where A_i and B_i are the water and salt permeability coefficients of the inner skin layer (i.e., selective skin), respectively. P_s is the hydraulic pressure within the support layer. According to Tang et al. (2011), different from the case of single-skinned HF membranes, it is possible to build a hydraulic pressure within the porous support because of the unbalanced water and salt transport through the two skin layers. The water and salt transport across the protective skin at position, r_o, can be quantified in the same way.

$$\xi_w = 2\pi r_o J_w(r_o) = 2\pi r_o A_o(\pi_{F/s} - \pi_{F,m} - P_s) \quad (36)$$

$$\xi_s = 2\pi r_o J_s(r_o) = 2\pi r_o B_o(C_{F/s} - C_{F,m}) \quad (37)$$

where A_o and B_o are the water and salt permeability coefficients of the outer skin layer (i.e., the protective skin), respectively. $C_{F/s}$ and $C_{F,m}$ are the surface salinity of the protective skin within the support layer and at the feed side, respectively. By combining Equations (34)–(37) to eliminate P_s and using the van't Hoff equation to correlate the salinity with the osmotic pressure, one can have

$$\xi_w\left(\frac{1}{2\pi r_i A_i} + \frac{1}{2\pi r_o A_o}\right) = iRT\xi_s\left(\frac{1}{2\pi r_i B_i} + \frac{1}{2\pi r_o B_o}\right) - \Delta P \quad (38)$$

i.e.,

$$\xi_w \frac{1}{2\pi r_i A} = iRT\xi_s \frac{1}{2\pi r_i B} - \Delta P \quad (39)$$

where A and B are the overall water and salt permeability coefficients of a double-skinned HF membrane, respectively. They can be calculated from the permeability coefficients of individual skin layers by assuming a resistance-in-series model. Experimentally, they can be determined via conventional RO method (Han et al., 2013; She et al., 2016; Sukitpaneenit and Chung, 2012).

$$\frac{1}{2\pi r_i A} = \frac{1}{2\pi r_i A_i} + \frac{1}{2\pi r_o A_o} \quad (40)$$

$$\frac{1}{2\pi r_i B} = \frac{1}{2\pi r_i B_i} + \frac{1}{2\pi r_o B_o} \quad (41)$$

We note that Equation (39) takes the identical form of Equation (16), which is used for the calculation of J_s from J_w in the single-skinned HF configuration.

The diffusion-convection transport of salt across the boundary layers and porous support can be written as the same as Equation (20). By rearranging Equation (20) to separate the variables, one obtains

$$\frac{dr}{r} = -\frac{2\pi D_s}{\xi_w}\frac{d(\xi_s + \xi_w C(r))}{\xi_s + \xi_w C(r)} \quad (42)$$

The BCs for the above differential equation are

$$\text{BC1}: C(r_i - \delta_D) = C_{D,b} \quad (43)$$

$$\mathrm{BC2}: C\left(r_i^-\right) = C_{D,m} \tag{44}$$

$$\mathrm{BC3}: C\left(r_i^+\right) = C_{D/s} \tag{45}$$

$$\mathrm{BC4}: C\left(r_o^-\right) = C_{F/s} \tag{46}$$

$$\mathrm{BC5}: C\left(r_o^+\right) = C_{F,m} \tag{47}$$

$$\mathrm{BC6}: C(r_o + \delta_F) = C_{F,b} \tag{48}$$

Integrating Equation (42) for the draw side (i.e., the boundary layer region) with BC1 and 2 yields Equation (49). Note that D_s here is simply the bulk diffusion coefficient of NaCl, D.

$$\ln(r)\Big|_{r_i - \delta_D}^{r_i} = -\frac{2\pi D}{\xi_w} ln(\xi_s + \xi_w C(r))\Big|_{C_{D,b}}^{C_{D,m}} \tag{49}$$

Equation (50) follows Equation (49)

$$\ln\frac{r_i}{r_i - \delta_D} = -\frac{2\pi D}{\xi_w} ln\frac{\xi_s + \xi_w C_{D,m}}{\xi_s + \xi_w C_{D,b}} \tag{50}$$

which can be rewritten as

$$\xi_s + \xi_w C_{D,m} = \left(\xi_s + \xi_w C_{D,b}\right)\left(\frac{r_i}{r_i - \delta_D}\right)^{-\frac{\xi_w}{2\pi D}} \tag{51}$$

On the feed side, a similar approach can be carried out with BC5 and 6. Equations (52)–(54) are analogous to Equations (49)–(51).

$$\ln(r)\Big|_{r_o}^{r_o + \delta_F} = -\frac{2\pi D}{\xi_w} ln(\xi_s + \xi_w C(r))\Big|_{C_{F,m}}^{C_{F,b}} \tag{52}$$

$$\ln\frac{r_o + \delta_F}{r_o} = -\frac{2\pi D}{\xi_w} ln\frac{\xi_s + \xi_w C_{F,b}}{\xi_s + \xi_w C_{F,m}} \tag{53}$$

$$\xi_s + \xi_w C_{F,m} = \left(\xi_s + \xi_w C_{F,b}\right)\left(\frac{r_o + \delta_F}{r_o}\right)^{\frac{\xi_w}{2\pi D}} \tag{54}$$

Within the support layer, integration of Equation (42) with BC3 and 4 gives Equation (55). Note that D_s now should be $D\varepsilon/\tau$ to count for the porosity and tortuosity of the porous support. Equations (55)–(57) are analogous to Equations (49)–(51).

$$\ln(r)\Big|_{r_i}^{r_o} = -\frac{2\pi D\varepsilon}{\xi_w \tau}\ln(\xi_s + \xi_w C(r))\Big|_{C_{D/s}}^{C_{F/s}} \quad (55)$$

$$\ln\frac{r_o}{r_i} = -\frac{2\pi D\varepsilon}{\xi_w \tau}\ln\frac{\xi_s + \xi_w C_{F/s}}{\xi_s + \xi_w C_{D/s}} \quad (56)$$

$$\xi_s + \xi_w C_{D/s} = \left(\xi_s + \xi_w C_{F/s}\right)\left(\frac{r_o}{r_i}\right)^{\frac{\xi_w \tau}{2\pi D\varepsilon}} \quad (57)$$

Combining Equations (35), (37), (51), (54), and (57) and eliminating the "internal" salinity terms (i.e., $C_{D,m}$, $C_{D/s}$, $C_{F/s}$, and $C_{F,m}$), which are hard to be determined, one obtains

$$\frac{\xi_s}{B_i 2\pi r_i} + \frac{\xi_s}{B_o 2\pi r_o}\left(\frac{r_o}{r_i}\right)^{\frac{\xi_w \tau}{2\pi D\varepsilon}} = \left(\frac{\xi_s}{\xi_w} + C_{D,b}\right)\left(\frac{r_i}{r_i - \delta_D}\right)^{-\frac{\xi_w}{2\pi D}} - \left(\frac{\xi_s}{\xi_w} + C_{F,b}\right)\left(\frac{r_o + \delta_F}{r_o}\right)^{\frac{\xi_w}{2\pi D}}\left(\frac{r_o}{r_i}\right)^{\frac{\xi_w \tau}{2\pi D\varepsilon}} \quad (58)$$

We note that Equation (58) only contains the measurable parameters such as the fiber dimensions of r_i and r_o, the salt permeability coefficients of the skin layers of B_i and B_o, and the bulk concentrations of the two solutions of $C_{D,b}$ and $C_{F,b}$. D is ~1.48×10^{-9} m^2 s^{-1} at 25 °C, which is obtained from the literature (Lobo, 1993). The tortuosity/porosity ratio, τ/ε, can be characterized using the reverse osmosis–forward osmosis (RO–FO) method (Cheng et al., 2016; Chou et al., 2013; Li et al., 2015; Song et al., 2013; Yip et al., 2011). The boundary layer thicknesses (δ_D and δ_F) on the two sides of HF can also be calculated via $\delta_D = D/k_D$ and $\delta_F = D/k_F$, respectively. If we substitute ξ_s with ξ_w using Equation (38) and assume the van't Hoff equation to be valid, one could finally reach the expression for ξ_w

$$\xi_w = A\left\{\frac{\pi_{D,b}\left(\frac{r_i}{r_i - D/k_D}\right)^{-\frac{\xi_w}{2\pi D}} - \pi_{F,b}\left(\frac{r_o + D/k_F}{r_o}\right)^{\frac{\xi_w}{2\pi D}}\left(\frac{r_o}{r_i}\right)^{\frac{\xi_w \tau}{2\pi D\varepsilon}}}{\frac{B}{B_i} + \frac{B r_i}{B_o r_o}\left(\frac{r_o}{r_i}\right)^{\frac{\xi_w \tau}{2\pi D\varepsilon}} + \frac{2\pi r_i B}{\xi_w}\left[\left(\frac{r_o + D/k_F}{r_o}\right)^{\frac{\xi_w}{2\pi D}}\left(\frac{r_o}{r_i}\right)^{\frac{\xi_w \tau}{2\pi D\varepsilon}} - \left(\frac{r_i}{r_i - D/k_D}\right)^{-\frac{\xi_w}{2\pi D}}\right]} - \Delta P\right\} \quad (59)$$

Equation (59) can be further rewritten by replacing the linear flux term (ξ_w) with the areal flux term (J_w) as

$$J_w = A\left\{\frac{\pi_{D,b}\left(\frac{r_i}{r_i - D/k_D}\right)^{-\frac{J_w r_i}{D}} - \pi_{F,b}\left(\frac{r_o + D/k_F}{r_o}\right)^{\frac{J_w r_i}{D}}\left(\frac{r_o}{r_i}\right)^{\frac{J_w r_i \tau}{D\varepsilon}}}{\frac{B}{B_i} + \frac{B r_i}{B_o r_o}\left(\frac{r_o}{r_i}\right)^{\frac{J_w r_i \tau}{D\varepsilon}} + \frac{B}{J_w}\left[\left(\frac{r_o + D/k_F}{r_o}\right)^{\frac{J_w r_i}{D}}\left(\frac{r_o}{r_i}\right)^{\frac{J_w r_i \tau}{D\varepsilon}} - \left(\frac{r_i}{r_i - D/k_D}\right)^{-\frac{J_w r_i}{D}}\right]} - \Delta P\right\} \quad (60)$$

4.2.4 Summary of the Mass Transfer Equations of J_w

Table 4.2 provides a general expression of J_w to summarize the mass transfer equations for different membrane configurations. The components of this

TABLE 4.2

Comparison of mass transfer equations to predict the water flux[†] in PRO, including the geometric/structural characteristics, detrimental effects (ECP, ICP, and reverse salt diffusion), and resistance raised by the skin layer(s) for various membrane configurations

$$J_w = A\left\{\frac{\pi_{D,b}F_{ECP,D} - \pi_{F,b}F_{ECP,F}F_{ICP}}{\Omega + \frac{B}{J_w}\left[F_{ECP,F}F_{ICP} - F_{ECP,D}\right]} - \Delta P\right\}$$

	FS	Inner-Selective HF	Double-Skinned (draw-on-lumen) HF[‡]	Outer-Selective HF	Double-Skinned (draw-on-shell) HF[‡]
Characteristic geometric parameters	t_s	r_i and r_o $(r_o - r_i = t_s)$			
Factor of ECP on the draw side, $F_{ECP,D}$	$\exp\left(-\frac{J_w}{k_D}\right)$	$\left(\frac{r_i}{r_i - D/k_D}\right)^{\frac{-J_w r_i}{D}}$		$\left(\frac{r_o + D/k_D}{r_o}\right)^{\frac{-J_w r_o}{D}}$	
Factor of ICP, F_{ICP}	$\exp\left(\frac{J_w t_s \tau}{D\varepsilon}\right)$	$\left(\frac{r_o}{r_i}\right)^{\frac{J_w r_i \tau}{D\varepsilon}}$		$\left(\frac{r_o}{r_i}\right)^{\frac{J_w r_o \tau}{D\varepsilon}}$	
Factor of ECP on the feed side, $F_{ECP,F}$	$\exp\left(\frac{J_w}{k_F}\right)$	$\left(\frac{r_o + D/k_F}{r_o}\right)^{\frac{J_w r_i}{D}}$		$\left(\frac{r_i}{r_i - D/k_F}\right)^{\frac{J_w r_o}{D}}$	
Resistance, Ω	1		$\frac{B}{B_i} + \frac{B r_i}{B_o r_o}F_{ICP}$	1	$\frac{B}{B_o} + \frac{B r_o}{B_i r_i}F_{ICP}$
Structural parameter, S	$\frac{\tau}{\varepsilon}t_s$	$\frac{\tau}{\varepsilon}r_i \ln\left(\frac{r_o}{r_i}\right)$		$\frac{\tau}{\varepsilon}r_o \ln\left(\frac{r_o}{r_i}\right)$	

[†] Water flux (J_w) is always calculated based on the area of the selective skin (facing the draw solution).

[‡] For double-skinned HF.

draw-on-lumen: $\frac{1}{A} = \frac{1}{A_i} + \frac{r_i}{A_o r_o}$; $\frac{1}{B} = \frac{1}{B_i} + \frac{r_i}{B_o r_o}$

draw-on-shell: $\frac{1}{A} = \frac{r_o}{A_i r_i} + \frac{1}{A_o}$; $\frac{1}{B} = \frac{r_o}{B_i r_i} + \frac{1}{B_o}$

generalized expression have been tabulated as geometric/structural characteristics, detrimental factors (i.e., F) to quantify different concentration polarization effects, and resistance raised by the skin layer(s) for easy comparison. It should be noted that J_w is always calculated based on the area of the skin layer on the draw side (i.e., the selective skin). The first difference between FS and HF configurations is the parameter used to characterize the geometry. Since the thickness of the skin layer is negligible compared to that of the support layer, a FS membrane only needs one geometric parameter to describe its dimension, which is the support layer thickness (t_s). In contrast, the geometric parameters of a HF membrane comprise both the inner radius (r_i) and outer radius (r_o).

The FS and HF models also differ in the mathematical forms of detrimental factors quantifying ECP on the draw and feed sides as well as ICP (i.e., $F_{ECP,D}$, $F_{ECP,F}$, and F_{ICP}) as functions of geometric parameters. For a FS membrane, F_{ICP} is proportional to the natural exponential of t_s, while $F_{ECP,D}$ and $F_{ECP,F}$ are irrelevant to t_s. On the other hand, all the detrimental factors for a HF configuration take the power form with both r_i and r_o involved, because the cross-section periphery changes along the radial direction. Nonetheless, for both FS and HF models, $F_{ECP,D}$ incorporates the J_w term with a minus sign in front, indicating it is a dilutive concentration polarization where $C_{D,b}$ is brought down to $C_{D,m}$. There is no such minus sign associated with J_w for $F_{ECP,F}$ and F_{ICP} because they concentrate $C_{F,b}$ to $C_{D/s}$. One may expect when the radii of a HF membrane (i.e., r_i and r_o) approach infinity, the HF and FS models converge as the HF curvature effect vanishes. Mathematically, this can be proved using the definition of exponential function first given by Euler (Maor, 1994).

$$\exp(x) = \lim_{n \to \infty} \left(1 + \frac{x}{n}\right)^n \tag{61}$$

For the inner-selective HF model:

$$F_{ECP,D}\colon \lim_{r_i \to \infty} \left(\frac{r_i}{r_i - \frac{D}{k_D}} \right)^{\frac{-J_w r_i}{D}} = \lim_{r_i \to \infty} \left(1 + \frac{-\frac{J_w}{k_D}}{\frac{J_w r_i}{D}} \right)^{\frac{J_w r_i}{D}} = \exp\left(-\frac{J_w}{k_D} \right) \tag{62}$$

$$F_{ICP}\colon \lim_{r_i \to \infty} \left(\frac{r_o}{r_i} \right)^{\frac{J_w r_i \tau}{D\varepsilon}} = \lim_{r_i \to \infty} \left(1 + \frac{\frac{J_w t_s \tau}{D\varepsilon}}{\frac{J_w r_i \tau}{D\varepsilon}} \right)^{\frac{J_w r_i \tau}{D\varepsilon}} = \exp\left(\frac{J_w t_s \tau}{D\varepsilon} \right) \tag{63}$$

$$F_{\mathrm{ECP,F}}\colon \lim_{r_o\to\infty}\left(\frac{r_o+\frac{D}{k_F}}{r_o}\right)^{\frac{J_w r_i}{D}}=\lim_{r_o\to\infty}\left(1+\frac{\frac{J_w}{k_F}}{\frac{J_w r_o}{D}}\right)^{\frac{J_w r_o}{D}-\frac{J_w t_s}{D}}=\exp\left(\frac{J_w}{k_F}\right) \tag{64}$$

For the outer-selective HF model:

$$F_{\mathrm{ECP},D}\colon \lim_{r_o\to\infty}\left(\frac{r_o+\frac{D}{k_D}}{r_o}\right)^{\frac{-J_w r_o}{D}}=\lim_{r_o\to\infty}\left(1+\frac{\frac{J_w}{k_D}}{\frac{J_w r_o}{D}}\right)^{\frac{-J_w r_o}{D}}=\exp\left(-\frac{J_w}{k_D}\right) \tag{65}$$

$$F_{\mathrm{ICP}}\colon \lim_{r_i\to\infty}\left(\frac{r_o}{r_i}\right)^{\frac{J_w r_o \tau}{D\varepsilon}}=\lim_{r_i\to\infty}\left(1+\frac{\frac{J_w t_s \tau}{D\varepsilon}}{\frac{J_w r_i \tau}{D\varepsilon}}\right)^{\frac{J_w r_i \tau}{D\varepsilon}+\frac{J_w t_s \tau}{D\varepsilon}}=\exp\left(\frac{J_w t_s \tau}{D\varepsilon}\right) \tag{66}$$

$$F_{\mathrm{ECP,F}}\colon \lim_{r_i\to\infty}\left(\frac{r_i}{r_i-\frac{D}{k_F}}\right)^{\frac{J_w r_o}{D}}=\lim_{r_i\to\infty}\left(1+\frac{-\frac{J_w}{k_F}}{\frac{J_w r_i}{D}}\right)^{-\left(\frac{J_w r_i}{D}+\frac{J_w t_s}{D}\right)}=\exp\left(\frac{J_w}{k_F}\right) \tag{67}$$

The structural parameter (S), quantifying the effective diffusion length ($t_s\tau/\varepsilon$) of solute transport across the support layer, is well defined for the FS model but not for the HF model. In order to obtain a comparable expression of S, we take the natural logarithm of F_{ICP} for both FS and HF models.

$$\text{For FS: } \ln\left(\exp\left(\frac{J_w t_s \tau}{D\varepsilon}\right)\right)=\frac{J_w t_s \tau}{D\varepsilon} \tag{68}$$

$$\text{For inner-selective HF: } \ln\left(\left(\frac{r_o}{r_i}\right)^{\frac{J_w r_i \tau}{D\varepsilon}}\right)=\frac{J_w r_i \tau}{D\varepsilon}\ln\left(\frac{r_o}{r_i}\right) \tag{69}$$

$$\text{For outer-selective HF: } \ln\left(\left(\frac{r_o}{r_i}\right)^{\frac{J_w r_o \tau}{D\varepsilon}}\right)=\frac{J_w r_o \tau}{D\varepsilon}\ln\left(\frac{r_o}{r_i}\right) \tag{70}$$

Equating Equations (69) and (70) to Equation (68) respectively yields

$$\text{For inner-selective HF: } t_s^{\mathrm{eq}}=r_i\ln\left(\frac{r_o}{r_i}\right) \tag{71}$$

$$\text{For outer-selective HF : } t_s^{\mathrm{eq}}=r_o\ln\left(\frac{r_o}{r_i}\right) \tag{72}$$

where the terms shown in Equations (71) and (72) were named as the equivalent membrane thickness by Sivertsen et al. (2012). It means that the FS model can also be employed for HF membranes if an equivalent thickness of the fiber is used in the calculation of S.

The difference between the single- and double-skinned configurations reflects on the value of resistance, Ω, located at the denominator of $\Delta\pi_{\text{eff}}$. It has the same effect as the reverse salt diffusion term (i.e., $\frac{B}{J_w}\left[F_{\text{ECP},F}F_{\text{ICP}} - F_{\text{ECP},D}\right]$) to reduce the effective osmotic driving force. For single-skinned (including both FS and HF) membranes, it equals 1. In comparison, it involves the salt permeability coefficients of skin layers (i.e., B_i and B_o), fiber radii (i.e., r_i and r_o), and F_{ICP} for the double-skinned HF membranes. Mathematically, if one takes the limit of their Ω

$$\lim_{r_i \to \infty}\left(\frac{B}{B_i} + \frac{Br_i}{B_o r_o}F_{\text{ICP}}\right) = \lim_{r_i \to \infty}\left(\frac{B}{B_o} + \frac{Br_o}{B_i r_i}F_{\text{ICP}}\right) = \frac{B}{B_D} + \frac{B}{B_F}\exp\left(\frac{J_w S}{D}\right) \tag{73}$$

where B_D and B_F are the salt permeability coefficients of skin layers facing the draw and feed solutions, respectively. Note that Equation (73) appears as the resistance term of the double-skinned FS model [Equation (20) of Tang et al., (2011)] derived by Tang et al. For the special case where the protective skin does not exist, Equation (73) reduces to 1 as $B_F \to \infty$ and $B = B_D$, reducing to the single-skinned model.

4.3 Results and Discussion of Model Simulations

4.3.1 The Single-Skinned Membrane Configuration

4.3.1.1 Comparison between the FS and HF Models

A simulation study has been carried out to quantify the disparity between the FS and HF models when describing a membrane with the same support layer thickness but various curvatures (i.e., the same $t_s = r_o - r_i$ but different pairs of r_o and r_i). The detailed parameters used in the simulation are summarized in Table 4.3, and the key results are shown in Figures 4.2 and 4.3. As PRO is a pressurized process, the fibers have to meet certain dimension criteria in order to sustain the high pressure (1) expansion for the inner-selective HF or (2) compaction for the outer-selective HF. The corresponding burst pressure and collapse pressure are proportional to $\frac{t_s}{2r_o}$ and $\left(\frac{t_s}{2r_o}\right)^3$ for the inner- and outer-selective HF configurations, respectively (Li and Chung, 2014; Sivertsen et al., 2013). Hence, three kinds of virtual HF membranes are chosen

TABLE 4.3
Parameters employed to simulate the PPD data used in Figures 4.2–4.5

Outer radius, r_o (μm)	300	500	700
Inner radius, r_i (μm)	100	300	500
Support layer thickness, t_s (μm)		200	
Tortuosity/porosity, τ/ε		2–10	
Water permeability coefficient, A (LMH/bar)		0.5–5	
Salt permeability coefficient, B (LMH)		$B = 0.0133A^3$ (Yip and Elimelech, 2011)	
Draw concentration (M)		1	
Feed concentration (mM)		10	
Boundary layer thickness (μm)		0–90	

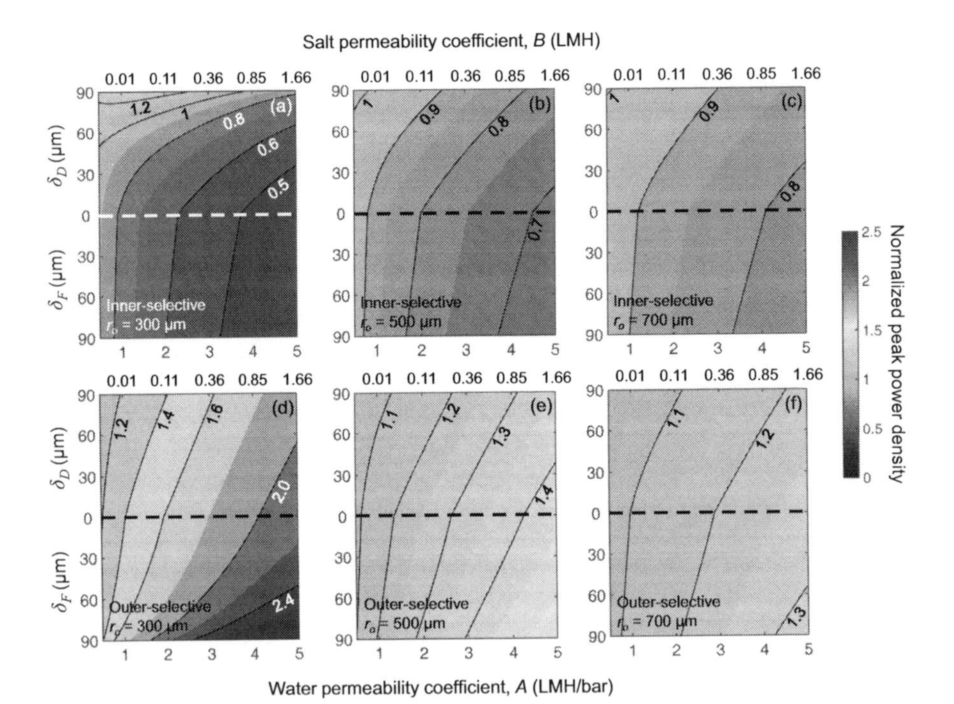

FIGURE 4.2
Normalized PPD (NPPD) of the FS model to (a–c) the inner-selective HF model and (d–f) the outer-selective HF model of various geometric parameters as functions of water permeability coefficient, A, salt permeability coefficient, B, boundary layer thickness on the feed side, δ_F, and the draw side, δ_D. B is correlated with A based on the permeability–selectivity trade-off relation (Yip and Elimelech, 2011). The dashed horizontal line indicates a boundary layer thickness of 0 μm (i.e., no ECP). The tortuosity/porosity ratio is assigned to 5.

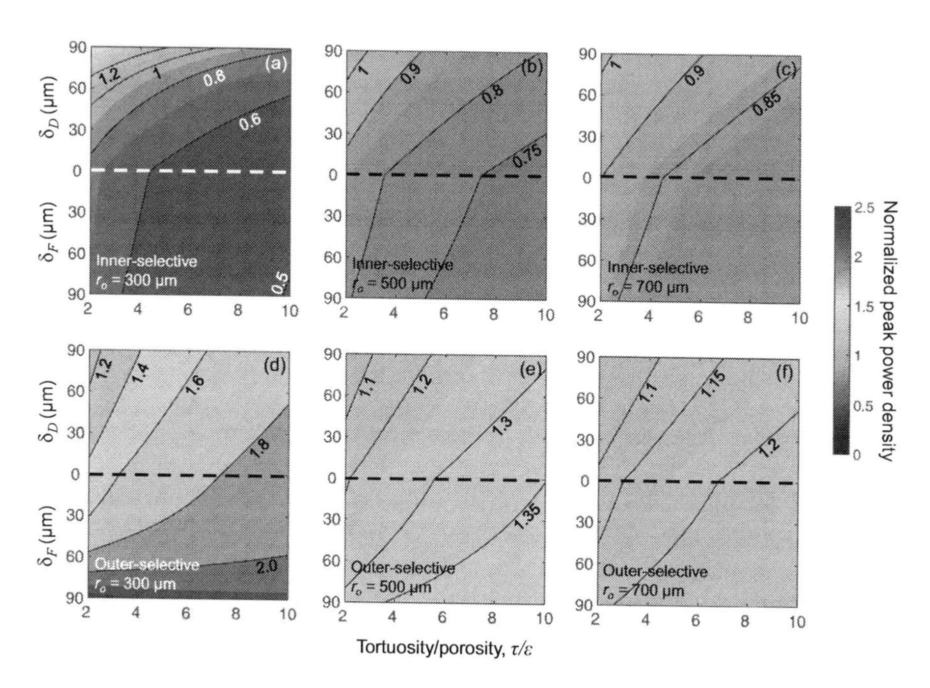

FIGURE 4.3
NPPD of the FS model to (a–c) the inner-selective HF model and (d – f) the outer-selective HF model of various geometric parameters as functions of tortuosity/porosity ratio, τ/ε, boundary layer thickness on the feed side, δ_F, and the draw side, δ_D. The dashed horizontal line indicates a boundary layer thickness of 0 μm (i.e., no ECP). The water and salt permeability coefficients are assigned to 2.5 LMH/bar and 0.21 LMH, respectively.

with different r_o of 300, 500, and 700 μm but at a fixed t_s of 200 μm, covering a wide range of $\frac{t_s}{2r_o}$ of reported HF PRO membranes (Table 4.1). In addition, the ranges of τ/ε and A match with those of PRO membranes (Table 4.1). B is calculated using the permeability–selectivity trade-off relation ($B = 0.0133A^3$) given by Yip and Elimelech (2011). To study the ECP effect on the draw side, δ_D changes from 0 to 90 μm and δ_F is fixed at 0. While δ_F varies by keeping δ_D as 0 when the feed side is concerned.

Figure 4.2 illustrates the NPPD of the FS model to the HF model [Equation (74)] for various geometric parameters as functions of A, B, δ_F, and δ_D.

$$\text{NPPD} = \frac{\text{peak PD}_{\text{FS}}}{\text{peak PD}_{\text{HF}}} \tag{74}$$

The FS model ignores the curvature effect and assumes t_s = 200 μm and τ/ε = 5 for all simulations in Figure 4.2(a) to (f). An arbitrary point in the contour plot indicates the scenario corresponding to the selective skin with A and B values reading from the bottom and top horizontal axes, respectively, and the ECP effect shown as δ_F and δ_D of bottom and top vertical axes, respectively. Generally, the PPD calculated from the FS model is underestimated for the inner-selective HF configuration [i.e., NPPD < 1 for Figure 4.2(a)–(c)] while overestimated for the outer-selective HF configuration [i.e., NPPD > 1 for Figure 4.2(d)–(f)]. As r_o grows, the HF and FS models tend to converge so that NPPD approaches to 1, which is consistent with the analyses in Chapter 4.2.4.

Since $F_{ECP,D}$ plays a more important role in determining the mass transport than $F_{ECP,F}$, a severer change of NPPD occurs when it moves away from the dashed horizontal line (i.e., no ECP) in the δ_D region compared to the δ_F region. Similar implications have been reported elsewhere (Xiong et al., 2017). The only exception is Figure 4.2(d), where NPPD seems to have more pronounced change in the δ_F region. This is attributed to the large curvature of the boundary layer on the feed side bounded by the inner circumference of HF, competing with the aforementioned $F_{ECP,D}$ factor. Since J_w and J_s are expressed as the areal fluxes, they are constant in the FS configuration but variable in the HF configuration. As a result, for Figure 4.2(a) to (f), an increase in A and B leads to both higher J_w and J_s, amplifying the difference between the FS and HF models. Similar trends of contour plots are observed in Figure 4.3, where τ/ε ranges from 2 to 10 with A and B assigned to 2.5 LMH/bar and 0.21 LMH, respectively. This can be understood that a larger τ/ε gives higher resistance of a membrane at a certain thickness, enlarging the difference between the FS and HF configurations.

4.3.1.2 Can the FS Model Apply for HF Membranes?

As discussed in Chapter 4.2.4, the FS model is more familiar to the PRO community and more convenient to be used, as it has four key parameters (A, B, t_s, and τ/ε), whereas the HF model has five parameters (A, B, r_i, r_o, and τ/ε). However, if the HF and FS models offer quite different results as shown in Figures 4.2 and 4.3, then the question is whether the classic FS model is still applicable to describe the mass transport of HF membranes as many researchers have practiced (Cheng et al., 2017; Chou et al., 2013; Li and Chung, 2014; Sun and Chung, 2013; Wan and Chung, 2015; Xiong et al., 2016; Zhang and Chung, 2016)? The short answer is yes under laboratory testing conditions.

To characterize A, B, and S (or τ/ε), the conventional approach is based on the RO–FO method, which can provide good predictions of PRO performance (Cheng et al., 2016; Chou et al., 2013; Li et al., 2015; Song et al., 2013; Yip et al., 2011). This method consists of two steps: (1) measuring water permeability and salt rejection under the RO mode to determine the transport properties (A and B) of the selective skin; and (2) testing the membrane under the FO mode using different NaCl concentrations (e.g., 0.5, 1, 1.5, and 2 M) and DI water as draw and feed solutions, respectively. After obtaining the corresponding water flux (J_w^{FO}), one can calculate S using Equation (75) derived from the FS model by assuming that there is no ECP effect (Bui et al., 2015; Cheng et al., 2016; Li et al., 2015; Yip et al., 2011).

$$S = \frac{\tau}{\varepsilon} t_s = \frac{D}{J_w^{\mathrm{FO}}} \ln \frac{A\pi_D + B}{J_w^{\mathrm{FO}} + B} \tag{75}$$

For the validation of applying the FS model to HF membranes, we employ the HF model to simulate J_w^{FO} using the virtual HF membranes of various curvatures presented in Table 4.3 and then calculate the PRO performance using these flux data with the FS model. Figures 4.4 and 4.5 display the normalized values of PPD predicted using the FS model to the virtual one simulated by the HF model. Surprisingly, two models are identical when the boundary layer thickness equals 0 (i.e., no ECP), even if the HF curvature is large (e.g., r_o = 300 μm). The relevant root causes can be examined through the characterization protocol of A, B, and S (or τ/ε). Theoretically, A and B are the intrinsic properties of the selective skin and should not be varied due to the curvature of HF membranes, since the scale of mass transport across the selective skin is orders of magnitude smaller than the geometric scale of the support layer (Lin, 2016; Xiong et al., 2016). In order to get the expression of S specifically for the HF configuration, one can derive it from the mass transfer equations under FO [refer to Equations (6) and (9) of Lin (2016)].

$$\text{For inner} - \text{selective HF: } S = \frac{\tau}{\varepsilon} r_i \ln\left(\frac{r_o}{r_i}\right) = \frac{D}{J_w^{\mathrm{FO}}} \ln \frac{A\pi_D + B}{J_w^{\mathrm{FO}} + B} \tag{76}$$

$$\text{For outer} - \text{selective HF: } S = \frac{\tau}{\varepsilon} r_o \ln\left(\frac{r_o}{r_i}\right) = \frac{D}{J_w^{\mathrm{FO}}} \ln \frac{A\pi_D + B}{J_w^{\mathrm{FO}} + B} \tag{77}$$

Because Equations (76) and (77) have the same mathematical form on the right hand side as that of Equation (75), an equal value of S is expected if A, B, and J_w^{FO} are predetermined using the RO–FO method. In other words, the resultant structural parameter itself does not tell the geometric difference of membranes. Note that the S values calculated via Equations (76) and (77) are bonded with the term of t_s^{eq} (i.e., $r_i \ln(\frac{r_o}{r_i})$ and $r_o \ln(\frac{r_o}{r_i})$), which guarantees the two models converge at no ECP effect.

As the boundary layer starts to build on the draw and feed sides, for inner-selective HF membranes, the FS model gives reasonable predictions within the δ_F region but fails when δ_D is large (60–90 μm) at r_o = 300 μm

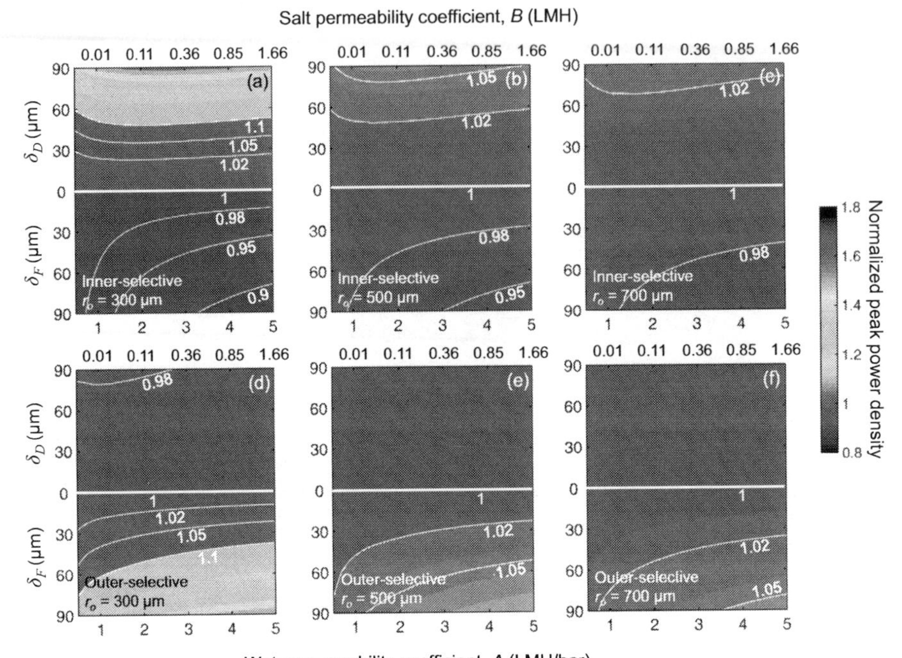

FIGURE 4.4
NPPD by dividing the predictions from the FS model to those from (a–c) inner-selective HF model and (d–f) outer-selective HF model of various geometric parameters as functions of water permeability coefficient, A, salt permeability coefficient, B, boundary layer thickness on the feed side, δ_F, and the draw side, δ_D. B is correlated with A based on the permeability–selectivity trade-off relation (Yip and Elimelech, 2011). The tortuosity/porosity ratio is assigned to 5.

[Figures 4.4(a) and 4.5(a)]. In contrast, the FS model gives reasonable predictions within the δ_D region but fails when δ_F is large (60–90 μm) at r_o = 300 μm for outer-selective HF membranes [Figures 4.4(d) and 4.5(d)]. This difference is caused by the large curvature bounded by the inner circumference of HF. It means that even if a small amount of water transports across the membrane, it will result in severe dilution or concentration in the region of δ_D or δ_F, respectively. Nonetheless, in lab-scale PRO tests, a high ratio of solution flow rate to membrane area on both feed and draw sides is normally employed to minimize the dilution effect on the observed water flux. Thus, the reported data are close to the membrane "true" properties (Straub et al., 2016; Zhang and Chung, 2016). In other words, most lab-scale PRO tests are run at conditions where F_{ICP} strongly dominates, so that the FS model is applicable for HF membranes. Even if $F_{ECP,D}$ is considered in the case of the outer-selective HF configuration

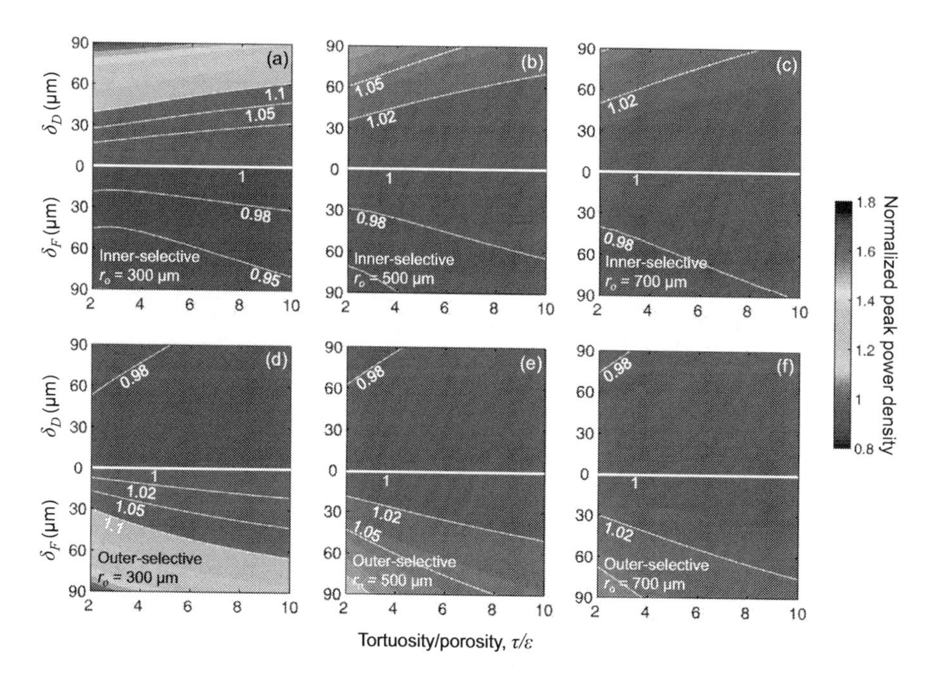

FIGURE 4.5
NPPD by dividing the predictions from the FS model to those from (a–c) inner-selective HF model and (d–f) outer-selective HF model of various geometric parameters as functions of tortuosity/porosity ratio, τ/ε, boundary layer thickness on the feed side, δ_F, and the draw side, δ_D. The water and salt permeability coefficients are assigned to 2.5 LMH/bar and 0.21 LMH, respectively.

(Cheng et al., 2017), the FS model still does a good prediction with a less than 2% deviation, as shown in Figure 4.4(d)–(f) and Figure 4.5(d)–(f). However, one should bear in mind that the right choice of mathematical model corresponding to the membrane configuration is always encouraged, as it best describes the real situation.

4.3.1.3 Structural Parameter (S) of FS and HF Membranes

To further illustrate the structural parameter of FS and HF membranes, Figure 4.6(a) shows the contour plot of S from literature-reported membranes (Table 4.1), which were experimentally determined via the RO–FO method as described in Chapter 4.3.1.2. The horizontal axis shows the curvature factor, which is defined as the ratio of the equivalent membrane thickness (i.e., $r_i\ln(\frac{r_o}{r_i})$ and $r_o\ln(\frac{r_o}{r_i})$ for inner- and outer-selective HFs, respectively) to the support layer thickness (t_s). The r_i, r_o, and t_s values can be gotten from the reported data or scanning electron microscopic

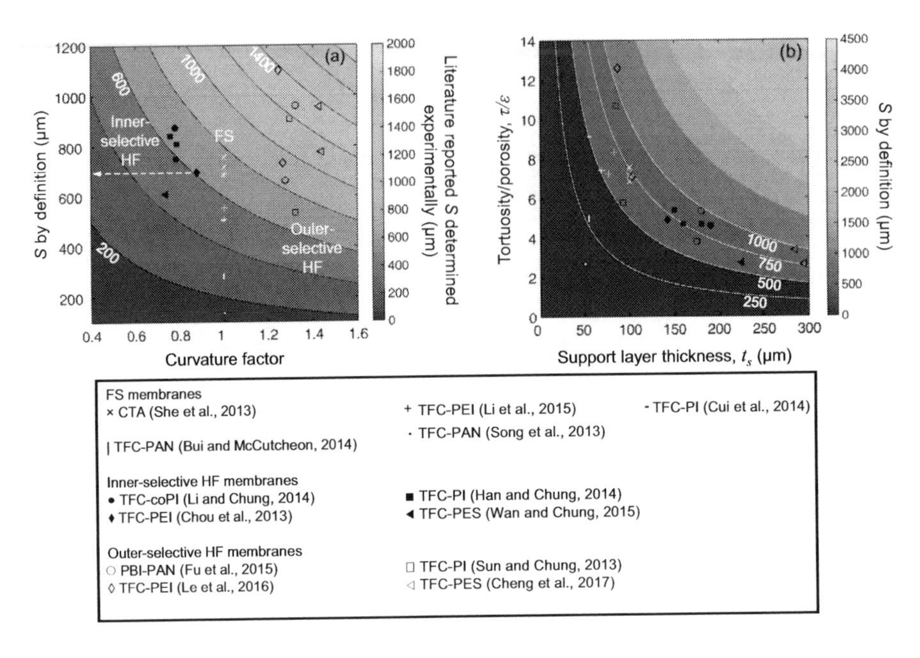

FIGURE 4.6
Contour plots of (a) structural parameter, S, from literature determined experimentally (Table 4.1) as a function of curvature factor [1 for FS, $r_i/t_s \times \ln(r_o/r_i) < 1$ for inner-selective HF and $r_o/t_s \times \ln(r_o/r_i) > 1$ for outer-selective HF] and S by definition as tortuosity/porosity ratio times support layer thickness, $\tau/\varepsilon \times t_s$; and (b) S by definition as $\tau/\varepsilon \times t_s$ as a function of t_s and τ/ε.

pictures in the literature. For FS membranes, the curvature factor is equal to 1, which means S obtained from the RO–FO method truly represents the value of $\tau/\varepsilon \times t_S$. However, it is smaller than 1 for inner-selective HF membranes and greater than 1 for outer-selective HF membranes, respectively (Figure 4.7).

If we remove the influence of curvature factor on the experimentally determined structural parameter, what will happen? The left vertical axis shows the resultant S (by definition as $\tau/\varepsilon \times t_S$) by dividing the experimental value (i.e., the elevation axis on the right vertical side of Figure 4.6(a)) to its corresponding curvature factor. A comparison between the left vertical and elevation values indicates the experimentally determined S is underestimated for inner-selective HF membranes and overestimated for outer-selective HF membranes. For example, in Figure 4.6(a), the literature reported that structural parameter of the inner-selective HF membrane (Chou et al., 2013) falls near the contour line of 600. If one assesses it using the left vertical axis of S by definition, it transfers to a value of around 700 (shown as the white arrow in Figure 4.6(a)). The deviation of S due to the curvature factor of HF

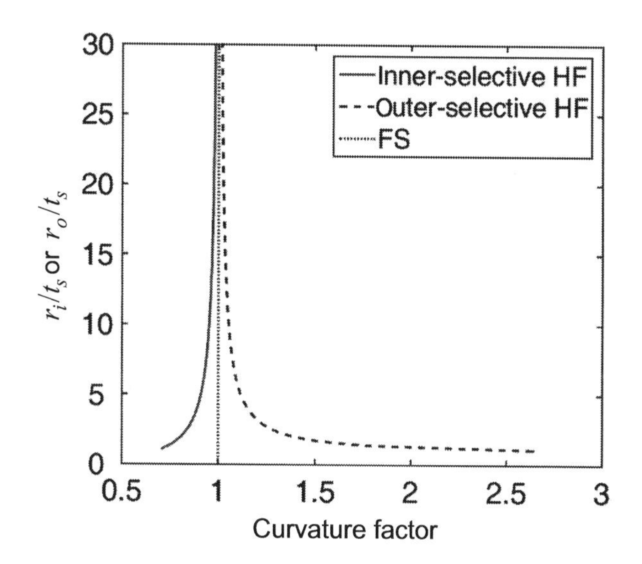

FIGURE 4.7
Normalized inner radius (r_i/t_s) for inner-selective HF membranes or normalized outer radius (r_o/t_s) for outer-selective HF membranes vs. curvature factor.

membranes can also be used to explain why generally the FS model under predicts the PPD than the HF model (i.e., NPPD < 1) for inner-selective HF membranes, while it over predicts the PPD than the HF model (i.e., NPPD > 1) for outer-selective HF membranes in Figures 4.2 and 4.3. Thus, in order to have a fair comparison of structural parameter among FS, inner- and outer-selective HF membranes, S determined experimentally via the RO–FO method should be adjusted using the curvature factor to yield the form of $\tau/\varepsilon \times t_s$.

Figure 4.6(b) re-plots the structural parameters defined as $\tau/\varepsilon \times t_s$ of those membranes in Figure 4.6(a) as a function of t_s and τ/ε. It can be seen that most membranes have S values between 500 and 1000 μm. In general, FS membranes have thinner support layers (50–100 μm) as feed spacers are used to reinforce the mechanical stability. While HF membranes have much thicker walls (100–300 μm) due to their self-supported nature. The τ/ε value of FS and HF membranes spreads over 2 to 10. It is known that the membrane with smaller values of t_s and τ/ε is preferable if it can withstand the high-pressure operations in PRO processes. Based on this criterion, the thin-film nanofiber composite membrane developed by Song et al. shows a low S of 135 μm and a reasonably high ΔP over 15 bar (Song et al., 2013), representing the state-of-the-art substrate for FS PRO membranes.

4.3.2 The Double-Skinned Membrane Configuration

4.3.2.1 Transport Properties of the Skin Layers

For the double-skinned FS configuration, Tang et al. pioneered its modeling work and conducted a detailed analysis on how the transport properties of skin layers influence the overall mass transport for FO processes (Tang et al., 2011). However, no HF PRO membrane of such a configuration has been studied before. Recently, Han et al. developed a novel TFC HF membrane with double-polyamide skin layers on a polyethersulfone (PES) substrate (named as *d*TFC-PES). It has a draw-on-lumen configuration where the selective and protective skins are on the inner and outer surfaces, respectively. By using a real wastewater retentate as the feed solution, *d*TFC-PES shows superior antifouling performance and flux recovery via simple cleaning processes because its outer polyamide protective skin has an outstanding rejection to the foulants (Han et al., 2017).

To further understand the transport properties (A and B) of the skin layers of *d*TFC-PES, the mass transfer equations derived in Chapter 4.2.3 are employed. Table 4.4 lists the specifications of dimensions and transport properties of *d*TFC-PES. Since the HF support was spun from the typical PES dope solution formulated by our group (Cheng et al., 2016; Wan and Chung, 2015) and different interfacial polymerization techniques of Wan and Chung (2015) and Cheng et al. (2016) were applied to deposit the polyamide skins on the inner and outer surfaces separately, the A and B values of the inner and outer skin layers of *d*TFC-PES can be taken from the measurements of Wan and Chung (2015) and Cheng et al. (2016)

TABLE 4.4

Specifications of dimensions and transport properties (water permeability and salt permeability coefficients) of the double-skinned HF membrane reported in Han et al., (2017)

	Diameter (μm)	Water Permeability Coefficient (LMH per bar)	Salt Permeability Coefficient (LMH)
Inner circumference	525^{*}	$A_i = 3.5^{\dagger}$	$B_i = 0.28^{\dagger}$
Outer circumference	1125^{*}	$A_o = 1.42^{\ddagger}(4.55^{\S})$	$B_o = 0.028^{\ddagger}(1.25^{\S})$
Overall	–	$A = 1.63^{\Vert}$ $(1.5 \pm 0.2^{*})$	$B = 0.05^{\Vert}$ (0.02^{*})

† Taken from Wan and Chung (2015) (measured at 5 bar).
‡ Taken from TFC-II of Cheng et al. (2016) (measured at 1 bar).
‖ Calculated by $\frac{1}{A} = \frac{1}{A_i} + \frac{r_i}{A_o r_o}$; $\frac{1}{B} = \frac{1}{B_i} + \frac{r_i}{B_o r_o}$ (draw-on-lumen configuration).
* Reported values in Han et al. (2017) (measured at 1 bar).
§ Fitting results at 15 bar based on the model of draw-on-lumen configuration using the reported power density and $B_o = 0.0133 A_o^{\ 3}$ Yip and Elimelech (2011).

correspondingly. By applying Equations (40) and (41), the overall A and B are calculated to be 1.63 LMH/bar and 0.05 LMH, respectively. These values agree with the experimental ones (Table 4.4). However, if one calculates the power density of *d*TFC-PES using Equations (17) and (60) (without considering $F_{ECP,D}$ and $F_{ECP,F}$ based on its testing conditions), it is 4.9 W m^{-2} at 15 bar using 1 M NaCl and DI water as feeds, which is below 10.7 W m^{-2} as reported in Han et al. (2017). One possible reason for this discrepancy is due to the ignorance of the change of transport properties in the skin layers, as *d*TFC-PES undergoes expansion in a PRO process. Previous studies showed that, within the safe operating range, A is almost constant while B slightly grows with increasing ΔP for the inner-selective skin (Gai et al., 2016; Wan and Chung, 2015). In contrast, the outer-protective skin may deform since it locates at the opposite direction of the applied hydraulic pressure. Thus, to count for this effect, its transport properties should be characterized at 15 bar for the calculation of power density. However, these data are not available as only 1 bar was employed for the measurements of A and B (Han et al., 2017). If one applies the reported power density of 10.7 W m^{-2} and the permeability–selectivity trade-off relation (Yip and Elimelech, 2011) to fit Equations (17) and (60) simultaneously, A and B of the protective skin are found to be 4.55 LMH/bar and 1.25 LMH at 15 bar, respectively. Consistent with our hypothesis, the protective skin does show some deformation (i.e., larger A and B/A) compared to that at 1 bar (Table 4.4). To further validate the model-fitting values of A and B, the power density of *d*TFC-PES is calculated to be 10 W m^{-2} at 15 bar using 1 M NaCl and synthetic wastewater retentate [11 mM NaCl (Wan and Chung, 2015)] as feeds. This value agrees well with the experimental data of 9.8 W m^{-2} (Han et al., 2017).

4.3.2.2 Implications for Antifouling Features

Ideally, the protective skin of *d*TFC-PES formed by a polyamide layer should have almost 100% rejections to the foulants in wastewater retentate (Han et al., 2017; Wan and Chung, 2015). Thus these foulants will build up a cake on top of the outer skin layer rather than penetrating into the porous support (Hu et al., 2016; She et al., 2016). Meanwhile, the membrane transport properties (A and B) could be protected and remain almost unchanged. Figure 4.8 depicts the corresponding salt concentration profile across the membrane. For simplicity, ECP on both draw and feed sides are neglected. We note that the cake layer causes an additional transport resistance and should be taken into account when formulating the mass transfer equations (Nagy et al., 2016). In this way, the water flux of *d*TFC-PES is able to be calculated at different cake layer thicknesses, which can be further translated to the corresponding hydraulic resistance induced by the foulants, R_f, based on the

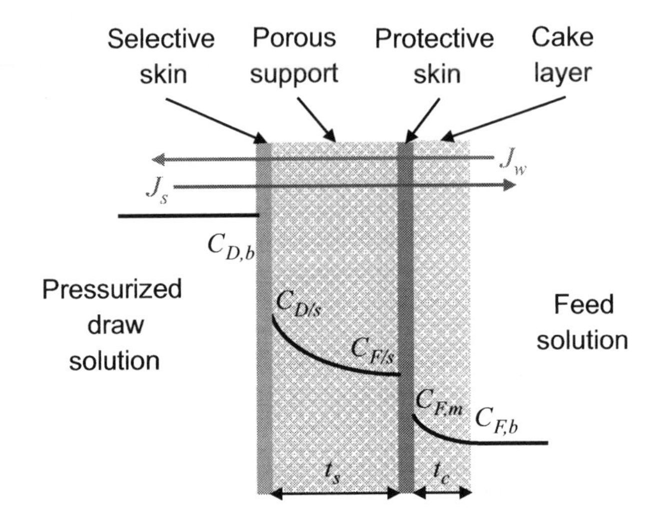

FIGURE 4.8
Schematics of the salt concentration profile across the double-skinned (draw-on-lumen) HF membrane including the cake layer built-up by the foulants in the feed solution. ECP on both draw and feed sides is neglected.

osmotic-resistance filtration model (She et al., 2016). To evaluate the antifouling features of *d*TFC-PES, a single-skinned HF membrane (retaining the inner-selective skin and HF support layer of *d*TFC-PES) is simulated for comparison. The detailed parameters and assumptions used in the modeling are provided as follows.

The same procedures of Chapter 4.2.3 can be used to derive the mass transfer equation of water flux in the presence of a fouling cake layer on a double-skinned (draw-on-lumen) HF membrane. Equation (42) serves as the governing equation. The corresponding BCs are depicted in Figure 4.8 without considering ECP on both draw and feed sides.

$$\text{BC1: } C\left(r_i^-\right) = C_{D,b} \tag{78}$$

$$\text{BC2: } C\left(r_i^+\right) = C_{D/s} \tag{79}$$

$$\text{BC3: } C\left(r_o^-\right) = C_{F/s} \tag{80}$$

$$\text{BC4: } C\left(r_o^+\right) = C_{F,m} \tag{81}$$

$$\text{BC5: } C(r_o + t_c) = C_{F,b} \tag{82}$$

where t_c is the thickness of the cake layer. The expression of J_w is obtained as

$$J_w = A\left\{\frac{\pi_{D,b} - \pi_{F,b}\left(\frac{r_o}{r_i}\right)^{\frac{J_w r_i \tau}{D\varepsilon}}\left(\frac{r_o+t_c}{r_o}\right)^{\frac{J_w r_i \tau_c}{D\varepsilon_c}}}{\frac{B}{B_i} + \frac{Br_i}{B_o r_o}\left(\frac{r_o}{r_i}\right)^{\frac{J_w r_i \tau}{D\varepsilon}} + \frac{B}{J_w}\left[\left(\frac{r_o}{r_i}\right)^{\frac{J_w r_i \tau}{D\varepsilon}}\left(\frac{r_o+t_c}{r_o}\right)^{\frac{J_w r_i \tau_c}{D\varepsilon_c}} - 1\right]} - \Delta P\right\} = A(\Delta\pi_{\text{eff}} - \Delta P)_c \tag{83}$$

where ε_c and τ_c are the porosity and tortuosity of the cake layer, respectively. The subscript c indicates the cake layer. Note that, although Equation (83) assumes no ECP, one can extend it to consider ECP on the draw or feed side by including $F_{ECP,D}$ or $F_{ECP,F}$, respectively, as shown in Table 4.2. Compared to the one without the cake layer (Equation (84)), Equation (83) has an additional term of $\left(\frac{r_o+t_c}{r_o}\right)^{\frac{J_w r_i \tau_c}{D\varepsilon_c}}$, which has a similar effect as F_{ICP}.

$$J_w = A\left\{\frac{\pi_{D,b} - \pi_{F,b}\left(\frac{r_o}{r_i}\right)^{\frac{J_w r_i \tau}{D\varepsilon}}}{\frac{B}{B_i} + \frac{Br_i}{B_o r_o}\left(\frac{r_o}{r_i}\right)^{\frac{J_w r_i \tau}{D\varepsilon}} + \frac{B}{J_w}\left[\left(\frac{r_o}{r_i}\right)^{\frac{J_w r_i \tau}{D\varepsilon}} - 1\right]} - \Delta P\right\} = A(\Delta\pi_{\text{eff}} - \Delta P) \tag{84}$$

By applying the osmotic-resistance filtration model (She et al., 2016) for the fouled membrane, the additional resistance brought by the cake layer can be lumped into the overall water permeability coefficient of the fouled membrane, A_f (Wan and Chung, 2015), where A_f is related to the feed water viscosity, μ, membrane hydraulic resistance, R_m, and hydraulic resistance induced by the foulants, R_f.

$$J_{w,f} = A_f(\Delta\pi_{\text{eff}} - \Delta P) \tag{85}$$

$$A_f = \frac{1}{\mu(R_m + R_f)} \tag{86}$$

Note that Equation (85) is applicable only if the cake layer thickness is much smaller than that of the porous support [e.g., $t_c = 3.53$ μm ≪ $t_s = 300$ μm for the *d*TFC-PES (Han et al., 2017)]. For the pristine membrane, which has no R_f, Equation (86) reduces to

$$A = \frac{1}{\mu R_m} \tag{87}$$

Combining Equations (83)–(87) and eliminating R_m and A_f, one can get the expression of R_f for the fouled membrane.

$$R_f = \frac{1}{\mu A}\left\{\frac{(\Delta\pi_{\text{eff}} - \Delta P)}{(\Delta\pi_{\text{eff}} - \Delta P)_c} - 1\right\} \tag{88}$$

For the properties of the fouled cake layer, ε_c can be estimated as 0.77 from the measurement of an alginate organic fouling layer in a FO process [see Figure 3A of (Xie et al., 2015)], while τ_c is calculated using the correlation proposed for fine-grained sediments (Boudreau, 1996).

$$\tau_c = 1 - \ln\left(\varepsilon_c^2\right) \tag{89}$$

In such a way, the tortuosity/porosity ratio (τ_c/ε_c) of the cake layer is approximated as 2.

On the other hand, the water flux of a fouled single-skinned (inner-selective) HF membrane is predicted via Equation (90).

$$J_{w,f} = A_f\left\{\frac{\pi_{D,b} - \pi_{F,b}\left(\frac{r_o}{r_i}\right)^{\frac{J_{w,f} r_i \tau}{D\varepsilon}}}{1 + \frac{B}{J_{w,f}}\left[\left(\frac{r_o}{r_i}\right)^{\frac{J_{w,f} r_i \tau}{D\varepsilon}} - 1\right]} - \Delta P\right\} \tag{90}$$

where A_f and B can be obtained by interpolating the experimental data presented in Figures 9a, 10a, and 11a of Wan and Chung (2015). B is founded almost as a constant of 0.33 LMH while A_f is significantly affected when fouling occurs. Figure 4.9 plots the A_f value versus R_f at ΔP = 15 bar. A nice fitting is achieved by using the classic pore constriction model ($A_f = 3.5 \times (1 + 2.1359 \times 10^{-15} R_f)^{-2}$) (Xiong et al., 2016; Zeman and Zydney, 1996).

Figure 4.10 compares the water flux of fouled *d*TFC-PES and virtual single-skinned HF membrane as functions of R_f at 15 bar using 1 M NaCl and synthetic wastewater retentate as feeds. Although *d*TFC-PES starts with a lower initial water flux compared to the single-skinned one due to the presence of additional protective skin, it shows a minimal flux decline when fouling occurs. On the other hand, a substantial reduction in water flux is observed for the virtual single-skinned HF membrane as the foulants easily penetrate and get stuck underneath the selective skin (Mi and Elimelech, 2008; She et al., 2016), ending with an even lower water flux. It can also be seen that the R_f value of Figure 4.10(b) is three orders of magnitude larger than that of Figure 4.10(a), confirming much severer fouling existing in the virtual single-skinned HF membrane. Considering the curves of Figure 4.10(a) and (b) have

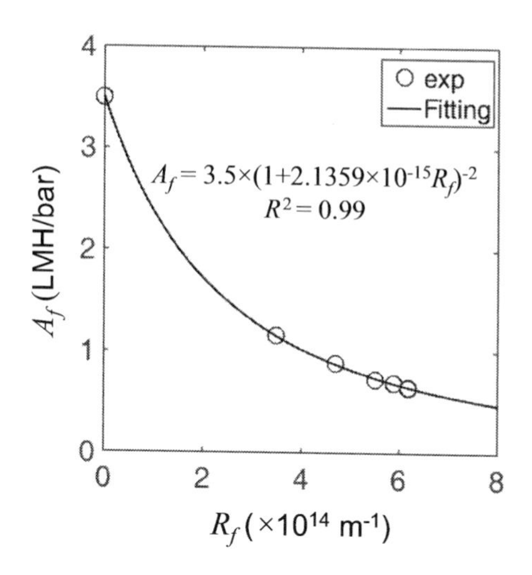

FIGURE 4.9
Water permeability coefficient of the fouled membrane, A_f, versus hydraulic resistance induced by the foulants, R_f, at ΔP = 15 bar. Experimental data (exp) are taken from Wan and Chung (2015), which are fitted by the classic pore constriction model (fitting).

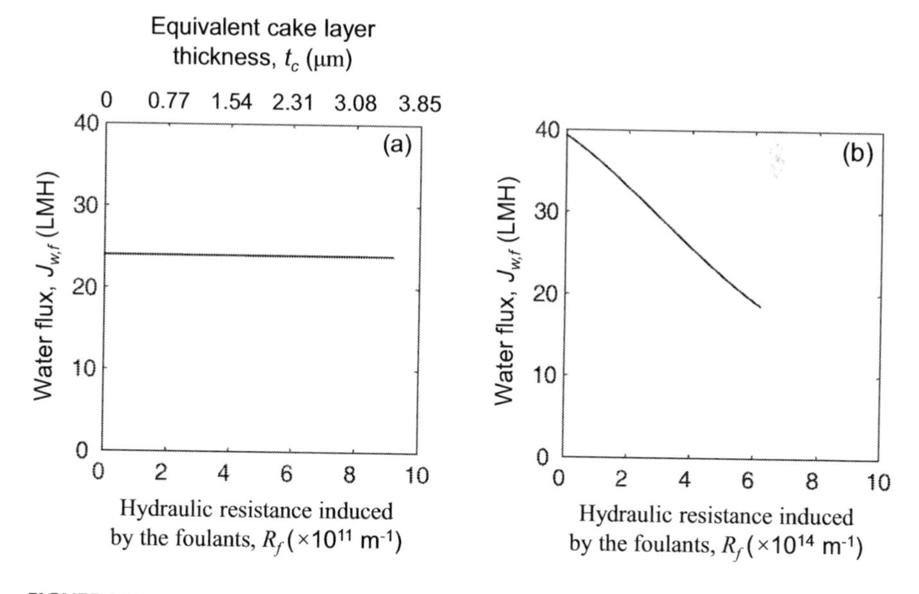

FIGURE 4.10
Comparison of water flux of fouled membranes, $J_{w,f}$, as functions of hydraulic resistance induced by the foulants, R_f, for (a) double-skinned (draw-on-lumen) and (b) single-skinned (inner-selective) HF configurations at ΔP = 15 bar using 1 M NaCl and synthetic wastewater retentate as feeds. The equivalent cake layer thickness, t_c, in the top horizontal axis of (a) is correlated using the osmotic-resistance filtration model (She et al., 2016).

timescales of 250 and 0.5 h, respectively [based on their corresponding testing conditions reported (Han et al., 2017; Wan and Chung, 2015)], *d*TFC-PES is superior in antifouling performance and promising for real applications. More importantly, its associated external fouling layer shows great reversibility, which can be more effectively removed by the developed cleaning strategies compared to the single-skinned configuration (Han et al., 2017; She et al., 2016). It has been demonstrated that physically flushing the fouled *d*TFC-PES membrane surface by either freshwater or commercial cleaner Genesol 704 can efficiently restore the water flux back to 87% or 98% of its initial level, respectively (Han et al., 2017).

4.4 Conclusions

In this chapter, a set of analytical PRO models has been derived to describe the mass transport of various membrane configurations, including FS, single-, and double-skinned HFs. This provides a universal platform to evaluate the existing state-of-the-art PRO membranes. The following conclusions can be drawn from this chapter:

(1) For the single-skinned configuration, dramatic differences exist between the FS and HF models when describing a membrane with the same support layer thickness but various curvatures. However, this does not negate the validity of applying the classic FS model to quantify the mass transport across HF membranes. Their structural parameters determined via conventional RO–FO method have "automatically" taken the curvature effect into account (i.e., bonded with the term of equivalent membrane thickness). Compared to the virtual ones simulated using the HF model, the FS model gives identical results of PPD if no ECP effect is considered. This is the scenario often encountered under laboratory testing conditions due to high feed flowrates applied. Nonetheless, in order to have a fair comparison of structural parameter of different membrane configurations, the experimentally determined S should be adjusted using the curvature factor to yield the form of $\tau/\varepsilon \times t_S$.

(2) For the double-skinned HF membrane with the draw-on-lumen configuration, the outer protective skin is found to deform to some degrees (i.e., larger A and B/A) by the expansion of HF under high-pressure PRO processes. Even so, this membrane experimentally shows superior antifouling performance than other reported ones in the literature. By modeling the foulant build-up as an external

cake layer on top of the protective skin, one can mathematically confirm the advantages of double-skinned configuration in terms of minimal flux decline and lower transport resistance when fouling occurs.

References

Achilli, A., Cath, T.Y., Childress, A.E., 2009. Power generation with pressure retarded osmosis: An experimental and theoretical investigation. *J. Membr. Sci.* 343, 42–52.

Boudreau, B.P., 1996. The diffusive tortuosity of fine-grained ulithified sediments. *Geochim. Cosmochim. Acta* 60, 3139–3142.

Bui, N.N., Arena, J.T., McCutcheon, J.R., 2015. Proper accounting of mass transfer resistances in forward osmosis: Improving the accuracy of model predictions of structural parameter. *J. Membr. Sci.* 492, 289–302.

Bui, N.N., McCutcheon, J.R., 2014. Nanofiber supported thin-film composite membrane for pressure-retarded osmosis. *Environ. Sci. Technol.* 48, 4129–4136.

Cheng, Z.L., Li, X., Feng, Y., Wan, C.F., Chung, T.S., 2017. Tuning water content in polymer dopes to boost the performance of outer-selective thin-film composite (TFC) hollow fiber membranes for osmotic power generation. *J. Membr. Sci.* 524, 97–107.

Cheng, Z.L., Li, X., Liu, Y.D., Chung, T.S., 2016. Robust outer-selective thin-film composite polyethersulfone hollow fiber membranes with low reverse salt flux for renewable salinity-gradient energy generation. *J. Membr. Sci.* 506, 119–129.

Chou, S., Wang, R., Fane, A.G., 2013. Robust and high performance hollow fiber membranes for energy harvesting from salinity gradients by pressure retarded osmosis. *J. Membr. Sci.* 448, 44–54.

Chung, T.S., Luo, L., Wan, C.F., Cui, Y., Amy, G., 2015. What is next for forward osmosis (FO) and pressure retarded osmosis (PRO). *Sep. Purif. Technol.* 156, 856–860.

Cui, Y., Liu, X.Y., Chung, T.S., 2014. Enhanced osmotic energy generation from salinity gradients by modifying thin film composite membranes. *Chem. Eng. J.* 242, 195–203.

Fu, F.J., Zhang, S., Chung, T.S., 2015. Sandwich-structured hollow fiber membranes for osmotic power generation. *Desalination* 376, 73–81.

Gai, W., Li, X., Xiong, J.Y., Wan, C.F., Chung, T.S., 2016. Evolution of micro-deformation in inner-selective thin film composite hollow fiber membranes and its implications for osmotic power generation. *J. Membr. Sci.* 516, 104–112.

Gerstandt, K., Peinemann, K.V., Skilhagen, S.E., Thorsen, T., Holt, T., 2008. Membrane processes in energy supply for an osmotic power plant. *Desalination* 224, 64–70.

Han, G., Cheng, Z.L., Chung, T.S., 2017. Thin-film composite (TFC) hollow fiber membrane with double-polyamide active layers for internal concentration polarization and fouling mitigation in osmotic processes. *J. Membr. Sci.* 523, 497–504.

Han, G., Chung, T.S., 2014. Robust and high performance pressure retarded osmosis hollow fiber membranes for osmotic power generation. *AIChE J.* 60, 1107–1119.

Han, G., Wang, P., Chung, T.S., 2013. Highly robust thin-film composite pressure retarded osmosis (PRO) hollow fiber membranes with high power densities for renewable salinity-gradient energy generation. *Environ. Sci. Technol.* 47, 8070–8077.

Han, G., Zhang, S., Li, X., Chung, T.S., 2015. Progress in pressure retarded osmosis (PRO) membranes for osmotic power generation. *Prog. Polym. Sci.* 51, 1–27.

Hu, M., Zheng, S., Mi, B., 2016. Organic fouling of graphene oxide membranes and its implications for membrane fouling control in engineered osmosis. *Environ. Sci. Technol.* 50, 685–693.

Ingole, P.G., Kim, K.H., Park, C.H., Choi, W.K., Lee, H.K., 2014. Preparation, modification and characterization of polymeric hollow fiber membranes for pressure-retarded osmosis. *RSC Adv.* 4, 51430–51439.

Kumano, A., Marui, K., Terashima, Y., 2016. Hollow fiber type PRO module and its characteristics. *Desalination* 389, 149–154.

Kurihara, M., Sakai, H., Tanioka, A., Tomioka, H., 2016. Role of pressure-retarded osmosis (PRO) in the mega-ton water project. *Desalin. Water Treat.* 57, 26518–26528.

Le, N.L., Bettahalli, N.M.S., Nunes, S.P., Chung, T.S., 2016. Outer-selective thin film composite (TFC) hollow fiber membranes for osmotic power generation. *J. Membr. Sci.* 505, 157–166.

Lee, J., Kim, S., 2016. Predicting power density of pressure retarded osmosis (PRO) membranes using a new characterization method based on a single PRO test. *Desalination* 389, 224–234.

Lee, K.L., Baker, R.W., Lonsdale, H.K., 1981. Membranes for power generation by pressure-retarded osmosis. *J. Membr. Sci.* 8, 141–171.

Li, X., Cai, T., Chung, T.S., 2014. Anti-fouling behavior of hyperbranched polyglycerol-grafted poly(ether sulfone) hollow fiber membranes for osmotic power generation. *Environ. Sci. Technol.* 48, 9898–9907.

Li, X., Chung, T.S., 2014. Thin-film composite P84 co-polyimide hollow fiber membranes for osmotic power generation. *Appl. Energy* 114, 600–610.

Li, Y., Wang, R., Qi, S., Tang, C.Y., 2015. Structural stability and mass transfer properties of pressure retarded osmosis (PRO) membrane under high operating pressures. *J. Membr. Sci.* 488, 143–153.

Lin, S., 2016. Mass transfer in forward osmosis with hollow fiber membranes. *J. Membr. Sci.* 514, 176–185.

Lobo, V.M.M., 1993. Mutual diffusion coefficients in aqueous electrolyte solutions. *Pure Appl. Chem.* 65, 2613–2640.

Maor, E., 1994. *e: The Story of a Number*, Princeton Univeristy Press, Princeton, NJ.

Mi, B., Elimelech, M., 2008. Chemical and physical aspects of organic fouling of forward osmosis membranes. *J. Membr. Sci.* 320, 292–302.

Mulder, M., 1996. *Basic Principles of Membrane Technology*, 2nd ed., Springer Netherlands, Dordrecht.

Nagy, E., Dudás, J., Hegedüs, I., 2016. Improvement of the energy generation by pressure retarded osmosis. *Energy* 116, 1323–1333.

She, Q., Hou, D., Liu, J., Tan, K.H., Tang, C.Y., 2013. Effect of feed spacer induced membrane deformation on the performance of pressure retarded osmosis (PRO): Implications for PRO process operation. *J. Membr. Sci.* 445, 170–182.

She, Q., Wang, R., Fane, A.G., Tang, C.Y., 2016. Membrane fouling in osmotically driven membrane processes: A review. *J. Membr. Sci.* 499, 201–233.

Sivertsen, E., Holt, T., Thelin, W., Brekke, G., 2012. Modelling mass transport in hollow fibre membranes used for pressure retarded osmosis. *J. Membr. Sci.* 417–418, 69–79.

Sivertsen, E., Holt, T., Thelin, W., Brekke, G., 2013. Pressure retarded osmosis efficiency for different hollow fibre membrane module flow configurations. *Desalination* 312, 107–123.

Song, X., Liu, Z., Sun, D.D., 2013. Energy recovery from concentrated seawater brine by thin-film nanofiber composite pressure retarded osmosis membranes with high power density. *Energy Environ. Sci.* 6, 1199–1210.

Straub, A.P., Deshmukh, A., Elimelech, M., 2016. Pressure-retarded osmosis for power generation from salinity gradients: is it viable? *Energy Environ. Sci.* 9, 31–48.

Sukitpaneenit, P., Chung, T.S., 2012. High performance thin-film composite forward osmosis hollow fiber membranes with macrovoid-free and highly porous structure for sustainable water production. *Environ. Sci. Technol.* 46, 7358–7365.

Sun, S.P., Chung, T.S., 2013. Outer-selective pressure-retarded osmosis hollow fiber membranes from vacuum-assisted interfacial polymerization for osmotic power generation. *Environ. Sci. Technol.* 47, 13167–13174.

Tang, C.Y., She, Q., Lay, W.C.L., Wang, R., Field, R., Fane, A.G., 2011. Modeling double-skinned FO membranes. *Desalination* 283, 178–186.

Thorsen, T., Holt, T., 2009. The potential for power production from salinity gradients by pressure retarded osmosis. *J. Membr. Sci.* 335, 103–110.

Wan, C.F., Chung, T.S., 2015. Osmotic power generation by pressure retarded osmosis using seawater brine as the draw solution and wastewater retentate as the feed. *J. Membr. Sci.* 479, 148–158.

Xie, M., Lee, J., Nghiem, L.D., Elimelech, M., 2015. Role of pressure in organic fouling in forward osmosis and reverse osmosis. *J. Membr. Sci.* 493, 748–754.

Xiong, J.Y., Cai, D.J., Chong, Q.Y., Lee, S.H., Chung, T.S., 2017. Osmotic power generation by inner selective hollow fiber membranes: An investigation of thermodynamics, mass transfer, and module scale modelling. *J. Membr. Sci.* 526, 417–428.

Xiong, J.Y., Cheng, Z.L., Wan, C.F., Chen, S.C., Chung, T.S., 2016. Analysis of flux reduction behaviors of PRO hollow fiber membranes: Experiments, mechanisms, and implications. *J. Membr. Sci.* 505, 1–14.

Yip, N.Y., Elimelech, M., 2011. Performance limiting effects in power generation from salinity gradients by pressure retarded osmosis. *Environ. Sci. Technol.* 45, 10273–10282.

Yip, N.Y., Tiraferri, A., Phillip, W.A., Schiffman, J.D., Hoover, L.A., Kim, Y.C., Elimelech, M., 2011. Thin-film composite pressure retarded osmosis membranes for sustainable power generation from salinity gradients. *Environ. Sci. Technol.* 45, 4360–4369.

Zeman, L.J., Zydney, A.L., 1996. *Microfiltration and Ultrafiltration: Principles and Applications*, M. Dekker, New York.

Zhang, S., Chung, T.S., 2013. Minimizing the instant and accumulative effects of salt permeability to sustain ultrahigh osmotic power density. *Environ. Sci. Technol.* 47, 10085–10092.

Zhang, S., Chung, T.S., 2016. Osmotic power production from seawater brine by hollow fiber membrane modules: Net power output and optimum operating conditions. *AIChE J.* 62, 1216–1225.

Zhang, S., Sukitpaneenit, P., Chung, T.S., 2014. Design of robust hollow fiber membranes with high power density for osmotic energy production. *Chem. Eng. J.* 241, 457–465.

5

Analysis of Flux Reduction Behaviors in PRO

Jun Ying Xiong
School of Life Science and Chemical Technology
Ngee Ann Polytechnic
Singapore

Zhen Lei Cheng and Chun Feng Wan
Department of Chemical and Biomolecular Engineering
National University of Singapore
Singapore

CONTENTS

5.1 Introduction

Pressure retarded osmosis (PRO) is a promising technology to harvest renewable osmotic energy using a semipermeable membrane. Compared to many fossil fuel-based energy sources, PRO is a green and sustainable technology friendly to environment because no greenhouse gas and chemicals are released during the harvest of osmotic power. However, in spite of the advantages over other traditional energy sources, harvesting osmotic energy from PRO processes is still in an infant stage because power density and water flux are severely reduced by many limiting factors, such as internal concentration polarization (ICP), external concentration polarization (ECP), reverse salt permeation, and membrane fouling (Chen et al., 2016; Han et al., 2015; Kim et al., 2015; She et al., 2012; Wan and Chung, 2015; Zhang and Chung, 2013; Zhang et al., 2014). Even though many breakthroughs have been achieved in recent years on PRO membranes with impressive mechanical strength and power density (Han et al., 2015), membrane performance deteriorates sharply in real applications (Han et al., 2015). As a result, the net energy drops to a level only marginally higher or even lower than the economically feasible value (Wan and Chung, 2015). Therefore, it is of great importance to investigate the underlying mechanisms responsible for the significant reduction of water flux in order to overcome the limitations and design innovative PRO membranes for real practices.

This chapter focuses on the elucidation of various underlying mechanisms responsible for the flux reduction behaviors in PRO. Three cases are examined. First, the flux reduction behaviors under a fixed bulk salinity gradient will be discussed. Figure 5.1 illustrates a schematic diagram of salt concentration across a PRO membrane at steady state. A fixed bulk salinity gradient means that the bulk salinities in the bulk draw solution ($C_{D,b}$) and in the bulk feed solution ($C_{F,b}$) are fixed. Both inner-selective and outer-selective thin film composite (TFC) hollow fiber membranes (Figure 5.2) are employed to examine how the fundamental internal factors (such as the surface salinity of the selective layer at the feed side ($C_{F\prime m}$) and its components) interact with one another under the fixed bulk salinity gradient, resulting in various behaviors of external performance indexes such as water flux, reverse salt flux, and power density. Second, the flux reduction behaviors under a growing bulk salinity gradient (i.e., an increasing $C_{D,b}$–$C_{F,b}$) will be investigated since in real operations, the bulk feed salinity ($C_{F,b}$) is always increasing due to the water permeation to the draw solution side and the accumulated reverse salt flux along PRO modules. Third, the flux reduction behaviors due to scaling will be elaborated. An advanced nucleation theory will be employed to elucidate the dynamic scaling process by visualizing how the multiple fundamental factors (such as local supersaturation, nucleation rate, and nuclei size) evolve and interplay with one another in various membrane regimes during the whole scaling process. Finally, the flux reduction behaviors in PRO modules will be examined. With an in-depth understanding of various underlying mechanisms responsible for the flux reduction behaviors in PRO, this study will surely provide useful insights to design more suitable TFC hollow fiber membranes and to operate them with enhanced water flux so that the PRO process may become more promising in the near future.

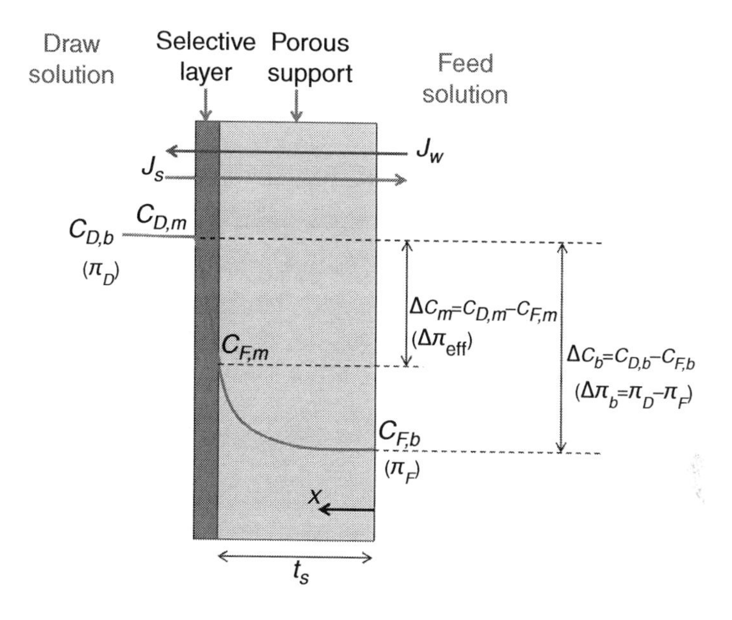

FIGURE 5.1
A schematic diagram of salt concentration across a TFC PRO membrane at steady state. Osmotic pressures corresponding to salt concentrations at different locations are illustrated in the brackets near the concentrations. In this diagram, the salt concentration on the surface of the selective layer ($C_{D,m}$) is nearly equal to the concentration of the draw solution in the bulk ($C_{D,b}$) due to the negligible ECP.

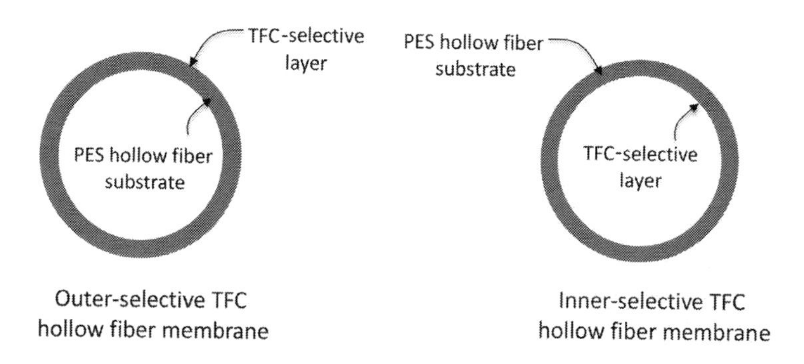

FIGURE 5.2
A schematic illustration of the inner- and outer-selective TFC hollow fiber membranes.

5.2 Flux Reduction Behaviors under a Fixed Bulk Salinity Gradient

In general, ICP coupled with undesirable reverse salt flux accounts for the water flux decline in PRO processes when the feed pair contains negligible fouling tendency. Prior works have indicated a direct link between intrinsic membrane properties such as water permeability (A), reverse salt permeability (B), structural parameter (S), and operation conditions (for example, operation pressure ΔP) and external performance indexes such as water flux (J_w), reverse salt flux (J_s), and power density (Yip et al., 2011; Zhang and Chung, 2013). However, although most prior works have observed direct links among the intrinsic membrane properties, operation conditions, and external performance, the work to systematically elaborate the complicated interactions among them is still absent. In this section, we choose two TFC hollow fiber membranes recently developed in our group, an inner-selective TFC membrane (Wan and Chung, 2015) and an outer-selective TFC membrane (Cheng et al., 2016), to in-depth bridge the intrinsic membrane properties, operation conditions, and external performance by elucidating the interactions among some internal factors and illustrating their effects on the external performance. The internal factors include the surface salinity of the selective layer at the feed side ($C_{F,m}$) and its ICP contribution and Js contribution, while the external performance comprises water flux (J_w), reverse salt flux (J_s), and power density. Table 5.1 summarizes the inner diameter (ID), outside diameter (OD), water permeability (A), reverse salt permeability (B), and structural parameter (S) of the inner- and outer-selective membranes. It is worth noting that A and B values of the outer-selective membrane are smaller compared to those of the inner-selective membrane. As shown in Figure 5.1, for the ease of the illustration, one-dimensional coordinate (i.e., radial direction) is used for the discussions of the flux reduction behaviors.

Assuming ECP in the draw solution side is negligible due to efficient mixing, the effects of ICP within the porous support and the reverse salt flux

TABLE 5.1

The inner diameter (ID), outside diameter (OD), pure water permeability (A), salt permeability (B), and structural parameter (S) of the outer-selective and inner-selective TFC hollow fiber membranes used in this chapter

TFC Membrane	ID/OD (mm)	A (LMH/bar)	B (LMH)	S (μm)
Outer-selective	0.53/1.1	1.42	0.028	996
Inner-selective	0.58/1.0	3.50	0.30	450

across the membrane are the main causes resulting in a smaller effective osmotic driving force ($\Delta\pi_{eff}$) than the bulk osmotic driving force ($\Delta\pi_b$). Water flux can be computed as follows (Lee et al., 1981; Lonsdale, 1973):

$$J_w = A(\Delta\pi_{\text{eff}} - \Delta P) = A\left(\pi_{D,m} - \pi_{F,m} - \Delta P\right) \tag{1}$$

where $\pi_{D,m}$ and $\pi_{F,m}$ are the osmotic pressures of the draw and feed solutions facing the selective layer surfaces, respectively. The salt flux across the selective layer (J_s) can be written as (Lee et al., 1981)

$$J_s = B\left(C_{D,m} - C_{F,m}\right) \tag{2}$$

where $C_{D,m}$ and $C_{F,m}$ are the salt concentrations of the draw and feed solutions at the selective layer surfaces, respectively.

The salt concentration of the feed at the selective layer surface ($C_{F,m}$) can be expressed as follows (Xiong et al., 2016):

$$C_{F,m} = \underbrace{C_{F,b}\overset{\text{ICP factor}}{\boxed{\exp\left(\frac{J_w S}{D}\right)}}}_{\text{ICP contribution}} + \underbrace{\frac{J_s}{J_w}\overset{\text{Equivalent ICP factor}}{\boxed{\left[\exp\left(\frac{J_w S}{D}\right) - 1\right]}}}_{J_s \text{ contribution}} \tag{3}$$

Eq. 3 indicates that $C_{F,m}$ comprises two terms: the first term represents the contributions from the internal factors (i.e., concentrative ICP and others) and is referred to as the ICP contribution. The second term stands for the contributions from the reverse salt permeation across the selective layer and is hereafter called as the J_s contribution. J_s/J_w is the specific reverse salt flux. As J_s/J_w has a concentration unit, it can be considered as an equivalent concentration induced by the reverse salt permeation. The percentage of the J_s contribution to $C_{F,m}$ is defined as J_s-contribution %:

$$J_s - \text{contribution } \% = \frac{J_s \text{ contribution}}{C_{F,m}} \times 100\% \tag{4}$$

To study the relationship between the internal factors and their effects on the external performance, a 0.81 M NaCl aqueous solution was used as the draw solution, which has the similar osmolality as the first-stage RO brine of an RO plant. A 0.011-M NaCl aqueous solution was used as the feed, which has the similar osmolality as the wastewater retentate obtained from a wastewater plant (Wan and Chung, 2015). Figure 5.3 (1a) to (2f) show various internal factors and external performance

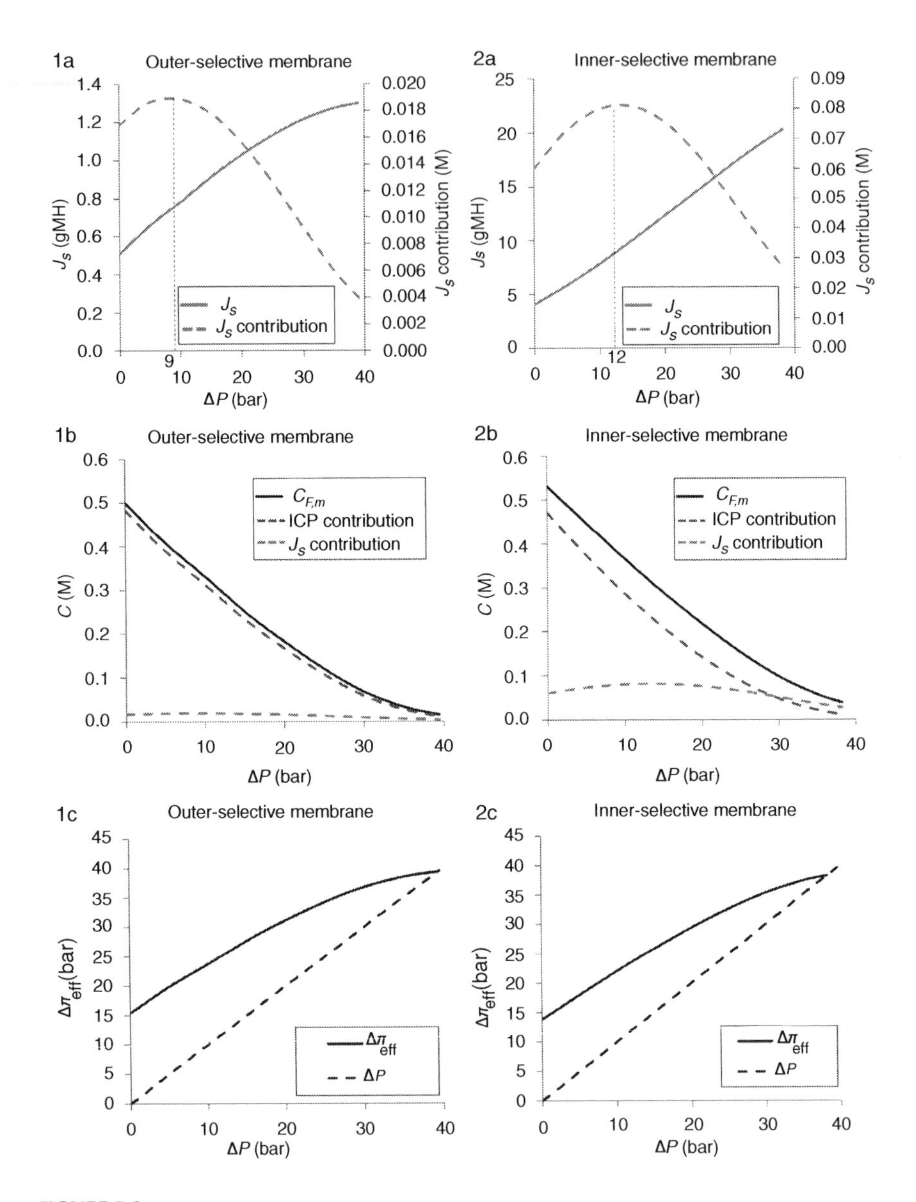

FIGURE 5.3
Various variables in the PRO model as a function of operation pressure (ΔP) under the fixed bulk salinity gradient (draw solution: 0.81 M NaCl versus feed: 0.011 M NaCl). Series 1a to 1f and 2a to 2f correspond to the outer-selective membrane and the inner-selective membrane, respectively. ideal: simulation results without considering concentration polarization and reverse salt permeation; model: simulation results with ICP and reverse salt permeation taken into account; and exp: experimental data.

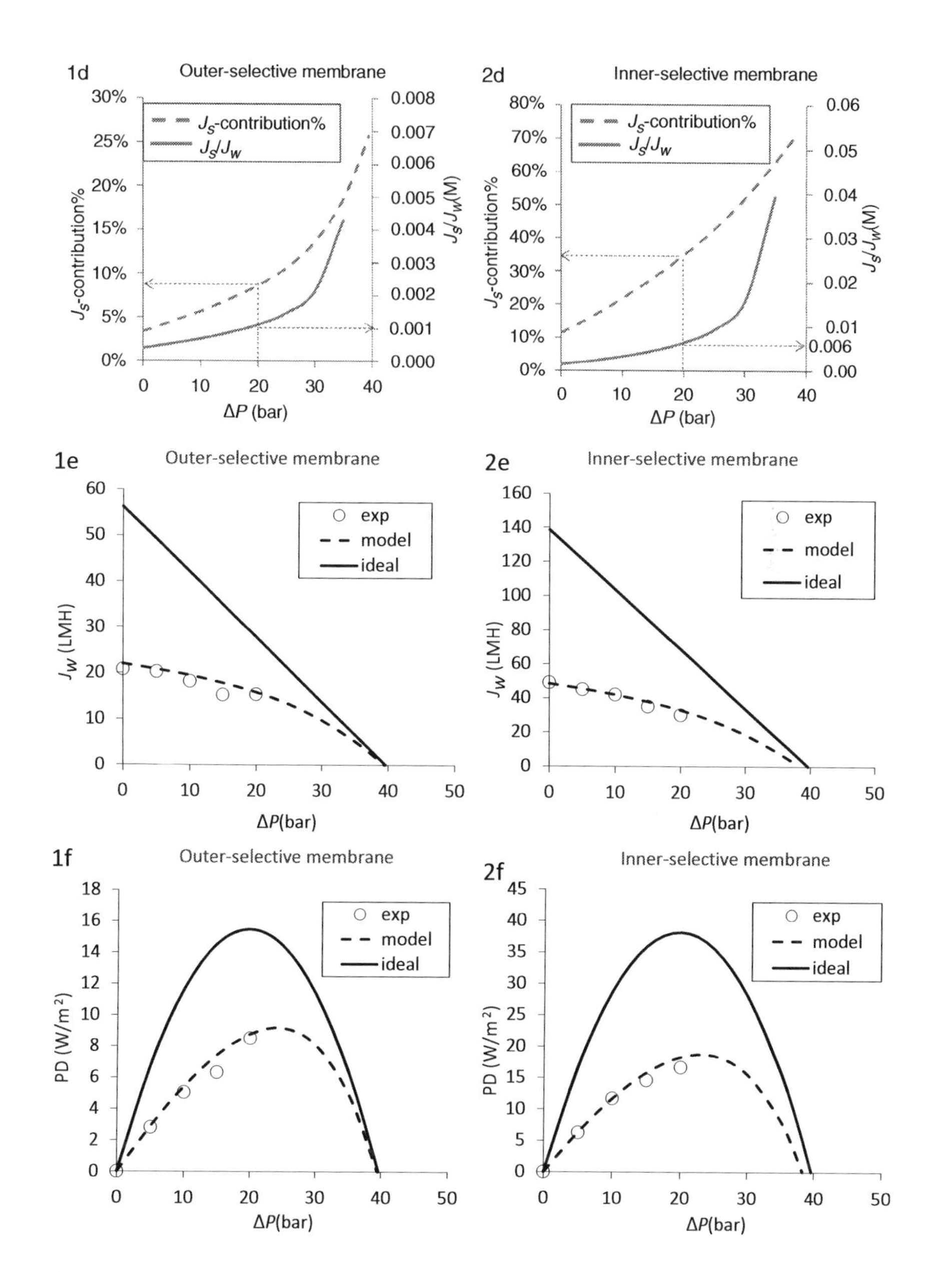

FIGURE 5.3
(continued)

indices as a function of operation pressure based on theoretical simulations.

As illustrated in Figure 5.3 (1a) and (2a), reverse salt fluxes (J_s) of both membranes increase with operation pressure (ΔP) under the fixed bulk salinity gradient. However, the inner-selective membrane has J_s values about ten times higher than the outer-selective one due to its higher B value. Interestingly, for both membranes, the J_s contribution exhibits a convex trend and reaches a peak at a certain operation pressure of less than 15 bar. To explain the trend, let us use the outer-selective membrane as an example. As defined in Eq. 3, the J_s contribution is a product of the J_s/J_w ratio and the equivalent ICP factor. Since J_s increases rapidly with ΔP in the range of 0–9 bar as shown in Figure 5.3 (1a), while J_w decreases relatively slowly at the low ΔP range (Figure 5.3 (1e)), this results in a slow reduction of the equivalent ICP factor. Therefore, the growth of J_s/J_w overwhelms the reduction of the equivalent ICP factor, leading to a peak in the J_s contribution profile at 9 bar. When ΔP is higher than 9 bar, the J_s contribution starts to decay due to the dominant reduction of the equivalent ICP factor.

Based on Eq. 1, the water flux at a given ΔP is determined by the effective osmotic pressure difference ($\Delta\pi_{\mathrm{eff}}$) and water permeability (A). As $\Delta\pi_{\mathrm{eff}}$ is proportional to the salinity gradient across the selective layer (i.e., $\Delta C_m = C_{D,m} - C_{F,m}$), it is important to investigate the behaviors of $C_{F,m}$ and its component factors defined in Eq. 3. As plotted in Figure 5.3 (1b) and (2b), the ICP contribution of both membranes monotonically decrease in the entire ΔP range, while the J_s contribution exhibits a convex trend. As the reduction of the ICP contribution is dominant, $C_{F,m}$ exhibits a monotonic decline with ΔP. Consequently, $\Delta\pi_{\mathrm{eff}}$ increases with an increase in ΔP (Figure 5.3 (1c) and (2c)), so does J_s because of Eq. 2. Since the inner-selective membrane has a much higher B value than the outer-selective one, the former has a more significant J_s contribution than the latter, as illustrated in Figure 5.3 (1b) versus (2b). As a consequence, the effects of reverse salt permeation on PRO performance are more severe for the inner-selective membrane. For example, the inner-selective membrane at 20 bar has a J_s/J_w value about five times higher than the outer-selective membrane. The former also has a J_s-contribution % more than three times higher than the latter.

Comparing Figure 5.3 (1b) and (2b), one can find that, although the inner-selective membrane has a much larger J_s contribution, its $C_{F,m}$ is only slightly higher than that of the outer-selective membrane. Consequently, as shown in Figure 5.3 (1c) versus (2c), its $\Delta\pi_{\mathrm{eff}}$ curve is only slightly below the curve of the outer-selective membrane. This is due to the fact that the effect of its large B is effectively offset by its small S value (i.e., small ICP factor and the equivalent ICP factor). As J_w is determined by $\Delta\pi_{\mathrm{eff}}$ and A, the inner-selective membrane still outperforms the outer-selective membrane in terms of J_w and power density because the former has a higher

A value than the latter, as evidenced when comparing Figure 5.3 (1e) versus (2e) and (1f) versus (2f).

In summary, the complicated relationship between various external performance and internal factors can be clearly illustrated in Figure 5.4. In the scenario of the fixed bulk salinity gradient, it is $C_{F,m}$ that essentially serves as the key factor to determine ΔC_m when ECP of the draw solution side is negligible. $C_{F,m}$ incorporates two parts: ICP contribution and J_s contribution (Eq. 3). For ICP contribution, S serves as the dominating factor affecting the ICP effect. For J_s contribution, beside S, equivalent concentration J_s/J_w also plays an important role. Based on Figure 5.4, Eqs. 1, and 2, ultimately a PRO membrane with a higher A, lower B, and lower S are preferred to have a larger J_w and smaller J_s. However, as seen in Figure 5.3, increased J_w has a limit, as a larger J_w will also result in a larger ICP factor because it is inside the term exp (J_wS/D). Therefore, during the start-up of a PRO module, J_w will keep increasing until it reaches a certain level at which the increasing tendency of J_w is compressed and offset by the simultaneously increased ICP effect.

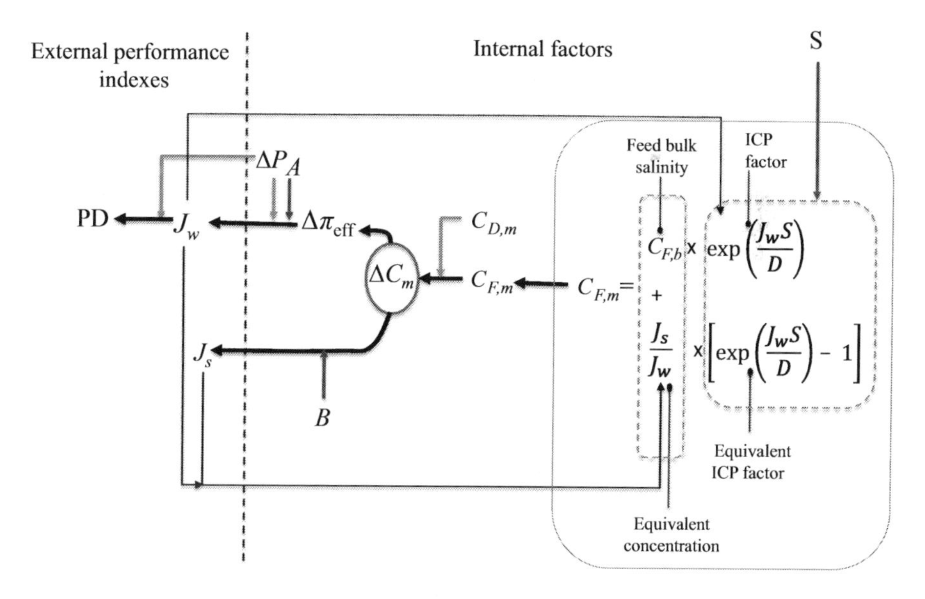

FIGURE 5.4
Schematic system diagram illustrating the relationship between external performance indexes and internal factors. PD is the power density. $\Delta\pi_{eff}$ is the effective osmotic pressure difference across the membrane. ΔC_m is the salinity gradient across the selective layer. $C_{D,m}$ is the surface salinity of the selective layer facing the draw solution. $C_{F,m}$ is the surface salinity of the selective layer facing the feed. $C_{F,b}$ is the bulk salinity of the feed.

5.3 Flux Reduction Behaviors under a Growing Bulk Feed Salinity

In reality, due to the water permeation into the draw solution and undesirable reverse flux to the feed, the bulk feed salinity ($C_{F,b}$) will increase along membrane modules. Therefore, it is meaningful to examine how internal factors response to this growing bulk feed salinity. In the following discussion, the draw solution has a fixed bulk salinity of 0.81 M NaCl and the operation pressure is set at 20 bar.

The relationship between reverse salt flux (J_s) and feed bulk salinity ($C_{F,b}$) is presented in Figure 5.5 (a). As the interface salinity ($C_{F,m}$) increases with the feed salinity ($C_{F,b}$) (Figure 5.5 (b)), the reverse salt flux (J_s) decreases with $C_{F,b}$.The inner-selective membrane exhibits a much higher J_s than the outer-selective membrane due to its large B value. Interestingly, as shown in Figure 5.5 (c), despite of the reducing J_s, the J_s/J_w ratio keeps growing with $C_{F,b}$ for both membranes. This trend can be explained by Eq. 5.

$$\frac{J_s}{J_w} = \frac{B\Delta C_m}{A(iRT\Delta C_m - \Delta P)} = \frac{B}{A\left(iRT - \frac{\Delta P}{\Delta C_m}\right)} \tag{5}$$

where ΔC_m is the salinity gradient across the selective layer of the membrane (i.e., $\Delta C_m = C_{D,m} - C_{F,m}$), i is the van't Hoff coefficient, R is the universal gas constant, and T is the absolute temperature. On one hand, as ΔC_m decreases with increasing $C_{F,b}$, J_s/J_w increases with $C_{F,b}$ because $(iRT\Delta C_m - \Delta P)$ in the denominator decreases faster than ΔC_m in the numerator. On the other hand, the growth rate of J_s/J_w depends on the B/A ratio, which can be considered as the reciprocal of the membrane selectivity. As the B/A ratio of the inner-selective membrane is much larger than that of the outer-selective membrane (i.e., 0.086 bar versus 0.020 bar), the inner-selective membrane presents a more rapid growth of J_s/J_w.

The trend of $C_{F,m}$ and its ICP contribution is illustrated in Figure 5.5 (b). Using the inner-selective membrane as an example, one can find that both $C_{F,m}$ (represented by line section $\overline{AC}$) and its ICP contribution (represented by $\overline{BC}$) increases monotonically with $C_{F,b}$. Meanwhile, the gap between the two curves (i.e., J_s contribution, represented by $\overline{AB}$) decreases with $C_{F,b}$, which is consistent with the trend of J_s-contribution % in Figure 5.5 (d). Compared to the inner-selective membrane, the curves of ICP contribution and $C_{F,m}$ of the outer-selective membrane are close to each other because its J_s contribution is effectively suppressed by its low B value. As shown in Figure 5.5 (d), the contribution of its J_s contribution to $C_{F,m}$ (i.e., J_s-contribution %) is less than 10% when $C_{F,b}$ is larger than 0.01 M.

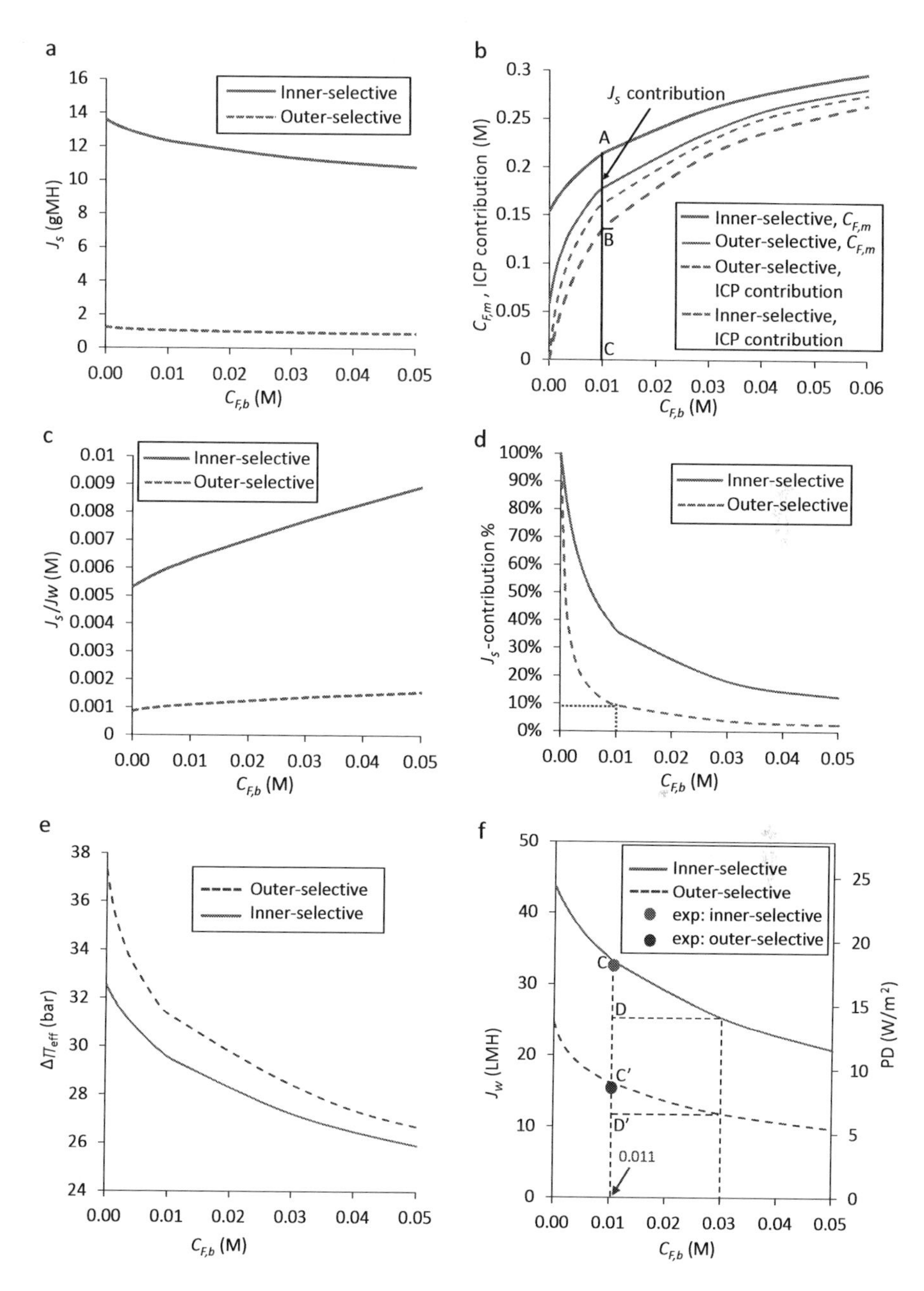

FIGURE 5.5
Comparison of various variables of inner- and outer-selective TFC hollow fiber membranes as a function of feed bulk salinity ($C_{F,b}$). The draw solution has a fixed bulk salinity of 0.81 M NaCl and the operation pressure is set at 20 bar. Solid and lines made up of longer dashes are the simulation results of the inner-selective membrane; solid and lines made up of shorter dashes are the simulation results of the outer-selective membrane. exp: experimental data.

Figure 5.5 (e) and (f) shows the effective driving force ($\Delta\pi_{eff}$) and water flux (J_w) as a function of $C_{F,b}$. For both membranes, $\Delta\pi_{eff}$ decreases rapidly with an increase in $C_{F,b}$ (Figure 5.5 (b)). As a result, water flux declines rapidly with an increase in $C_{F,b}$, especially in the low $C_{F,b}$ range up to 0.01 M. This implies that power density will drop drastically when employing low-salinity feeds such as RO-treated membrane bioreactor permeate as the feed in PRO. Since the feed has a salinity of 0.011 M in this study, the inner-selective membrane exhibits faster drops in both water flux and power density compared to the outer-selective membrane. For example, when $C_{F,b}$ increases from 0.011 to 0.03 M, the power density of the inner-selective membrane drops by 4.4 W/m^2 (represented by the line $\overline{CD}$) while that of the outer-selective membrane drops by only 2.2 W/m^2 (represented by the line $\overline{C'D'}$). In summary, since $C_{F,b}$ is typically ≥0.01 M in real applications, the effect of the ICP contribution on $C_{F,m}$ would become dominant with growing $C_{F,b}$, as illustrated in Figure 5.5 (b). This would result in fast declines in both water flux and power density (Figure 5.5 (f)).

5.4 Flux Reduction Behaviors Due to Scaling

When the feed pair contains scaling precursors, the performance of TFC hollow fiber membranes will not only drop due to the growing salinity in the feed but also deteriorate by severe scaling within the membrane (Chen et al., 2016; Zhang et al., 2014). In this section, the outer-selective TFC hollow fiber membrane is used to investigate the underlying mechanism of gypsum scaling. The detailed experiment conditions have been elaborated elsewhere (Xiong et al., 2016). Figure 5.6 (a) plots the weight loss profiles of the feed as a function of time under different bulk saturation indices (SIs). The scaling process can be divided into two stages: (1) an initial rapid decline stage with a slightly concave trend from their original starting points to points A, A′, and A″, and (2) then an approximately linear stage. Weight loss profiles can be further translated into water flux curves as illustrated in Figure 5.6 (b) and (c). Four classical flux decline models listed in Table 5.2 (Cheryan, 1986; Zeman et al., 1996) are employed to fit the water flux curves in order to examine their patterns and divide them into more detailed stages. As shown in Figure 5.6 (b), the best fitting is achieved by the modified pore blocking model. Therefore, as shown in Figure 5.6 (c), the water flux curves can hypothetically be divided into three stages: (1) an initial exponentially descending stage (for example, from the starting point up to point B), (2) a transition stage, and (3) a final pseudo-stable stage (for example, the profile after point A).

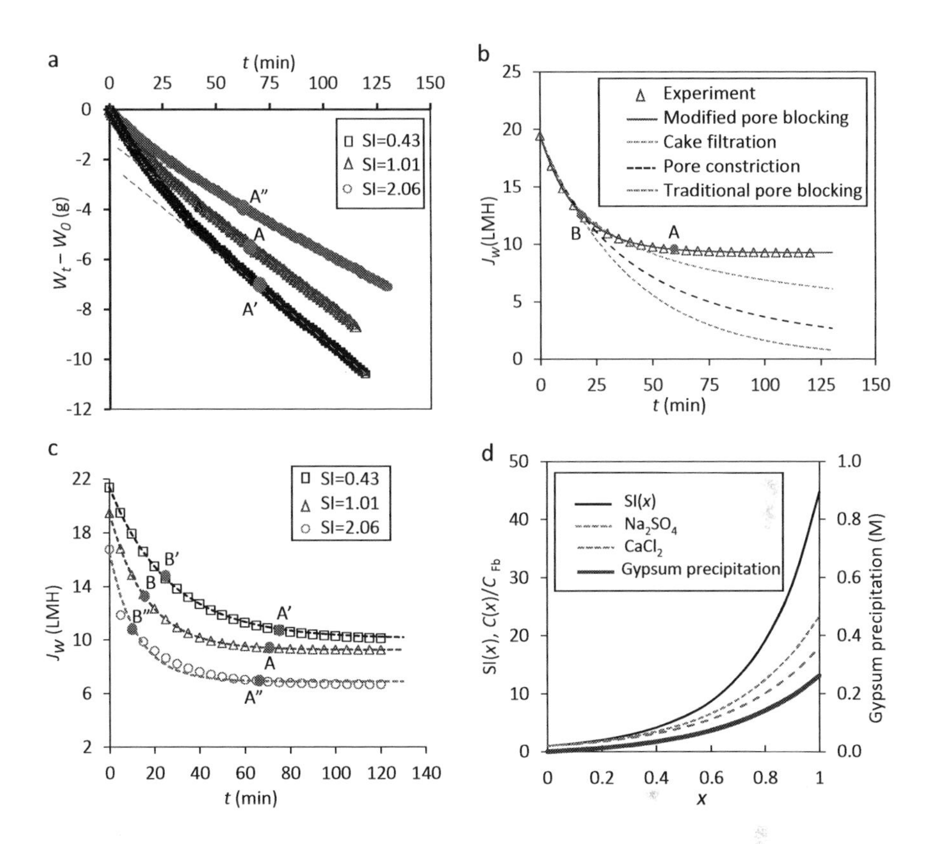

FIGURE 5.6
Investigation of PRO gypsum scaling at a fixed operation pressure (ΔP) of 10 bar. (a) Weight loss of the feed as a function of time for different bulk feed SIs. W_0 is the initial weight of the feed while W_t is the weight at time t. (b) Data fitting of the flux reduction curve for the bulk feed SI = 1.01 with four models. (c) Flux reduction curves for bulk SI = 0.43, 1.01, and 2.06. Separate symbols are experimental data. Dotted lines are the fitted curves using the modified pore blocking model. (d) Simulated SIs, dimensionless precursor concentrations for $CaCl_2$ and Na_2SO_4, and gypsum precipitation as a function of location inside the membrane porous support. x = 0 is the surface of the membrane porous support facing the bulk feed. x = 1 represents the interface between the porous support and the selective layer.

TABLE 5.2
Four classical models used in data fitting for the PRO gypsum scaling study

Classical Models	Model Equations	Parameters for Data Fitting
Cake filtration	$J_w = J_0(1+bt)^{-1/2}$	J_0 (initial water flux), b
Pore constriction	$J_w = J_0(1+bt)^{-2}$	J_0 (initial water flux), b
Traditional pore blocking	$J_w = ae^{-bt}$	a, b
Modified pore blocking	$J_w = J_\infty + ae^{-bt}$	J_∞ (limiting water flux), a, b

To examine the relationship between pore blocking and subsequent flux reduction behavior, the distributions of scaling precursors within the porous support are estimated based on Eq. 6 (Xiong et al., 2016) and plotted in Figure 5.6 (d).

$$C(x) = C_{F,b}\exp\left(J_w K \times \frac{x}{t_s}\right) + \frac{J_s}{J_w}\left[\exp\left(J_w K \times \frac{x}{t_s}\right) - 1\right] \quad (6)$$

where t_s is the thickness of the porous support and $C_{F,b}$ is the salt concentration of the bulk feed solution. K is defined as $\frac{\tau t_s}{D\varepsilon}$, referring to the resistance to the salt transport in the porous support, in which τ and ε are the tortuosity and porosity of the porous support, respectively, and D is the salt diffusivity in water. As shown in Figure 5.6 (d), the concentrations of scaling precursors grow relatively slowly at positions near the porous support surface facing the bulk feed (i.e., the locations near $x = 0$), and progressively increase when approaching to the interface between the porous support and the selective layer (i.e., the location near $x = t_s$). The local SI displays a similar trend accordingly. Based on the advanced nucleation theory (Liu, 2000; Sato et al., 2001), the relationship among nucleation barrier (ΔG^*), nucleation rate (J) (i.e., the population of critical nuclei generated per unit-volume and unit-time), and critical radius of nuclei (r_c) can be described as

$$\Delta G^* = \frac{16\pi r_{cf}^3 \Omega^3}{3[\kappa T \ln(1+\sigma)]^2} \quad (7)$$

$$J = B' \exp\left(-f\frac{\Delta G^*}{\kappa T}\right) \quad (8)$$

$$r_c = \frac{2\Omega r_{cf}}{\kappa T \ln(1+\sigma)} \quad (9)$$

where κ is the Boltzmann constant, T is the absolute temperature, r_{cf} is the surface-free energy between the nuclei and the mother phase, Ω is the volume of the growth unit, and B' is a kinetic parameter that is constant for a given system. f ($0 \leq f \leq 1$) is a factor describing the reduction of the nucleation barrier due to a foreign body. σ is the supersaturation, which can be well represented by SI (Sato et al., 2001) as follows:

$$\ln(1+\sigma) = \ln(\mathrm{IAP}/K_{sp}) = \ln(\mathrm{SI}) \quad (10)$$

where SI is the saturation index defined as IAP/K_{sp}, in which IAP is the product of the precursor ion activities and K_{sp} is the solubility constant of gypsum. Clearly, a higher local SI value leads to a higher local supersaturation and hence a lower nucleation barrier. This will promote a larger nucleation rate of the critical nuclei with a smaller size. As the local SI value changes with different locations inside the membrane, nucleation rate and nuclei size also change with different locations accordingly. With these insights in mind, Figure 5.7 schematically illustrates the possible mechanisms of flux decline in the scaling process.

Once scaling is initiated, scaling precursors such as Ca^{2+} and SO_4^{2-} will soon build-up inside the membrane porous support due to the effect of concentrative ICP. As illustrated in inset (I) of Figure 5.7, in the region just beneath the selective layer, a large population of small critical nuclei may be generated either on the selective layer through surface crystallization (i.e., heterogeneous nucleation) or from the mother phase through homogeneous-like nucleation. The nuclei in the mother phase may rapidly deposit underneath the selective layer due to the local water flux toward the selective layer induced by the draw solution. These deposited nuclei, together with the nuclei produced by surface crystallization, occupy a certain portion of the selective layer surface, leading to instant losses of water permeability (A) and water flux (J_w). Such surface coverage can be verified by DI water backwash performed at 15 bar immediately after

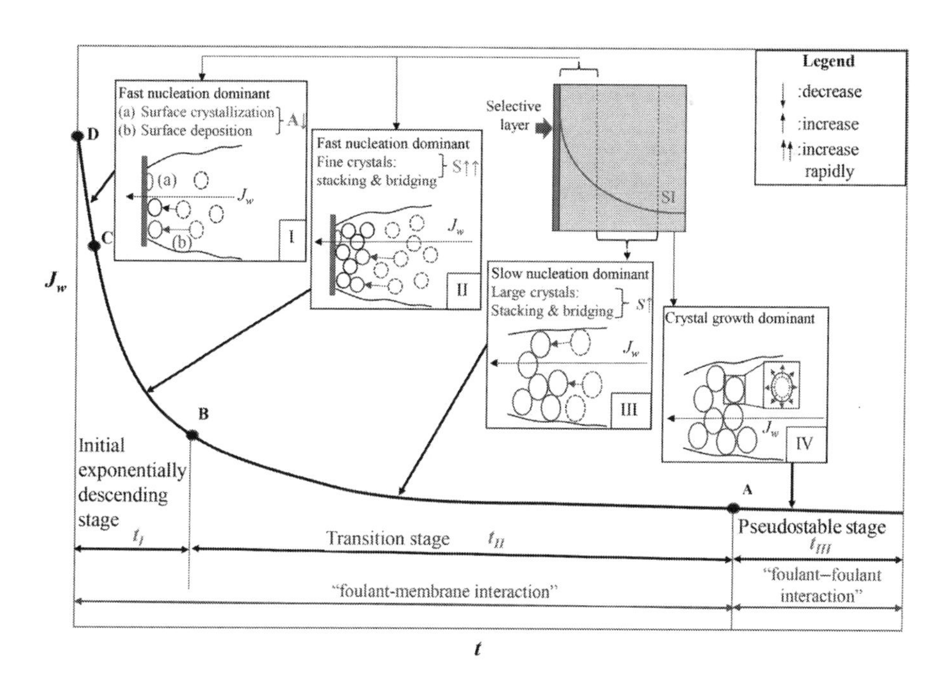

FIGURE 5.7
Schematic illustration of gypsum scaling mechanisms across the substrate wall.

every scaling test. If water flux can be quickly recovered after the backwash, surface crystallization and surface deposition will be less severe. On the contrary, if it takes a certain period of time to recover water flux, then the surface crystallization is more severe and the nuclei deposition is more compact. Experimental results indicate that an instant recovery of nearly 100% water flux can be achieved after backwash when using a feed bulk SI of 1.01 for the scaling tests. In contrast, the recovery of the initial water flux is only 85% if a feed bulk SI of 2.06 is employed; it takes 30 minutes to fully recover the water flux. Such different responses to the backwash may be due to the following reasons. Since the actual space available for the initial fast nucleation underneath the selective layer in the asymmetric membrane support is highly constricted, a higher feed bulk SI value would lead to a larger local supersaturation near the selective layer, which will result in nuclei with a smaller size. Therefore, a feed with a higher SI value would facilitate the deposition and form a more compact crystalline layer. Consequently, as shown in Figure 5.6 (c), one can observe a steeper initial flux reduction for the case of SI = 2.06 and less steep initial flux drops for SI = 0.43 and SI = 1.01.

After the rapid surface crystallization and nuclei deposition, quick stacking or bridging may occur in the porous substrate a bit far away from the selective layer, as illustrated in inset (II) of Figure 5.7. Owing to the high supersaturation in the mother phase, nucleation is rather fast in this regime and the nascent nuclei size is still small due to the low nucleation barrier at the high local supersaturation. These small crystals may quickly block the flow channels in the membrane porous support. Not only do they induce a quick growth of tortuosity (τ) but also a fast decline of porosity (ε). Therefore, the structural parameter (S) increases rapidly and the initial flux declines exponentially because of the drastically enhanced ICP. After a certain period of time t_I, fast nucleation near the selective layer is almost complete, growth of S is then dominated by the relatively slower nucleation with larger nuclei sizes occurring in the region further away from the selective layer, as illustrated by inset (III). Meanwhile, the previously mentioned regions adjacent to the selective layer (corresponding to insets (I) and (II)) enters the crystal growth stage. Compared to nucleation, crystal growth is relatively sluggish and provides less contributions to the growth of S. As illustrated in inset (III), because nucleation rate in the transition stage slows down, the water flux decline also slows down. Finally, when nucleation comes to the end for all the regions inside the porous support, crystal growth eventually takes over as the dominant role. A sluggish S growth coupling with a very slow water flux decline leads to an apparently pseudo-stable stage, as illustrated in inset (IV) in Figure 5.7.

Based on the above analysis, one may have a better understanding of Figure 5.6 (c). Among the three curves, the feed with a SI value of 2.06 has the fastest growth in structural parameter due to the largest local supersaturation inside the porous support. As a result, it has earlier turning points as compared to the other two stages at points B″ and A″. On the other hand, it also has the smallest gypsum nuclei size because it has the largest local supersaturation within the membrane. The combination of these two factors would result in the fouled membrane with the densest gypsum scaling inside the porous support, the largest *S* growth, and the lowest water flux compared to the other two cases. However, if the scaling process continues for a sufficiently long period of time, all three curves in Figure 5.6 (c) would merge together as reported in some prior works (Tang et al., 2009, 2011). This phenomenon implies that the initial exponentially declining stage and the following transition stage in Figure 5.7 may be termed as "foulant-membrane interaction." Similar to those pressurized filtration processes such as NF and RO (Tang et al., 2011, 2009), scalants interact with the membrane in these stages resulting in a significant change of membrane structural parameter through nucleation. The final pseudo-stable stage can be named as "foulant–foulant interaction" (Tang et al., 2009, 2011) because gypsum crystal growth is dominant in this stage. Nevertheless, the kinetics of scaling in PRO are still much different from those of the pressurize-driven filtration processes in nature.

5.5 Flux Reduction Behaviors in PRO Modules

In this section, a full model will be employed to elaborate the flux reduction behaviors in PRO hollow fiber modules. A schematic diagram of salt concentration across an inner-selective PRO hollow fiber membrane at steady state is illustrated in Figure 5.8. The schematic diagram of the simulation approach of the PRO hollow fiber membrane module is illustrated in Figure 5.9. Membrane parameters and module dimensions for simulation are given in Table 5.3. Development of model and a comprehensive analysis of fluid motion, mass transport, thermodynamics, and power generation during PRO processes have been elaborated elsewhere (Xiong et al., 2017). This section mainly aims to (1) elucidate the fundamental relationship among various membrane properties and operation parameters and (2) analyze their individual and combined impacts on PRO module performance.

5.5.1 ECP Effect of PRO Hollow Fiber Membranes

ECP effects on the feed and the draw solution sides commonly exist in all PRO operations. However, the understanding of their impacts on the performance of PRO hollow fiber membranes is still not sufficient. To

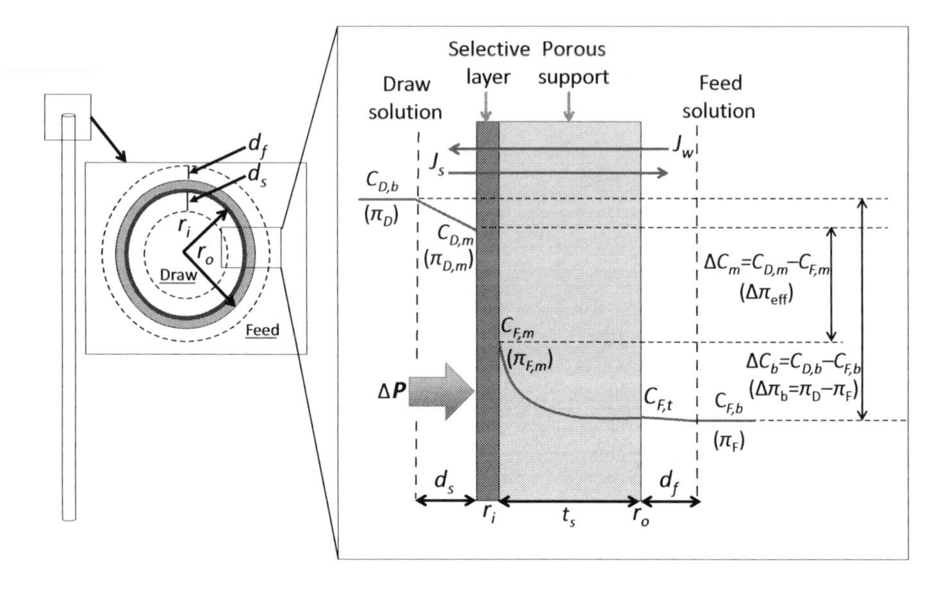

FIGURE 5.8
A schematic diagram of salt concentration across an inner-selective PRO hollow fiber membrane at steady state. Osmotic pressures corresponding to salt concentrations at different locations are illustrated in the brackets near the concentrations.

elaborate on the significance of ECP effects on both feed side and draw solution side, the ECP factors on the feed side and draw solution side can be defined in Eqs. 11 and 12, respectively (Xiong et al., 2017) as follows:

$$F_{\mathrm{ECP},f} = \left(\frac{r_0 + d_f}{r_0}\right)^{\frac{Jw(r_i)r_i}{D}} \tag{11}$$

$$F_{\mathrm{ECP},s} = \left(\frac{r_i - d_s}{r_i}\right)^{\frac{Jw(r_i)r_i}{D}} \tag{12}$$

where r_o is the outer radius of hollow fibers, r_i is the inner radius of hollow fibers, and D is the salt diffusivity in water. d_f and d_s are the boundary layer thicknesses of the ECP layer at the feed side and at the draw solution side, respectively. The profiles of water flux and power density versus operation pressure under various boundary thicknesses are plotted in Figure 5.10. It is very interesting to find that both water flux and power density decrease rapidly with an increase in the boundary thickness d_s on the draw solution side while they only slightly decrease with an increase in the boundary thickness d_f on the feed side. As ECP factors on both sides

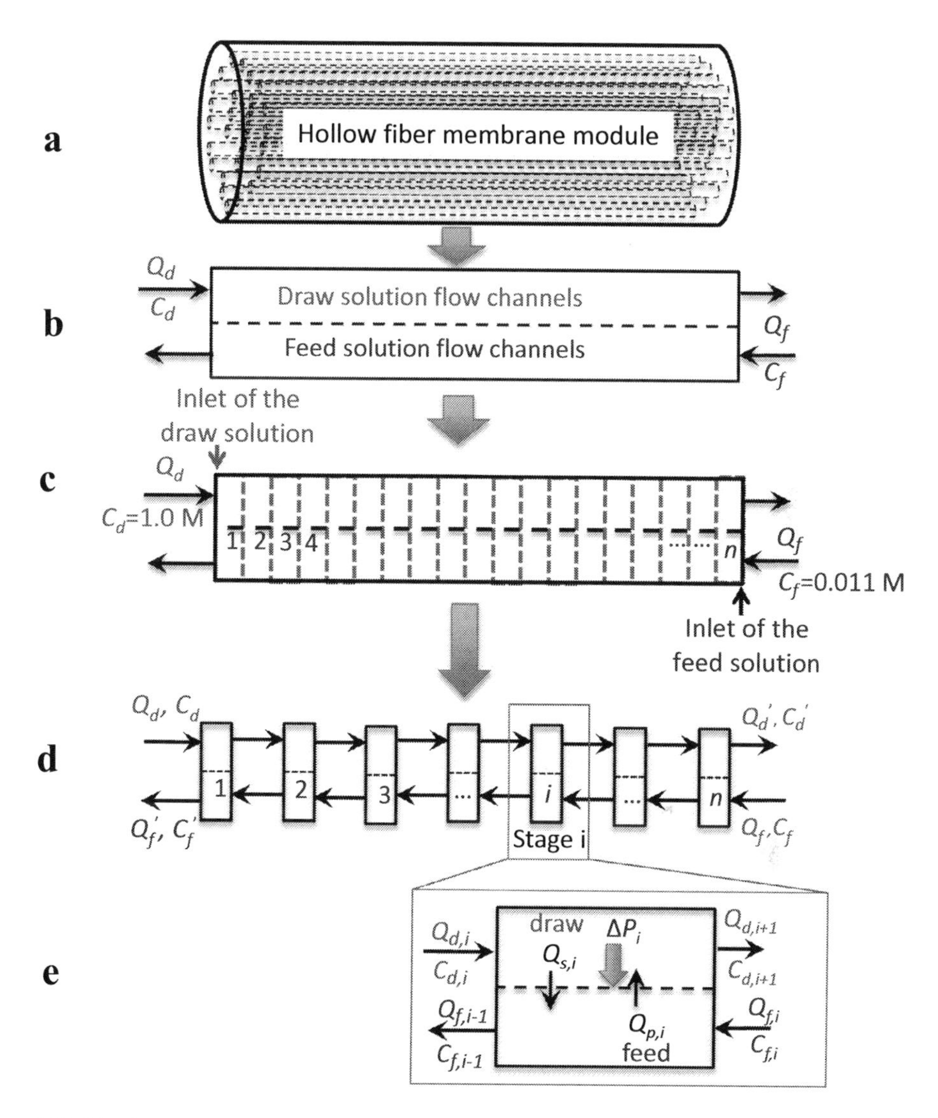

FIGURE 5.9
A schematic diagram of the simulation approach of the PRO hollow fiber membrane module.

have been defined in Eqs. 11 and 12, these two equations may provide the root causes why they respond differently with an increase in boundary thickness.

As shown in Eqs. 11 and 12, the ECP effect on the draw solution side is closely related to d_s and the inner radius of the hollow fiber membrane r_i, while the ECP effect on the feed side is closely associated with d_f and the

TABLE 5.3

Membrane parameters and module dimensions for simulation

Variable	Value
1. Membrane model parameters	
Water permeability, A (LMH/bar)	3.5
Salt permeability, B (LMH)	0.3
Membrane structural parameter, τ/ε (–)	2.8
Temperature, T (K)	298
Diffusion coefficient for NaCl (m^2/s)	1.61×10^{-9}
Draw solution concentration, $C_{D,b}$ (mol/L)	1.0
Feed water concentration, $C_{F,b}$ (mol/L)	0.011
Outer radius of fiber, r_o (mm)	0.50
Inner radius of fiber, r_i (mm)	0.29
2. Module parameters	
The draw solution inlet flow rate, Q_d (m^3/h)	20
The feed solution inlet flow, Q_f (m^3/h)	12
Transmembrane pressure at the sea water brine inlet, ΔP (bar)	20
The hollow fiber length, L (m)	2
Total number of hollow fibers, N	27,437
The module inner diameter, d_{module} (inch)	12
The module effective membrane area, A_m (m^2)	100
Number of stages, n	100

outer radius of the hollow fiber membrane r_o. For the TFC hollow fiber membrane used in this study, r_i is 0.29 mm, which is much smaller than r_o (i.e., 0.5 mm). Therefore, the variation of d_s in Eq. 12 has a much stronger impact on $F_{ECP,s}$ than that of d_f in Eq. 11 on $F_{ECP,f}$. In other words, the asymmetric geometry of the hollow fiber membrane results in very different behaviors of water flux and power density when d_s and d_f varies.

There are two implications from the above findings. First, based on Eqs. 11 and 12, ECP effects on both sides can be reduced by employing hollow fiber membranes with relatively larger inner and outer diameters. However, hollow fiber membranes with a smaller diameter are sometimes preferred to provide better mechanical strength, larger surface area, and higher packing density. Therefore, it is worthy of investigation in the future to design the hollow fiber membranes with the optimal inner and outer diameters for PRO applications. Second, the boundary thicknesses on both sides (i.e., d_s and d_f) are actually determined by the fluid dynamics within PRO modules. From the viewpoint of operating PRO modules made from inner-selective membranes, more energy is needed to pump

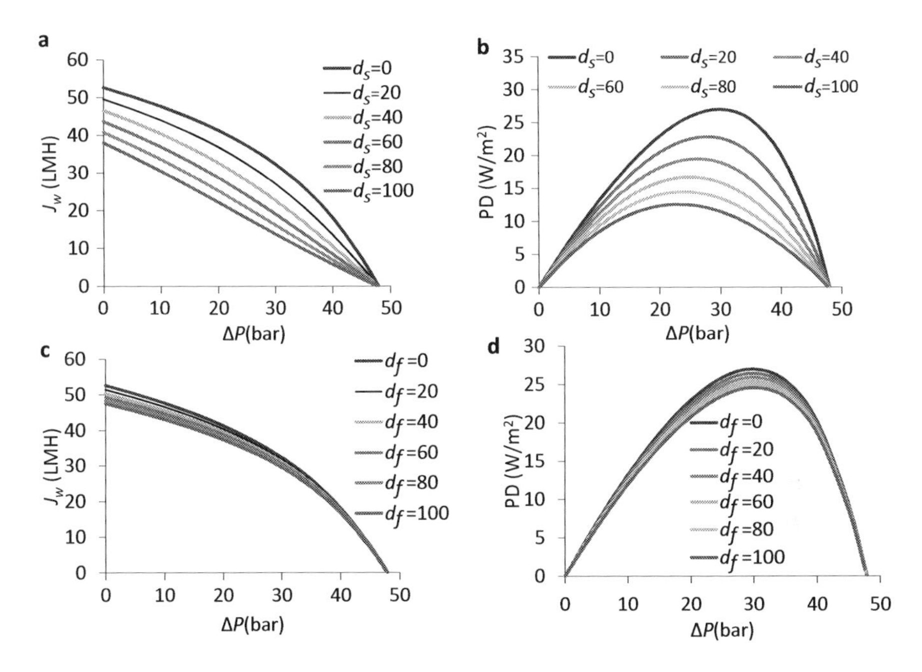

FIGURE 5.10
ECP effects on the draw solution side and the feed side. d_s and d_f are the thicknesses of the boundary layer on the draw solution side and the feed side, respectively. The unit of d_s and d_f is micrometer. The bulk feed salinity $C_{F,b}$ and the bulk draw solution salinity $C_{D,b}$ are fixed as 0.011 and 1.0 M, respectively. J_w is the water flux at the internal surface of the inner-selective hollow fiber membrane. When the boundary layer of one side changes from 0 to 100 μm, the boundary layer of the other side is set to be constant.

the draw solution in order to ensure a relatively larger draw solution flow rate and to reduce d_s so that the ECP effect on the draw solution side can be effectively suppressed.

5.5.2 Relationship between External Performance Indexes and Internal Factors at Membrane Level

To clearly understand the relationship between external performance indexes and internal factors, all the relevant mass transfer equations can be reorganized in the schematic system diagram in Figure 5.11 (Xiong et al., 2017). This membrane level analysis will serve as the basis for the module scale investigation in later sections. As shown in Figure 5.11, there are three groups of variables linking to one another in the system diagram, namely, membrane intrinsic properties, membrane structural parameters, and operational parameters. Some of the parameters can be combined to form functional factors, which play important roles in the

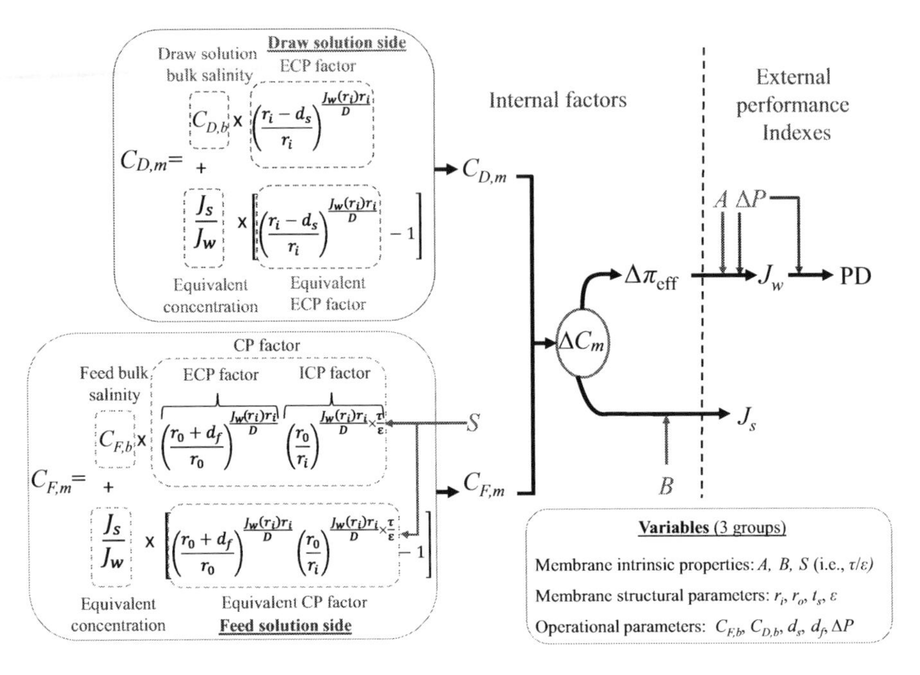

FIGURE 5.11
Schematic system diagram illustrating the relationship between external performance indexes and internal factors of inner-selective hollow fiber membranes.

mass transfer across the PRO membranes. These factors include ECP factors for both feed and draw solution sides, ICP factor for the porous support, and equivalent ECP or CP factors for both sides. As illustrated in Figure 5.11, both the water flux and reverse salt flux are ultimately linked to the salinity gradient ΔC_m across the PRO membrane, which is determined by the surface salinities $C_{D,m}$ and $C_{F,m}$.

$C_{D,m}$ and $C_{F,m}$ actually comprise two terms with two associated factors. For example, $C_{D,m}$ consists of (1) the bulk salinity of the draw solution $C_{D,b}$ multiplied by the ECP factor on the draw solution side; and (2) the equivalent concentration J_s/J_w (Xiong et al., 2016) multiplied by the equivalent ECP factor, which is due to the presence of the reverse salt flux. In addition, membrane intrinsic properties, membrane structural parameters, and operational parameters take effect at different locations in the system diagram, and result in various external performance behaviors. As investigated in the earlier section, the asymmetric nature of the hollow fiber configuration induces a significant ECP effect on the draw solution side. In fact, besides ECP factors on both sides, the hollow fiber dimensions such as r_i and r_o also impact the ICP factor and its equivalent CP factor on the feed side. Hollow fiber membranes with different combinations of r_i and r_o are expected to display distinct J_w and power density behaviors due to the different ICP effects as well.

5.5.3 Various Salinities along Flow Channels in PRO Modules

In the hollow fiber PRO modules, $C_{D,m}$, $C_{F,m}$, and $C_{F,t}$ illustrated in Figure 5.8 can be simplified as (Xiong et al., 2017):

$$C_{D,m} = C_{D,b}F_{\text{ECP},s} + \frac{J_s}{J_w}\left(F_{\text{ECP},s} - 1\right) \tag{13}$$

$$C_{F,m} = C_{F,t}F_{\text{ICP}} + \frac{J_s}{J_w}(F_{\text{ICP}} - 1) \tag{14}$$

$$C_{F,t} = C_{F,b}F_{\text{ECP},f} + \frac{J_s}{J_w}\left(F_{\text{ECP},f} - 1\right) \tag{15}$$

where F_{ICP} is the ICP factor of the porous support:

$$F_{\text{ICP}} = \left(\frac{r_0}{r_i}\right)^{\frac{J_w(r_i)r_i}{D_e}} \tag{16}$$

In addition, $F_{\text{ECP},f}$ and $F_{\text{ECP},s}$ are the ECP factors of the feed side and draw solution side defined in Eqs. 11 and 12, respectively.
Various salinities along the flow channels within the PRO module of a typical operation case are presented in Figure 5.12. As expected, $C_{F,b}$ and $C_{D,b}$ decrease with the normalized position due to the water permeation and reverse salt flux across the PRO membrane. Intermediate concentrations such as $C_{D,m}$, $C_{F,m}$, and $C_{F,t}$ follow the similar trends as $C_{D,b}$ and $C_{F,b}$ on their respective sides. In the counter-current mode, the local effective driving force for mass transfer, defined by the difference between $C_{D,m}$ and $C_{F,m}$ (or ΔC_m in Figure 5.11), remains relatively stable across the length of the entire module. In addition, as enlarged in the magnified insets in Figure 5.12, on the draw solution side, the significant drop from $C_{D,b}$ to $C_{D,m}$ is dominated by the dilutive ECP contribution (i.e., $C_{D,b}$ $F_{\text{ECP},s}$ in Eq. 13), while the J_s contribution (i.e., $\frac{J_s}{J_w}\left(F_{\text{ECP},s} - 1\right)$ in Eq. 13, due to the presence of J_s) is negligible. Based on Eq. 13, this is because the equivalent concentration J_s/J_w (~0.007–0.008 M in this study) caused by the presence of the reverse salt flux J_s is much smaller than $C_{D,b}$ (~0.9–1.0 M).
In contrast, the growth from $C_{F,b}$ to $C_{F,m}$ on the feed side is largely due to the ICP contribution (i.e., $C_{F,t}$ F_{ICP} in Eq. 14) and J_s contribution (i.e., $\frac{J_s}{J_w}(F_{\text{ICP}} - 1)$ in Eq. 14) inside the membrane porous support (Figure 5.12 (b)), while the ECP contribution (i.e., $C_{F,b}F_{\text{ECP},f}$ in Eq. 15) and J_s

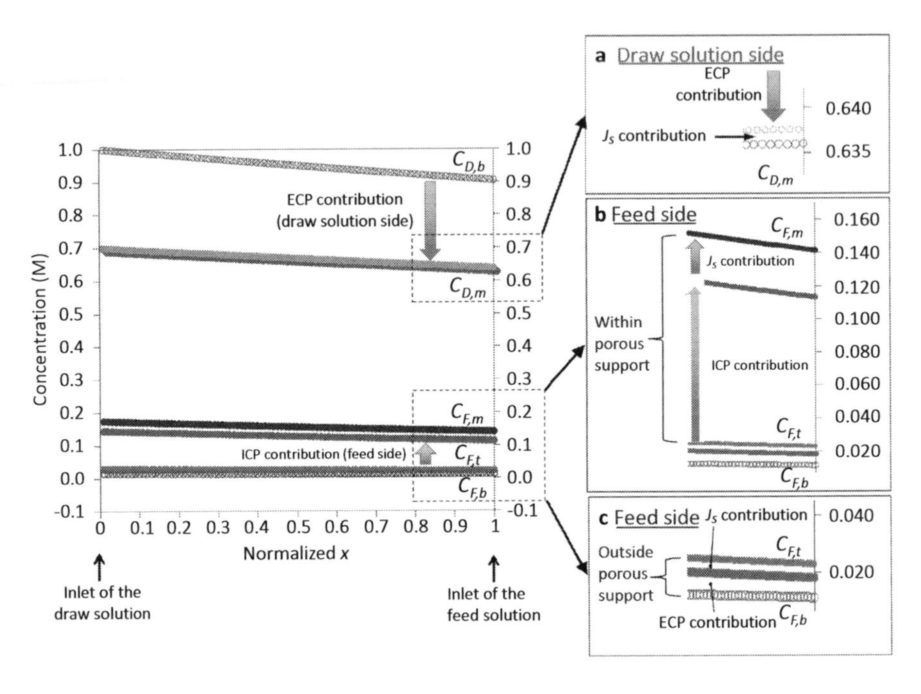

FIGURE 5.12
Various salinities along the PRO module (counter-current flow, conditions of incoming streams: Q_d=20 m^3/h, C_d=1.0 M, Q_f=12 m^3/h, C_f = 0.011M, multi-stage simulation, stage number n = 100).

contribution (i.e., $\frac{J_s}{J_w}(F_{ECP,f} - 1)$ in Eq. 15) outside the membrane support (Figure 5.12 (c)) are not so important. According to Eqs. 14 and 15, it is because the ICP factor F_{ICP} for the membrane support (~4.85–4.99) is much larger than the ECP factor $F_{ECP,f}$ outside the membrane support (~1.63–1.66). Moreover, within the membrane porous support, the ICP contribution (i.e., $C_{F,t}$ F_{ICP} in Eq. 14) is more significant than the J_s contribution (i.e., $\frac{J_s}{J_w}(F_{ICP} - 1)$ in Eq. 14); however, the latter is still nontrivial and therefore cannot be neglected. Referring to Eq. 14, this is because the equivalent concentration J_s/J_w (~0.007–0.008 M) is now comparable to $C_{F,b}$ (~0.011–0.015 M). In summary, the dilutive ECP effect is dominant on the draw solution side, while the concentrative ICP effect is dominant on the feed side. In addition, the J_s contribution on the feed side due to the reverse salt flux is still obvious even for the TFC hollow fiber membrane having a relatively small salt permeability (Xiong et al., 2016). Further efforts in the future to develop PRO membranes with a lower salt permeability is required to suppress the negative J_s contribution toward the PRO system performance.

5.5.4 Effect of Structural Parameter

Besides the operation pressure and ECP effects, the membrane's structural parameter is another important factor affecting the PRO module performance. As indicated in the equation of $C_{F,m}$ in Figure 5.11, the structural parameter S of hollow fibers in this study has the form of τ/ε, in which τ is the tortuosity and ε is the porosity of the porous support. Figure 5.13 shows that a reduction of structural parameter can significantly enhance water flux at both membrane and module levels because a smaller structural parameter can effectively reduce $C_{F,m}$ and enlarge the effective driving force $\Delta\pi_{\text{eff}}$ across the membrane. As a result, the water flux and power density are enhanced.

5.5.5 Significance of Various Factors on Module Scale Performance

To understand the significance of various factors on module scale performance, the effects of these factors are compared in Figure 5.14. Among all the factors, the ECP in the draw solution side and ICP in the feed side are

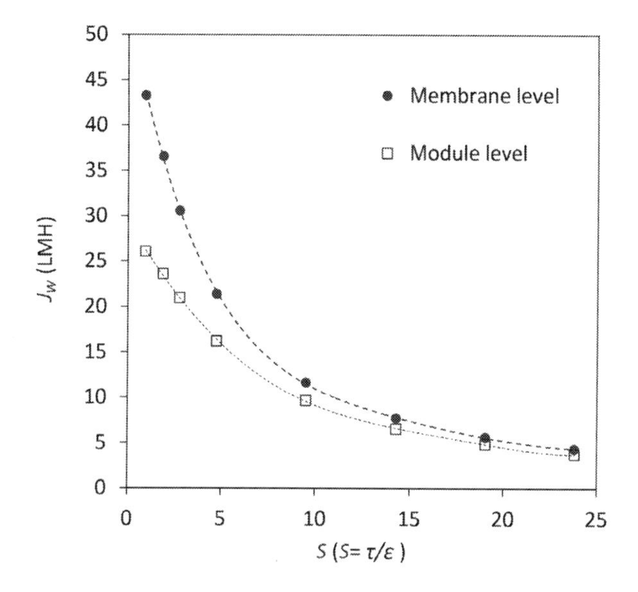

FIGURE 5.13
Water flux as a function of structural parameter S ($S = \tau/\varepsilon$) for the inner-selective membrane at the operation pressure of 20 bar. Water permeability A and salt permeability B remain unchanged. The dotted lines are the best polynomial fitting. Module level results are obtained from a typical case with $Q_d = 20\ \text{m}^3/\text{h}$, $C_d = 1.0$ M, $Q_f = 12\ \text{m}^3/\text{h}$, and $C_f = 0.011$ M. Membrane level results are obtained from the case with very large flow rates ($Q_d = 100\ \text{m}^3/\text{h}$ and $Q_f = 100\ \text{m}^3/\text{h}$), in which the bulk concentrations of the draw solution and the feed along the entire module actually remains constant ($C_{D,b} = 1.0$ M and $C_{F,b} = 0.011$ M).

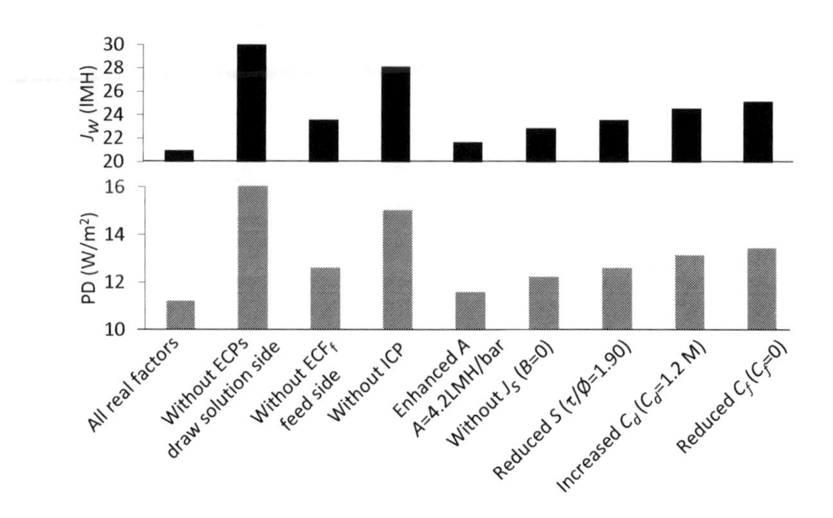

FIGURE 5.14
Factor analyses of module level performance for the typical case with Q_d = 20 m^3/h, C_d = 1.0 M, Q_f = 12 m^3/h, C_f = 0.011 M, and the operation pressure set at 20 bar.

the top two factors in determining module scale performance. To effectively mitigate the ECP effect on the draw solution side, a relatively large draw solution flow rate Q_d may be wanted to achieve a high power density. On the other hand, a small structural parameter S is required to mitigate the ICP effect. In addition, as investigated in some prior works (Xiong et al., 2016), ICP may also promote severe scaling within the porous support, resulting in more severe ICP and rapid water flux reduction. Therefore, more efforts are needed to find solutions in removing scaling precursors from the feed streams, and developing antiscaling hollow fiber membranes.

As indicated in Figure 5.14, the module performance can also be enhanced by increasing the salinity of the draw solution from 1.0 to 1.2 M or by further bringing down the feed salinity from 0.011 M to a fresh water level. However, although the feed salinity can possibly be reduced further, it would be more viable to explore the usage of draw solutions with higher salinities. Finally, although the membrane intrinsic properties such as water permeability A, salt permeability B, and membrane structural parameter S seem to have less significant effects on the module performance, their importance should not be neglected. In fact, since the TFC hollow fiber membrane used in this study has been well engineered to have a high A, low B, and rather small S, further moderate improvements on membrane's intrinsic properties may not noticeably enhance the module performance.

5.6 Conclusions

In summary, flux reduction behaviors can be very complex as multiple factors must be considered. A schematic system diagram illustrated in Figure 5.11 can serve as the platform to facilitate an overview of all the internal factors, external performance indexes, and interlinking relationships among them. This platform can be applied to scaling processes and explain the dynamic mechanisms of water flux reduction (Figure 5.7). The platform can also be employed at every stage of a PRO module to provide various salinities along the flow channels (Figure 5.12). Therefore, various flux reduction behaviors in PRO investigated in this chapter will surely provide useful insights in both PRO membrane development and operations of PRO modules.

Acknowledgements

This work is granted by the Singapore National Research Foundation under its Environmental & Water Research Programme and administered by the Public Utilities Board (PUB), Singapore's National Water Agency. It is funded under the projects entitled "Membrane Development for Osmotic Power Generation, Part 1. Materials Development and Membrane Fabrication" (1102-IRIS-11-01) and "Part 2, Module fabrication and system integration" (1102-IRIS-11-02). We also thank Professor Tai-Shung Chung for fruitful discussions and the improvement of writing.

References

Chen, S.C., Amy, G.L., Chung, T.S., 2016. Membrane fouling and anti-fouling strategies using RO retentate from a municipal water recycling plant as the feed for osmotic power generation, *Water Res.* 88, 144–155.

Cheng, Z.L., Li, X., Liu, Y., Chung, T.S., 2016. Robust outer-selective thin-film composite polyethersulfone hollow fiber membranes for osmotic power generation, *J. Membr. Sci.* 506, 119–129.

Cheryan, M., 1986. *Ultrafiltration handbook*, Technomic Pub. Co, Lancaster, PA.

Han, G., Zhang, S., Li, X., Chung, T.S., 2015. Progress in pressure retarded osmosis (PRO) membranes for osmotic power generation, *Prog. Polym. Sci.* 51, 1–27.

Kim, D.I., Kim, J., Shon, H.K., Hong, S., 2015. Pressure retarded osmosis (PRO) for integrating seawater desalination and wastewater reclamation: Energy consumption and fouling, *J. Membr. Sci.* 483, 34–41.

Lee, K., Baker, R., Lonsdale, H., 1981. Membranes for power generation by pressure-retarded osmosis, *J. Membr. Sci.* 8, 141–171.

Liu, X.Y., 2000. Heterogeneous nucleation or homogeneous nucleation? *J. Chem. Phys.* 112, 9949–9955.

Lonsdale, H.K., 1973. Recent advances in reverse osmosis membranes, *Desalination* 13, 317–332.

Sato, K., Furukawa, Y., Nakajima, K., 2001. *Advances in crystal growth research*, Elsevier Science, Amsterdam and New York.

She, Q., Jin, X., Tang, C.Y., 2012. Osmotic power production from salinity gradient resource by pressure retarded osmosis: Effects of operating conditions and reverse solute diffusion, *J. Membr. Sci.* 401–402, 262–273.

Tang, C.Y., Chong, T.H., Fane, A.G., 2011. Colloidal interactions and fouling of NF and RO membranes: A review, *Adv. Colloid Interface Sci.* 164, 126–143.

Tang, C.Y., Kwon, Y.-N., Leckie, J.O., 2009. The role of foulant–Foulant electrostatic interaction on limiting flux for RO and NF membranes during humic acid fouling—Theoretical basis, experimental evidence, and AFM interaction force measurement, *J. Membr. Sci.* 326, 526–532.

Wan, C.F., Chung, T.S., 2015. Osmotic power generation by pressure retarded osmosis using seawater brine as the draw solution and wastewater retentate as the feed, *J. Membr. Sci.* 479, 148–158.

Xiong, J.Y., Cai, D.J., Chong, Q.Y., Lee, S.H., Chung, T.S., 2017. Osmotic power generation by inner selective hollow fiber membranes: An investigation of thermodynamics, mass transfer, and module scale modelling, *J. Membr. Sci.* 526, 417–428.

Xiong, J.Y., Cheng, Z.L., Wan, C.F., Chen, S.C., Chung, T.S., 2016. Analysis of flux reduction behaviors of PRO hollow fiber membranes: Experiments, mechanisms, and implications, *J. Membr. Sci.* 505, 1–14.

Yip, N.Y., Tiraferri, A., Phillip, W.A., Schiffman, J.D., Hoover, L.A., Kim, Y.C., Elimelech, M., 2011. Thin-film composite pressure retarded osmosis membranes for sustainable power generation from salinity gradients, *Environ. Sci. Technol.* 45, 4360–4369.

Zeman, L.J., Zydney, A.L., NetLibrary, I., 1996. *Microfiltration and ultrafiltration: Principles and applications*, M. Dekker, New York.

Zhang, M., Hou, D., She, Q., Tang, C.Y., 2014. Gypsum scaling in pressure retarded osmosis: Experiments, mechanisms and implications, *Water Res.* 48, 387–395.

Zhang, S., Chung, T.S., 2013. Minimizing the instant and accumulative effects of salt permeability to sustain ultrahigh osmotic power density, *Environ. Sci. Technol.* 47, 10085–10092.

6

Development of Antifouling Pressure Retarded Osmosis Membranes

Wen Gang Huang, Xue Li, and Tao Cai

Key Laboratory of Biomedical Polymers of Ministry of Education
College of Chemistry and Molecular Science, Wuhan University
Wuhan, P. R. China

Wuhan University Shenzhen Research Institute
Shenzhen, P. R. China

CONTENTS

6.1 Introduction

Pressure retarded osmosis (PRO) providing sustainable use of salinity gradient from water resources has gained worldwide attention (Chung et al., 2012; Logan and Elimelech, 2012; Han et al., 2015; Kim et al., 2015; Touati and Tadeo, 2016). One drawback of PRO is that the water permeation flux is limited by fouling. Fouling blocks membrane pores, causes a severe performance loss, and shortens membranes' lifetime. Scaling of inorganic salts, deposition of organic compounds, and colonization of microorganisms often cause fouling. When wastewater is used as the feed, mixed foulants with bacteria and inorganic and organic compounds lead to severe pore clogging and irreversible flux reduction because of synergistic mechanisms, which make the fouling eradication a very difficult task (Liu and Mi, 2012).

The severity of fouling depends on the membrane orientation since the active layer of asymmetric membranes helps keep foulants away from internal pores. Consequently, fouling under the AL-FS (i.e., the active layer facing the feed solution) mode remains moderate and easily reversible, while membranes operating under the AL-DS (i.e., the active layer facing the draw solution) mode are prone to serious irreversible fouling. Unfortunately, PRO is operating under the AL-DS mode, and foulants in the feed are inevitably brought into naked pores by water permeation (Yip and Elimelech, 2013; Li et al., 2014; Fang et al., 2018). What's worse, the foulants accumulated inside the support are extremely difficult to wash out.

To mitigate PRO fouling, accomplishments in understanding the fouling mechanisms have been made. Thelin et al. (2013) proposed that some aspects related to PRO fouling are different from traditional pressure-driven processes. They focused on fouling caused by natural organic matter (NOM) and pointed out that the flux reduction as a result of NOM load was independent of the NOM concentration. Moreover, increased ions in the support layer exacerbated PRO fouling. She and coworkers (2013) also studied the organic fouling in PRO and examined its effects on power density. The appearance of a large number of divalent cations led to significant alginate fouling in the draw solution side. Chen and coworkers (2014) fabricated a polybenzimidazole-based hollow fiber membrane and studied its fouling propensity in PRO. They found that gypsum scaling was inhibited by the reverse magnesium chloride flux, while alginate fouling was slightly worsened by the reverse sodium chloride flux. The authors further investigated the PRO fouling under different operating pressures and reported that gypsum-alginate synergistic fouling was eliminated by high hydraulic pressures (Chen et al., 2015). Wan and Chung (2015) employed a real wastewater retentate (WWRe) as the feed and systematically investigated PRO fouling by using a state-of-the-art hollow fiber membrane. They identified fouling on the support layer induced by the wastewater effluent to have a significant role in PRO fouling; on the other hand, negligibly negative impacts were caused by the draw solution. She and coworkers (2017) investigated both organic and inorganic fouling in PRO and identified three fouling mechanisms: internal concentration polarization (ICP)–enhanced fouling, inter-molecular interaction–enhanced fouling, and pressure-independent limiting flux behavior.

The severity of fouling also depends on membrane surface properties, such as hydrophilicity or hydrophobicity, surface charge, morphology, and topology. For example, membranes prepared from hydrophobic materials may have high fouling tendency due to hydrophobic–hydrophobic interactions (Díez et al., 2018). Thus, altering membrane surface properties gives us an essential way to produce a low-fouling membrane. So far, researchers have developed a significant number of strategies for fouling control via chemistry approaches (Sofia et al., 1998; Li et al., 2008; Chang et al., 2011; Chen et al., 2013; Yang et al., 2015), but most studies were conducted to prevent fouling under the AL-FS mode. For example, hydrophilic polymers have been anchored on the active layer (Yang et al., 2011; Yu et al., 2014), or nanocomposite active layers have been developed (Nguyen et al., 2014; Ghanbari et al., 2015). Efforts to defend porous support layers against fouling in PRO and investigations of antifouling membranes under the AL-DS mode are recently emerging. In this chapter, we cover developments of antifouling PRO membranes via surface functionalization treatments. First, synthesis of hyperbranched polyglycerols (HPGs), preparation of HPG-grafted PRO membranes, and their fouling behaviors will be discussed, respectively. It is followed by a review of PRO membranes grafted of zwitterionic materials. Lastly, other antifouling approaches in PRO membranes, such as surface tethering of different agents and double-skinned structure will be summarized.

6.2 HPGs-Grafted Membranes for Pressure Retarded Osmosis

Polyglycerols have substantially hydrophilic characters and present good protein and bacterial repellences. They are nontoxic, chemically stable under basic or neutral conditions, ion-transportable, highly water soluble, and inexpensive (Herzberger et al., 2016). Their linear analogues, for example, poly(ethylene glycol) (PEG) or poly(ethylene oxide), has only one or two active sites providing limited functional groups and low anchoring capacity. By contrast, HPGs, a subclass of dendritic polyglycerols, have well-preserved ending group functionalities and offer multiple anchoring sites (Kainthan et al., 2006; Yeh et al., 2008; Khandare et al., 2012; Paulus et al., 2014), as illustrated in Figure 6.1. When applying HPGs on membranes, its size ranging from several to dozens of nanometers can give a broader shielding area to foulants.

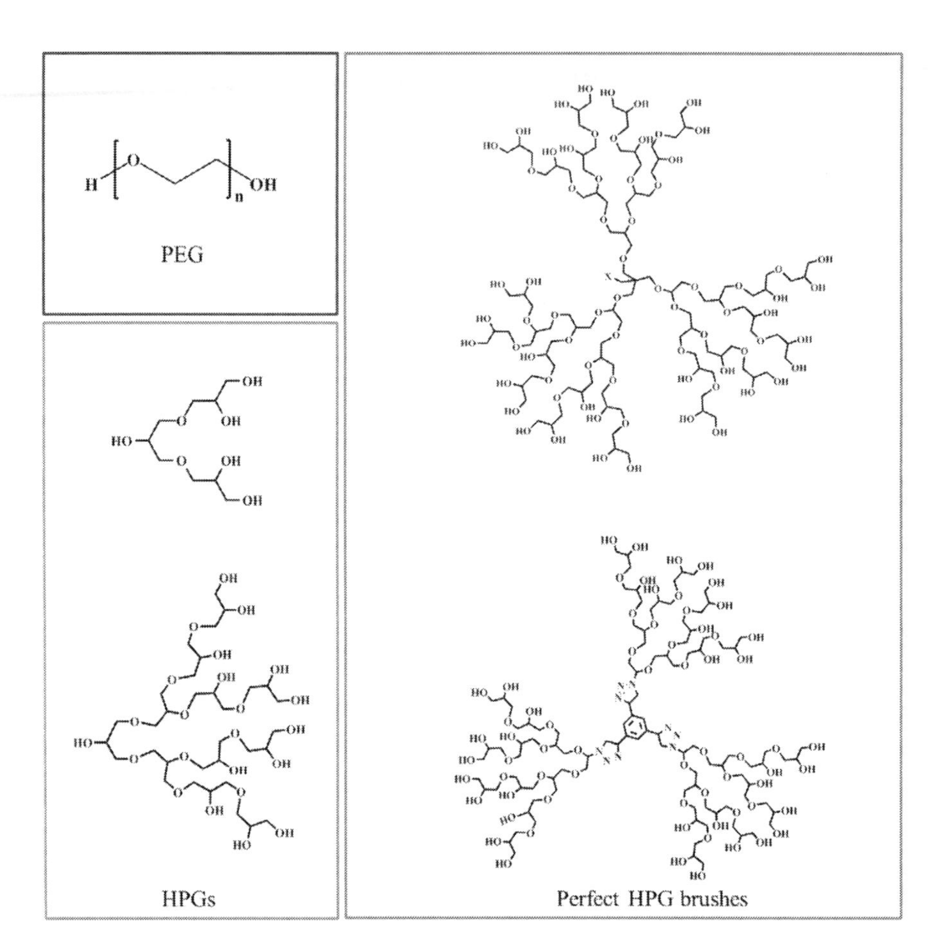

FIGURE 6.1
Examples of PEG and HPGs.

6.2.1 Synthesis Philosophy of HPGs

HPGs can be synthesized via three ways: bottom up (like growth from a seed to a tree), middle upon (like hybrid grafting of as-prepared precursors), and top down (like cuttings of undesired branches). Most HPGs are prepared through the first approach because of the diverse availability of monomers (Wilms et al., 2010; Cai et al., 2013). Both self-condensation of a monomer (e.g., one-monomer strategy) and polycondensation of a monomer pair (e.g., two-monomer strategy) can prepare HPGs from the bottom-up way. Among all synthesis strategies, self-polymerization of AB_2-type monomers (e.g., cyclic molecules) leads to HPGs with a controllable degree of branching and a narrow polydispersity index (PDI).

Glycidol is chosen to synthesize HPGs for antifouling PRO membranes by employing anionic ring-opening multibranching polymerization (AROMP) (Yan et al., 2000). The general reaction route is depicted in Scheme 6.1a. Glycidol with an attached hydroxyl group can be polymerized by the addition of nucleophilic initiators, for example, bis (2-hydroxyethyl)disulfide (BHEDS) or 1,1,1-tris(hydroxymethyl)ethane (TME), to generate branching points (Li et al., 2014, 2016; Zhang et al., 2018). 1,4-Polymerization involving a proton migration dominates the polymerization. Consequently, hydroxyl groups exist on both the polymer and the monomer. In the architectural construction, BHEDS or

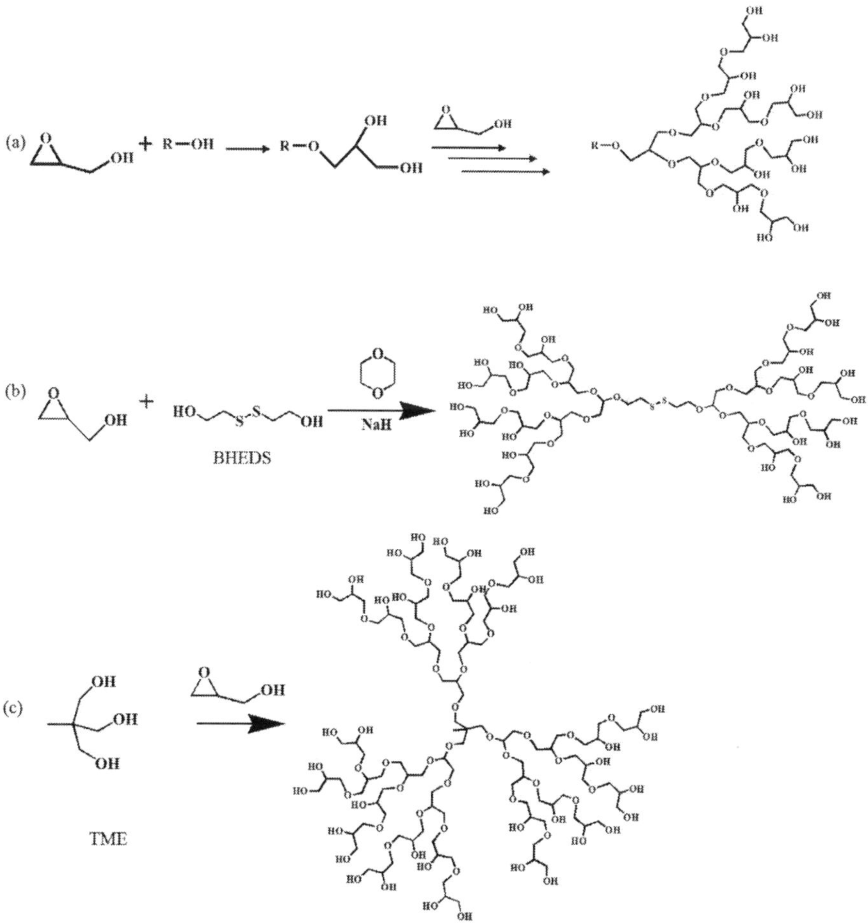

SCHEME 6.1
(a) A general reaction route of ring-opening polymerization of glycidol. (b) Synthesis route of HPG-S-S-HPG. (c) Synthesis route of a three-limb HPG.

TME, bearing two or three hydroxyl groups, can be used as the two-limb or three-limb initiator core. Properly adjusting (1) the addition speed of monomers and (2) the monomer-to-initiator ratio theoretically allows every newly produced glycerol unit to directly attach to the branched core, resulting in a perfect hyperbranched architecture (Figure 6.1). Hence, the key to a successful HPGs synthesis is the slow monomer addition technique to surpass the formation of a linear structure (Sunder et al., 1999).

Li and coworkers (2014) used BHEDS as the initiator to synthesize HPG for the PRO membrane modification. The HPG-S-S-HPG polymer with a molecular weight of 12,200 g mol^{-1} as shown in Scheme 6.1b was first synthesized from glycidol via anionic ring-opening multi-branching polymerization. Then a reducing agent, DL-1,4-dithiothreitol, cleaved of the disulfide bond in HPG-S-S-HPG and produced the final target, HPG-SH polymer (Figure 6.2a). Li et al. (2016) and Zhang et al. (2018) employed the three-OH-group initiator, TME, to make a three-limb HPG structure (Scheme 6.1c). The monomer-to-initiator ratio was fixed at 100:1, resulting in the HPG with a molecular weight of 5,700 g mol^{-1}. Then two negatively charged HPG derivatives, HPG-COOH and HPG-SO_4H (Figure 6.2b and 6.2c), were further synthesized from the three-limb precursor.

6.2.2 Grafting Approach of PRO Membranes

The state-of-the-art PRO membranes are thin-film composite (TFC) polyamide membranes based on polyethersulfone (PES) hollow fibers, whose configuration and structure can be referred to Chapter 4. Briefly, they

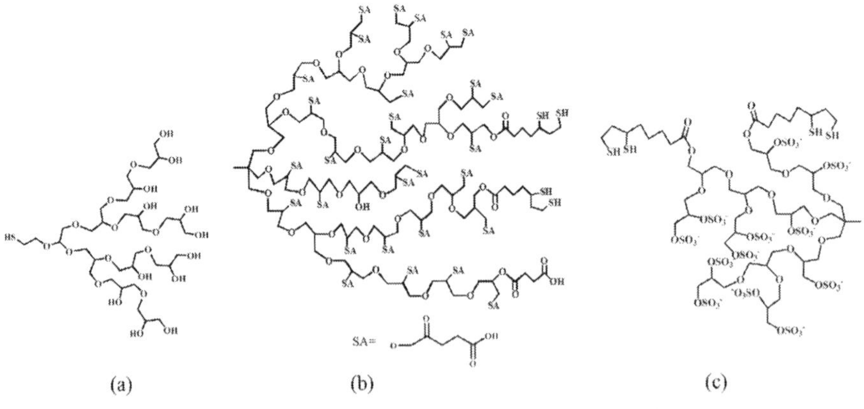

(a) (b) (c)

FIGURE 6.2
Three HPG polymers: (a) HPG-SH. (b) HPG-COOH. (c) HPG-SO_4H.

possess an asymmetric structure. Fouling on the thin film selective layer is relatively mild and reversible. On the other hand, PES substrates that directly face the feed are susceptible to foulants. So an effective surface modification must be carried out on the PES surface rather than on the polyamide dense-selective layer. PES supports are chemically inert in neutral conditions, therefore covalent grafting onto PES supports needs a bridge. In most cases, a polydopamine (PDA) gutter layer acts as a bridge between PES supports and grafting materials. The PDA coating has been widely implemented in various applications as it enables a strong linkage between the inert surface and functional polymers via Schiff base reactions or Michael additions (Liu et al., 2014). For example, thiol groups on grafting materials capable of easily tethering to PDA via Michael addition provide a simple approach of grafting reaction in mild conditions and aqueous solvents and keep the support structure robust and strong. Hence, HPGs with thiol ending groups are favorable in the grafting. A direct way of constructing an HPG-SH is the AROMP of glycidol from disulfide containing initiators, such as the synthetic route depicted in Scheme 6.1b. In the indirect construction of HPGs without thiol groups, a typical conversion of –OH ending groups to –SH ending groups employs *α*-lipoic acid to react with the HPGs in an esterification reaction, followed by a cleavage of disulfide bonds. Although thiol ending groups are available in three types of HPGs as illustrated in Figure 6.2, HPG-COOH and HPG-SO_4H (2b and 2c) converted from *α*-lipoic acid have pendent thiol ending groups and a higher grafting efficiency.

6.2.3 Antibacterial Adhesion, Antiprotein Adsorption, and Cytotoxicity of PRO Membranes

The resistance against bacteria and protein attachments is an important mark of membranes' antifouling behaviors. Usually, the pristine PES surface is seriously fouled due to the hydrophobic–hydrophobic interactions between foulants and the surface. In comparison, noncharged HPG-grafted PRO membranes exhibit good antimicrobial adhesion and antiprotein adsorption. For example, the HPG-SH (Figure 6.2a) grafted membrane displays a limited number of *Escherichia coli* (*E. coli*) and *Staphylococcus aureus* (*S. aureus*) attachments, while the pristine PES surface and PDA-coated surface are fully covered with bacterial colonization and biofilm matrix, as shown in Figure 6.3. A PEG-grafted membrane is also impervious to bacterial cells. Hydrophilic modifications form a hydrated barrier to protect the susceptible PES surface. Compared with PEG, HPGs are more efficient in providing an enthalpic and entropic fence, repelling most bacteria. In addition, the protein attaching tendency of the HPG-SH-grafted membrane is much lower than the PES supports. The pristine is easily covered with fluorescein-conjugated bovine serum albumin (BSA) after the protein exposure (Figure 6.4), where a green layer can be found. The green light emission is similar for the PDA-coated surface because its benzene rings induce π–π

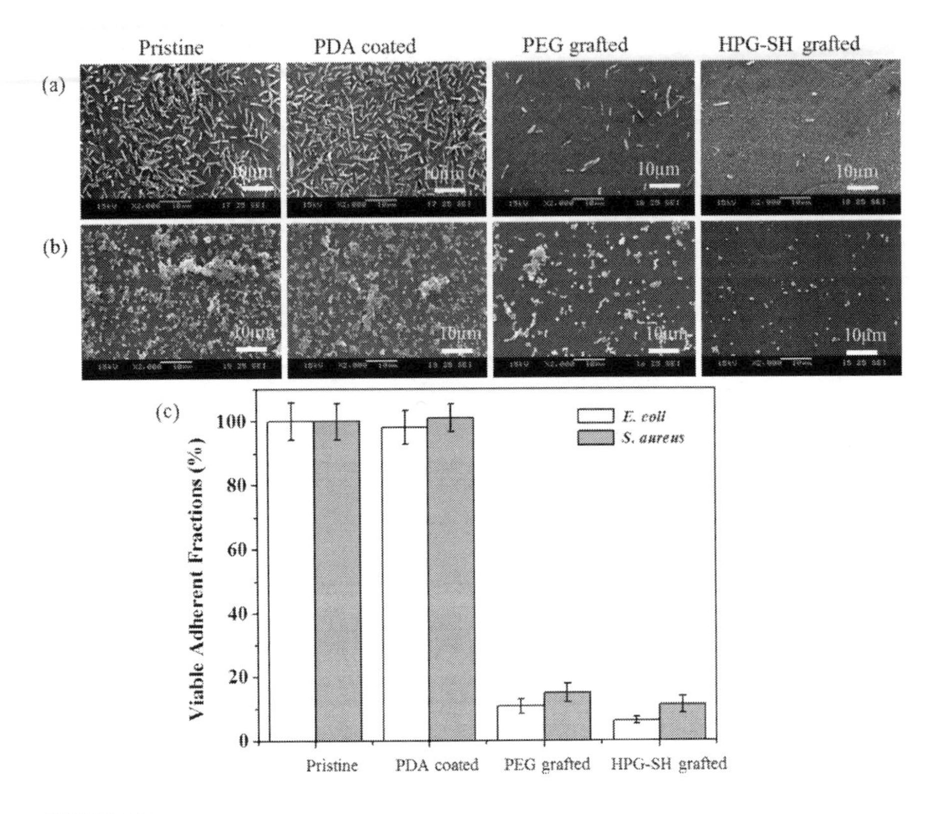

FIGURE 6.3
Antibacterial adhesion behavior of four membrane supports. (a) *Escherichia coli* (*E. coli*, ATCC DH5α); (b) *Staphylococcus aureus* (*S. aureus*, ATCC 25923); (c) Viable adherent fractions of *E. coli* and *S. aureus* cells in PBS (5×10^7 cells mL^{-1}) in contact with four hollow fiber membrane supports at 37°C for 4 h.

Adapted with permission from Li et al. (2014). Copyright (2014) American Chemical Society.

stacking to attract proteins. After the hydrophilic grafting, protein adsorption on the PEG-grafted surface considerably reduces. Among the four membranes, the HPG-SH grafting achieves the best antiadsorption capability.

Conversion of a portion of neutral –OH groups in HPGs to charged groups, for example, –COOH or $-SO_4H$, does not reduce their antibacterial and antiprotein abilities. On the contrary, the microbial repellences of both HPG-COOH- and HPG-SO_4H-grafted membranes seem to increase to some extent. The viable colonies of Gram-positive *S. aureus* or *Staphylococcus epidermidis* (*S. epidermidis*), or Gram-negative *E. coli* on HPGs-modified membranes are barely evident, as shown in Figure 6.5. Regarding the antiprotein capability, the HPG-SO_4H-grafted membrane has the smallest fluorescence intensity of BSA due to its strong electrostatic force of sulfonate (Figure 6.6) when

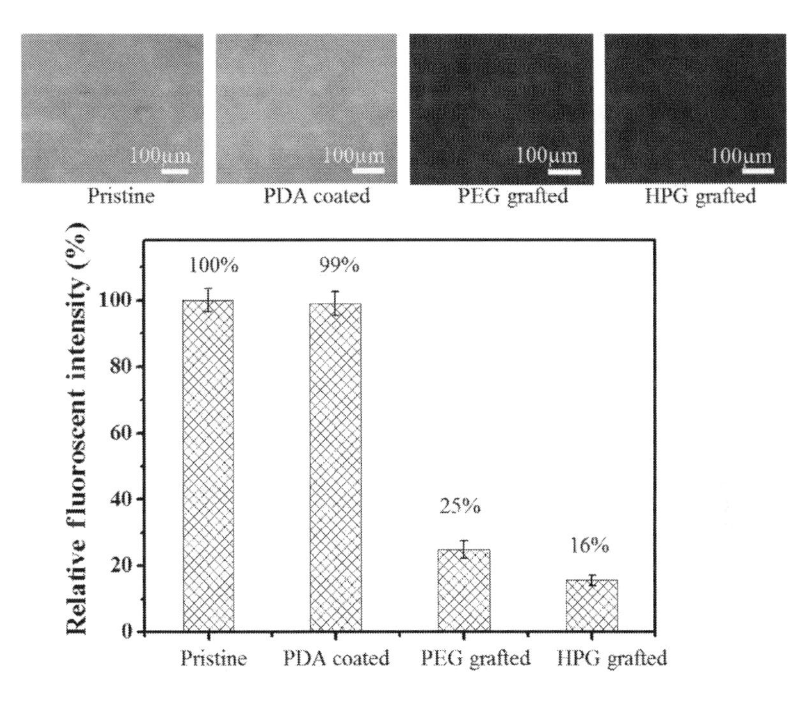

FIGURE 6.4
Relative fluorescence intensities and respective fluorescence microscopy images of the PES, PDA-coated, PEG-grafted, and HPG-grafted hollow fiber membrane supports after exposure to 0.5 mg mL^{-1} BSA-FITC solution for 1 h.

Adapted with permission from Li et al. (2014). Copyright (2014) American Chemical Society.

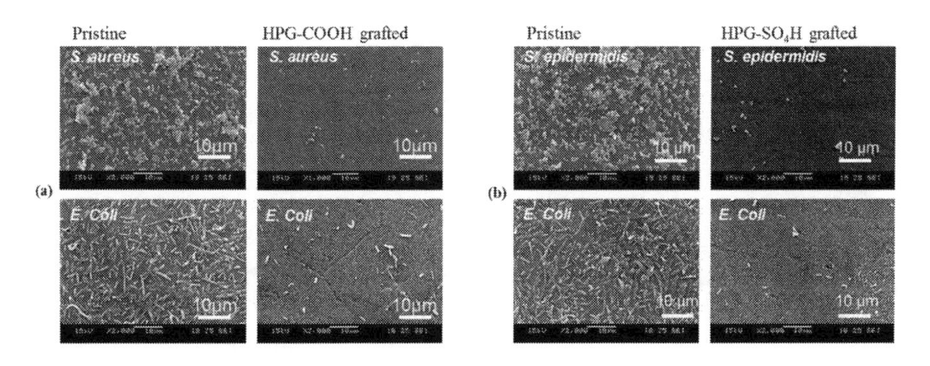

FIGURE 6.5
SEM images of outer surfaces of pristine and HPG-grafted membranes after exposure to *S. aureus*, *S. epidermidis*, and *E. coli*. The cell number was determined by the spread plate method.

Reprinted with permission from Elsevier. (a) Li et al. (2016); (b) Zhang et al. (2018).

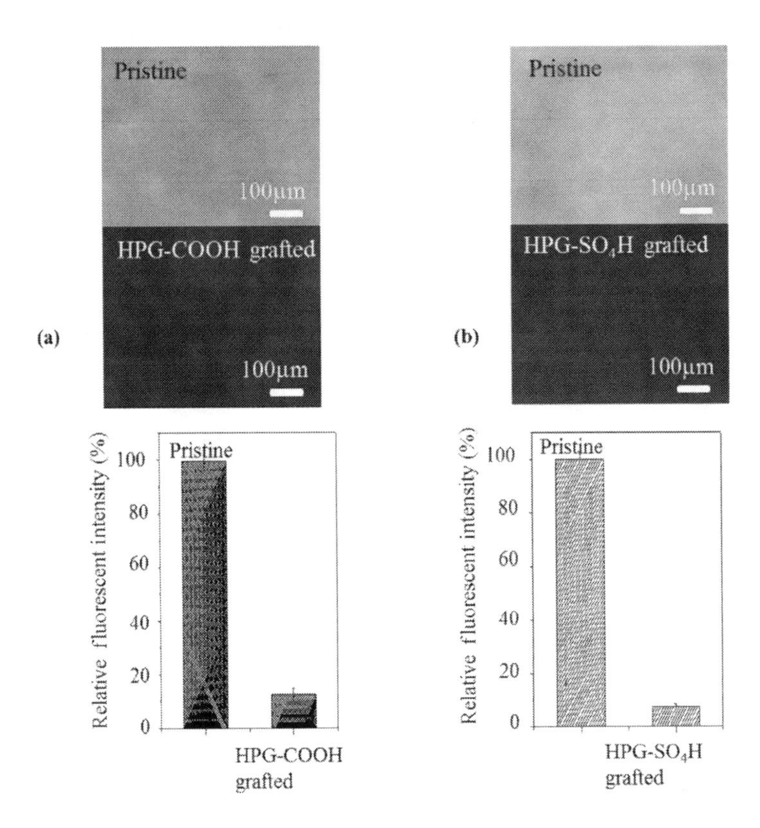

FIGURE 6.6
Relative fluorescence intensities and respective fluorescence microscopy images of support surfaces of different membranes after exposure to 0.5 mg/mL BSA solutions for 1 h.

Reprinted with permission from Elsevier. (a) Li et al. (2016); (b) Zhang et al. (2018).

comparing with other fouled membranes. Quantitatively, the HPG-SO_4 H-grafted membranes reduce BSA adsorption to 6% of the reference value, under the given testing conditions.

The saline stream pair utilized in PRO may be the source of water consumption and must be strictly nontoxic. Hence, the toxicity of modified membranes is of great importance in practical use. The *in vitro* experiments of viable cells in 3T3 fibroblasts or RAW macrophages culture media tell the cytotoxicity of membranes (Figure 6.7). A number of viable cells on PEG- and HPG-grafted membranes exceed more than 90% when compared with the pristine, which are coincident with the good biocompatibility of glycerol and its analogues. In all cases, PDA-coated membranes show marginally higher cytotoxicity to cells, probably due to the imine groups of PDA. This problem may be a potential drawback of bare PDA coatings.

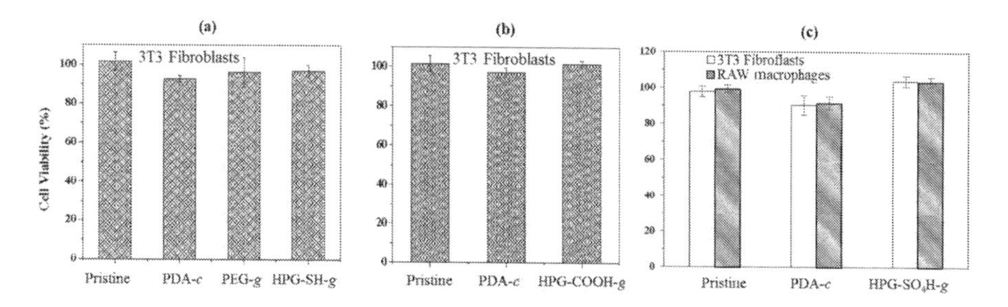

FIGURE 6.7
Cytotoxicity assays of membranes in 3T3 fibroblasts or RAW macrophages culture medium after 24 h of incubation. PDA-*c*: PDA coated; PEG-*g*: PEG grafted; HPG-SH-*g*: HPG-SH grafted; HPG-COOH-*g*: HPG-COOH grafted; HPG-SO_4H-*g*: HPG-SO_4H grafted.

Adapted with permission from (a) Li et al. (2014). Copyright (2014) American Chemical Society. Reprinted with permission from Elsevier. (b) Li et al. (2016); (c) Zhang et al. (2018).

6.2.4 Fouling in PRO Processes

The grafting of HPGs advances the hydrophilicity and wettability of PRO membranes and certainly improves the water permeation flux in PRO. Figure 6.8 displays the membrane performance of the pristine, PEG-grafted, and HPG-SH–grafted membranes using DI as the feed solution and 3.5% NaCl as the draw solution. The initial water flux at 0 bar of the HPG-SH-grafted membrane enhances to ~30 L m^{-1} h^{-1} (LMH), while the pristine bears an initial flux of only 20 LMH. Although the PEG grafting also affects membrane's hydrophilicity, the limited number of –OH group caps its flux enhancement. Thus, the PEG-grafted membrane only exhibits a marginal flux increase by 2–3 LMH.

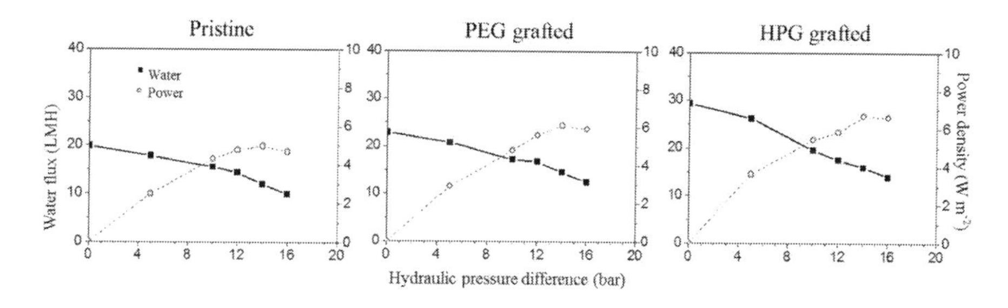

FIGURE 6.8
Membrane performance in PRO tests using DI as the feed solution and 3.5% NaCl as the draw solution.

Adapted with permission from Li et al. (2014). Copyright (2014) American Chemical Society.

To mimic the organic fouling in wastewater, PRO performance using a BSA solution as the feed was studied (Figure 6.9a). The pristine, PEG-grafted, and HPG-SH-grafted membranes behave differently. The former two suffer steep flux decreases immediately after being exposed to foulants, and the latter one shows an eased-off decline. The water flux reduction of the HPG-SH-grafted membrane is only around 10% of the initial value. In summary, the non-charged HPG-grafted membrane has an enhanced flux and an impressively low fouling susceptibility to organic compounds.

When municipal wastewater retentate (WWRe) containing mixed foulants is used as the feed, complex fouling of inorganic salts, organic compounds, and microorganisms severely deteriorate the membrane performance. A tremendous flux decrease of 60–80% can be found in the real wastewater testing of the pristine, as shown in Figure 6.9b and 6.9c, while the flux in previous BSA testing only reduces by 20–30% (Figure 6.9a). In the real wastewater case, the neutral –OH groups in HPG-SH lack strong repulsive forces to the charged foulants. Undoubtedly, charged ending groups such as –COOH and $-SO_4H$ are more favorable. The experimental results confirm that both HPG-COOH- and $HPG\text{-}SO_4H$-grafted membranes exhibit a much slower flux reduction in WWRe than the pristine (Figure 6.9b and 6.9c). Moreover, DI cleaning on charged HPG-grafted membrane is able to effectively remove the foulants deposited on the surface (Figure 6.10) and restore the performance to more than 90% (Figure 6.9c). However, the restoration is vulnerable because many foulants still clog the porous support (Chen et al., 2016; Li et al., 2017a). In contrast, cleaning with the aid of acid or ethylenediaminetetraacetate (EDTA) exhibits a more thorough washing and restores the flux effectively (Figure 6.10).

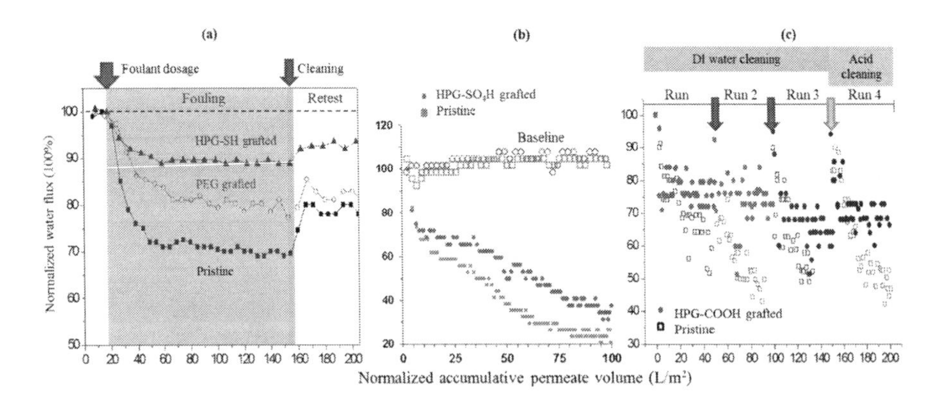

FIGURE 6.9
Membrane fouling performance in PRO tests. (a) 200 ppm BSA solution as the feed. (b and c) Municipal wastewater retentate as the feed.

Adapted with permission from (a) Li et al. (2014). Copyright (2014) American Chemical Society. Reprinted with permission from Elsevier. (b) Zhang et al. (2018); (c) Li et al. (2016).

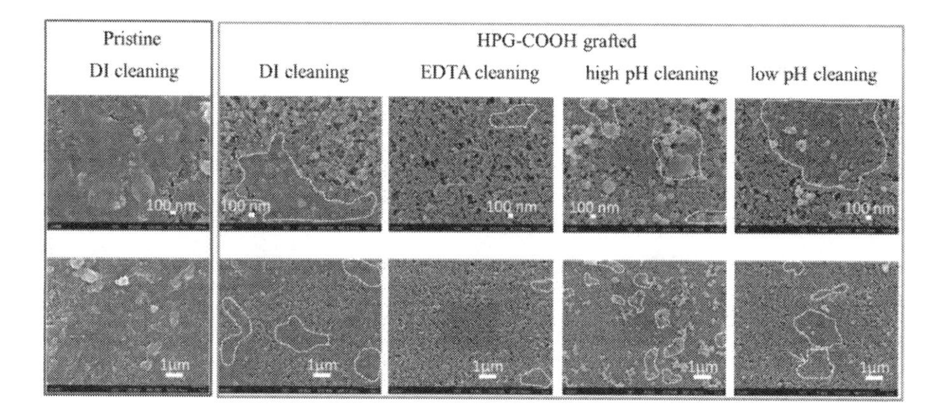

FIGURE 6.10
SEM detection on surfaces of membranes after different cleaning.

Reprinted with permission from Elsevier (Li et al., 2017a).

6.3 Zwitterionic Material-Grafted Membranes for Pressure Retarded Osmosis

Zwitterionic materials with both negative and positive charges in the same molecule build a robust hydration layer in aqueous solutions. Their externally neutral nature broadens the applications (Lowe and McCormick, 2002; Li et al., 2008; Jiang and Cao, 2010; Chang et al., 2011; Yang et al., 2011; Chen et al., 2013; Yu et al., 2014; Le et al., 2017). To take advantage of zwitterions, zwitterionic material-grafted PRO membranes are mainly prepared via the "grafting to" approach (Hansson et al., 2013) together with direct synthesis of well-designed structures.

6.3.1 Preparation Routes of PRO Membranes

Zwitterionic copolymers can be synthesized via free-radical polymerization using 2,2′-azobis(2-methylpropionitrile) (AIBN) as the initiator (Cai et al., 2016; Zhao et al., 2016; Han et al., 2018), as shown in Scheme 6.2a, 2b, 2c. In the first two cases, a disulfide-containing monomer, 2-methacryloyloxyethyl lipoate (MEL), is copolymerized to provide the grafting sites of –SH. The copolymer poly(MPC-*co*-AEMA), as shown in Scheme 6.2c bearing $-NH_2$ groups, enables a direct reaction with the imide group of the Torlon support. In the fourth case, thiol groups in thiolated-2-methacryloyloxyethyl phosphorylcholine (MPC-SH) facilitate its grafting to a PDA-coated surface (Scheme 6.2d). Different from these four membranes made by a "grafting to"

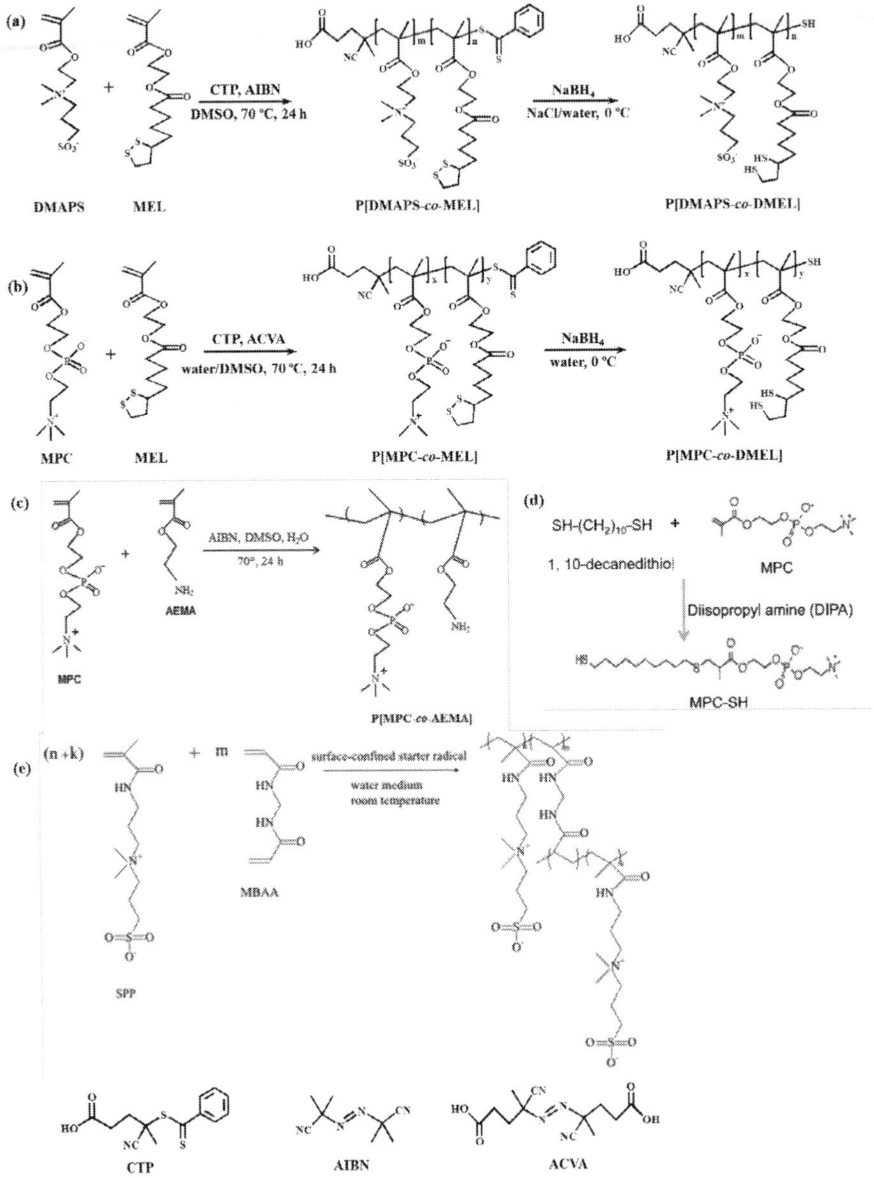

SCHEME 6.2
Synthesis routes of zwitterionic materials.

method, Le and coworkers (2017) use a "grafting from" method to graft zwitterionic copolymers on the polyetherimide (PEI) supports. A pre-anchored layer of poly((2-dimethylamino)ethyl methacrylate-*co*-butyl methacrylate) (poly(DMAEMA-*co*-BMA)) on the support provides initiating sites for the direct radical polymerization of zwitterionic monomers on the surface. Then the monomer 3-((3-methacryla-midopropyl)dimethyl ammonio)-propane-1-sulfonate (SPP) is crosslinked with *N,N*-methylenebisacrylamide (MBAA) (Scheme 6.2e). When "grafting from" a surface, a "bottom-up growth" is conducted. Comparatively, a well-prepared structure is covalently attached to the surface in a "grafting to" method. The "grafting from" method obviously leads to a higher grafting density with a poorer polydispersity index, and the major problem of "grafting to" is the low grafting efficiency due to the stereo hindrance of the pre-formed polymer. The low grafting rate can be solved by employing highly efficient "click" reactions, like thiol-ene chemistry (Hansson et al., 2013). Therefore, if the grafting materials have $-NH_2$ or –SH groups to "click" on the surface, the "grafting to" method is more promising to prepare well-designed antifouling membranes from an industrial point of view.

No matter what method is applied, phosphobetaine and sulfobetaine pendant groups are popular in designing zwitterionic materials because of their good hydrophilicity and stable chemistry.

6.3.2 Antibacterial Adhesion, Antiprotein Adsorption, and Cytotoxicity of PRO Membranes

Positive and negative charges of zwitterionic materials provide strong electrostatic attraction to water molecules. In aqueous solutions, the zwitterionic materials grafting takes up a thick hydration layer of water molecules, prevents foulants from close contact with the membrane surface, and substantially eliminates microbial and protein accumulation. Fluorescence microscopic images clearly show extensive protein attachments in the case of pristine membranes (Figure 6.11). Their hydrophobic–hydrophobic interaction with organic molecules intensely adsorbs BSA on the surface. On the contrary, fewer proteins are found on membranes with zwitterionic materials grafting. The microbial adhesion of pristine membranes is also evident, but zwitterionic material-grafted membranes, such as PES-*g*-P(DMAPS-*co*-DMEL) and PES-*g*-P(MPC-*co*-DMEL), are mostly free of bacteria and bacterial colonizations (Figure 6.12). Polyzwitterion-grafted PEI membranes prepared by the "grafting from" method (Scheme 6.2e) have a moderate antimicrobial ability. The modified PEI membranes show the relative cell viability in the range of 50–80%. It is worth noting that bacterial clusters are mostly prevented in all grafting cases. The hydration layer of zwitterionic materials exhibits strong repulsion against extracellular substances and biofilm formation. Besides, PES-*g*-P(DMAPS-*co*-DMEL) and PES-*g*-P(MPC-*co*-DMEL) membranes have been proven to be non-cytotoxic to live cells by

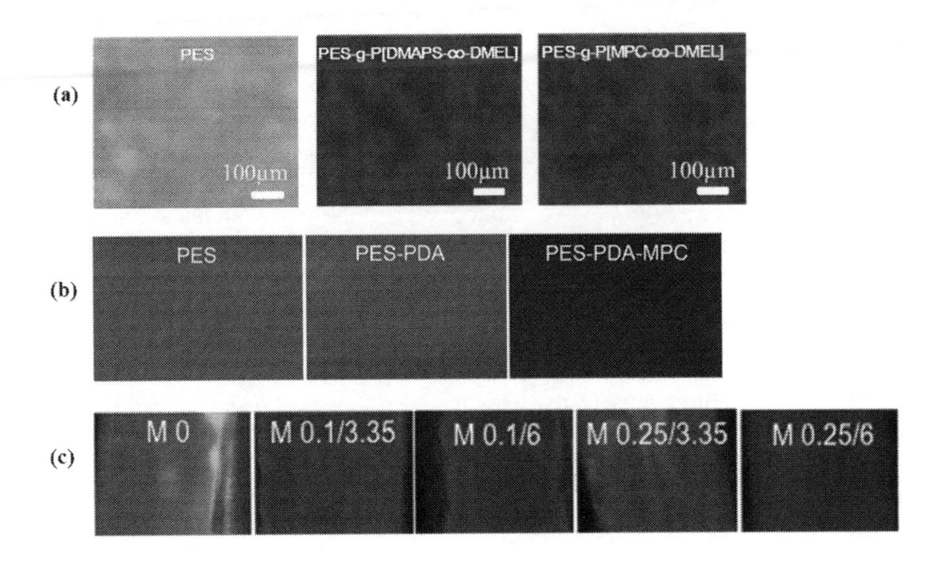

FIGURE 6.11
Fluorescence microscopic images of membranes. M0 is an unmodified PEI membrane, M0.1/3.35, M0.1/6, M0.25/3.35, M0.25/6 are polyzwitterion-grafted PEI membranes.

Reprinted with permission from Elsevier. (a) Cai et al. (2016); (b) Zhao et al. (2016); (c) Le et al. (2017).

cytotoxicity experiments as shown in Figure 6.13. The high biocompatibility of PES-*g*-P(DMAPS-*co*-DMEL) and PES-*g*-P(MPC-*co*-DMEL) membranes avoids any secondary contamination to the water source.

6.3.3 Fouling in PRO Processes

The fouling behaviors in PRO process are also examined for zwitterionic material-grafted membranes. BSA was chosen as the model foulant in the PRO tests of polyzwitterion-grafted PEI membranes (Figure 6.14a). Compared to the sharp flux drop of the pristine, the polyzwitterion-grafted PEI membranes show less serious fouling and better recovery. Their better antiprotein capabilities resulted from the hydrophilic surface.

Multiple fouling and cleaning cycles using WWRe as the feed provide more practical results. Multiple cycle performances of the PES-P(MPC-*co*-AEMA) membrane, PES-PDA-MPC membrane, PES-*g*-P(DMAPS-*co*-DMEL) membrane, and PES-*g*-P(MPC-*co*-DMEL) membrane were shown in Figure 6.14b, 6.14c, and 6.14d, respectively. Zwitterionic material-grafted membranes display quite stable and regenerable PRO performance. Quantitatively, the PES-P(MPC-*co*-AEMA), PES-*g*-P(DMAPS-*co*-DMEL), and PES-*g*-P(MPC-*co*-DMEL) membranes have a water flux of >70% of the initial value, under the given testing conditions.

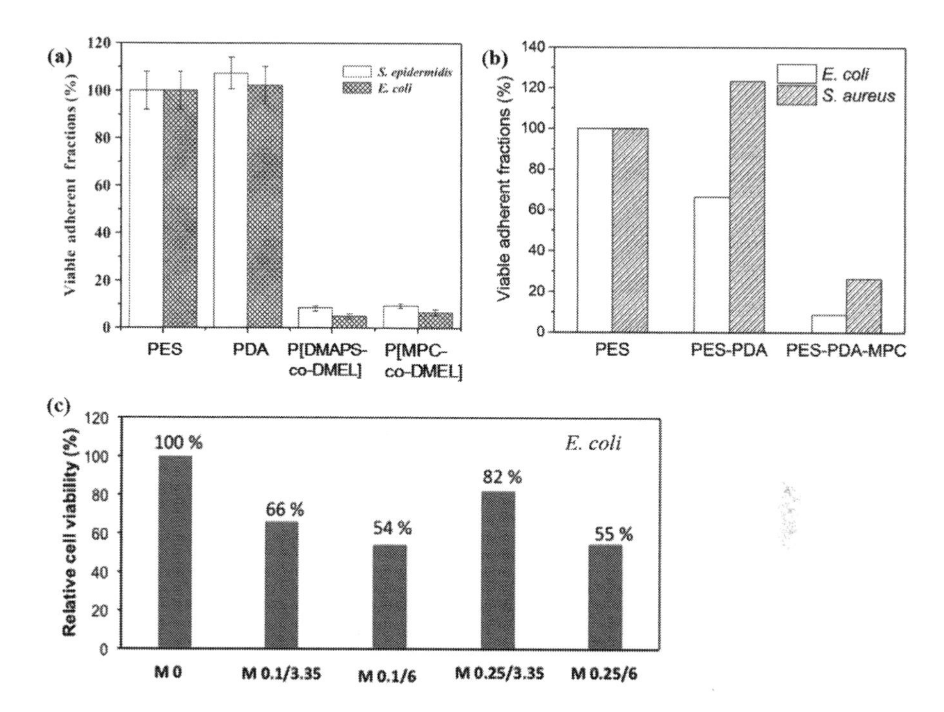

FIGURE 6.12
Bacterial growth inhibition.

Reprinted with permission from Elsevier. (a) Cai et al. (2016); (b) Zhao et al. (2016); (c) Le et al. (2017).

6.4 Other Surface-Modified Membranes for Pressure Retarded Osmosis

6.4.1 Membranes Modified by Different Agents

In addition to HPGs and zwitterionic materials grafting, there are some other surface-grafted PRO membranes made from commercially available agents, such as poly(vinyl alcohol) (PVA) and aminosilane. For example, PVA is a commonly used polymer with many active –OH groups as shown in Scheme 6.3.

Zhang and coworkers (2016b) grafted PVA onto the PDA-coated membranes with the aid of glutaraldehyde (GA) in an acidic environment. As shown in Scheme 6.3, GA is capable of reacting with the –OH groups and $-NH_2$ groups. The PVA molecules dispersed in the solution have a higher probability to participate in the reaction. Simultaneously a few GA connect PVA and PDA and link the PVA layer on top of the

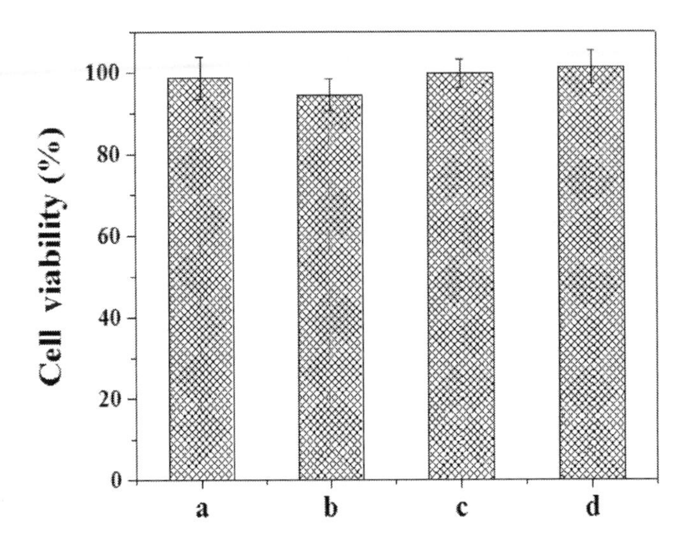

FIGURE 6.13
Cytotoxicity assays of (a) PES, (b) PDA, (c) PES-*g*-P(DMAPS-*co*-DMEL) and (d) PES-*g*-P(MPC-*co*-DMEL) hollow fibers in 3T3 fibroblasts culture medium after 24 h of incubation.

Reprinted with permission from Elsevier (Cai et al., 2016).

membrane. The more reactive sites on PDA, $-NH_2$ groups, become positive charged and hardly take part in the crosslinking because hydrochloride has pre-protonated these free amine groups. The PVA grafting on membranes results in a more hydrophilic surface (Figure 6.15a) and higher resistances to alginate fouling (Figure 6.15b and 6.15c). When sodium alginate is dosed in the feed, a mild improvement in antifouling performance is found for the PVA-grafted membrane (Figure 6.15b). This modified membrane acts better in complicated conditions. In a calcium-alginate complex circumstance (Figure 6.15c), 78% of the initial flux can be maintained for the PVA-grafted membrane. Although the calcium-alginate fouling complex always readily attaches on the membranes and forms irreversible deposition, simple cleaning on the PVA-grafted membrane can effectively restore the water flux up to 90% (Figure 6.15d).

Aminosilane is another commercial grafting agent to prepare surface-grafted PRO membranes. Its silicon center connects with both aminoalkyl group and alkoxy groups, as shown in Scheme 6.4. Alkoxy groups can be easily hydrolyzed and provide –OH groups for fouling control, while the aminoalkyl group serves as a grafting site. Zhang and coworkers (2016a) used 3-aminopropyltrimethoxysilane (APTMS) to modify the polyimide membrane surface (Scheme 6.4). Aminoalkyl groups of APTMS react with aromatic imide groups of polyimide and open the imide rings. PRO performance of the grafted membrane was investigated in a model wastewater

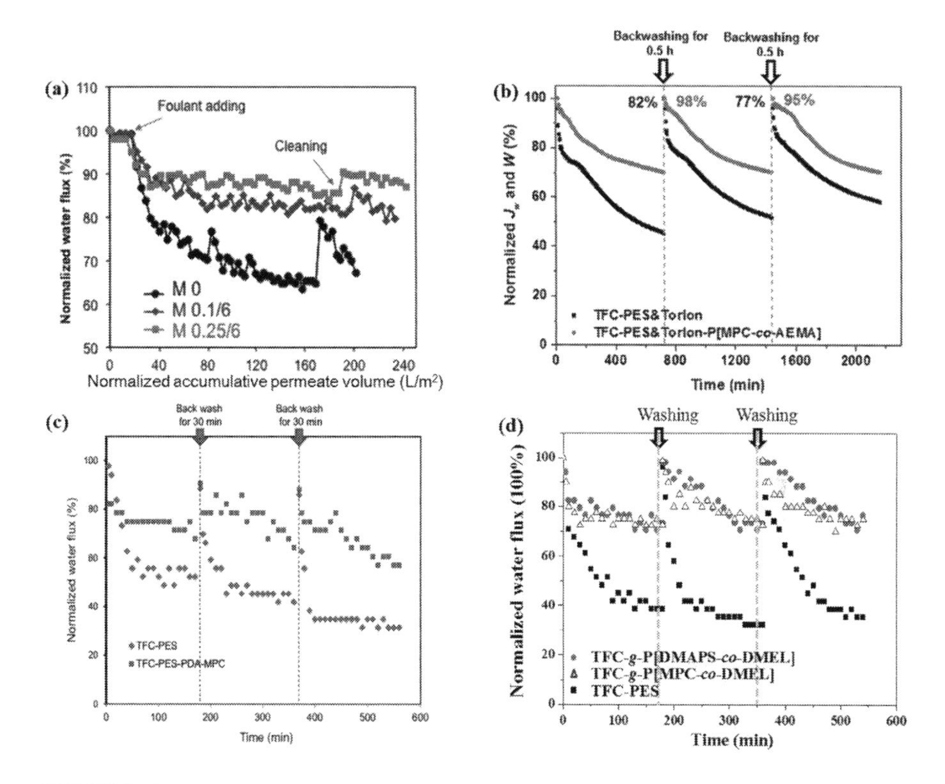

FIGURE 6.14
PRO performance in wastewater feeds. Reprinted with permission from Elsevier. (a) Le et al. (2017); (c) Zhao et al. (2016); (d) Cai et al. (2016).

Adapted with permission from (b) Han et al. (2018). Copyright (2018) American Chemical Society.

containing 10 mM NaCl, 3 mM $CaCl_2$, and 200 mg/L humic acid. The pristine possesses a very steep flux reduction and the grafted membrane has enhanced fouling resistances, as illustrated in Figure 6.16.

Carbon quantum dots (CQDs) with tunable hydrophilicity and ultra-fine particle size are a novel candidate for the PRO membrane modification. Zhao and coworkers (2017) developed CQD-modified PRO membranes for fouling control. They synthesized CQDs from citric acid, and then fixed CQDs on a PDA-coated PES support with the aid of *N*-(3-(dimethylamino) propyl)-*N′*-ethylcarbodiimide hydrochloride (EDC) and *N*-hydroxysuccinimide (NHS) via covalent bindings (Figure 6.17). The PRO performance in organic foulants was investigated for the CQD-modified membrane. It exhibited high resistances to organic fouling because the carboxyl groups on CQDs have strong repulsive forces to organic compounds. The water flux retains 82% of the initial value after 3 h of testing.

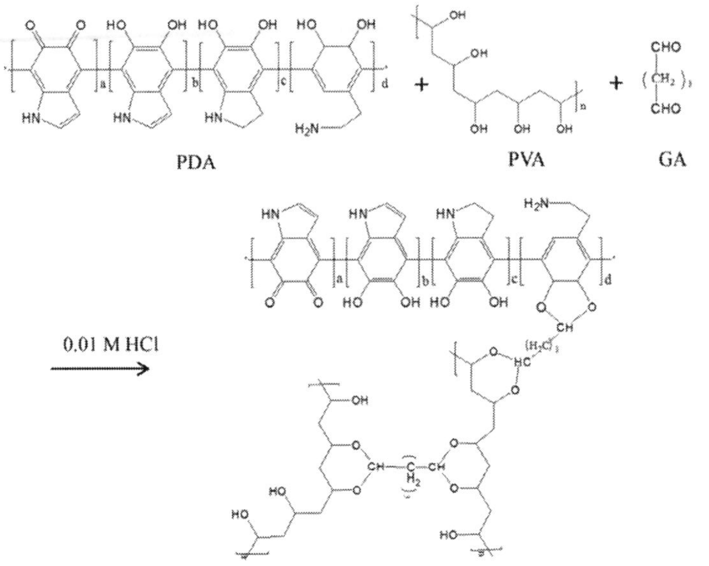

SCHEME 6.3
Possible reaction of PVA and PDA with the aid of GA.

Adapted with permission from Zhang et al. (2016b). Copyright (2016) American Chemical Society.

In addition to surface modification via covalent bindings, surface attachments by electrostatic interactions offer a feasible alternative. For instance, oppositely charged polyelectrolytes are also deposited on membrane supports to modify PRO membranes. Li and coworkers (2017b) deposited two polyelectrolytes, poly(allylamine) hydrochloride (PAH) and poly(acrylic acid) (PAA), on the polyetherimide support in a sequence of alternating charge. This is a typical layer-by-layer (LbL) adsorption. A large number of charged species can be employed in the LbL adsorption to produce a surface layer with different functions. Li and coworkers chose PAH and PAA because they have strong electrostatic repulsions to foulants. A charged PAH/PAA coating protects the substrate from organic foulants, such as BSA or alginate. The modified membrane suffers less severe organic fouling. However, the electrostatic attachments are not substantially stable especially when in some harsh operations.

6.4.2 Double-Skinned Membranes

Foulants accumulated inside membrane supports significantly lower the membrane performance. Researchers have pioneered double-skinned membranes to mitigate the internal accumulation of foulants (Wang et al., 2010; Zhang et al., 2010; Duong et al., 2014), where a secondary dense skin was

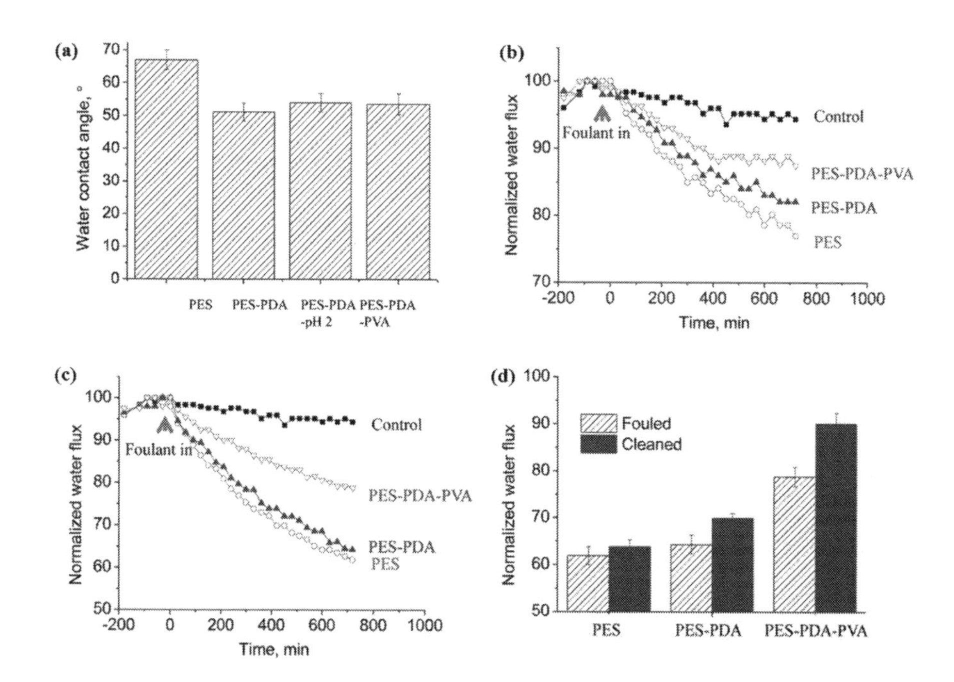

FIGURE 6.15

(a) Water contact angles of the PES and modified membranes. (b) Fouling performance of the PES and modified membranes in 200 ppm alginate; draw solution, 1 M NaCl. (c, d) Fouling performance of the PES and modified membranes in 200 ppm alginate +1.5 mM $CaCl_2$; draw solution, 1 M NaCl.

Adapted with permission from Zhang et al. (2016b). Copyright (2016) American Chemical Society.

synthesized to reject impurities in the feeds. Han and coworkers (2017) took advantages of double-skinned structure and developed a double-thin-film hollow fiber membrane for PRO. This hollow fiber membrane has a dense inner skin and a less dense outer skin both made by interfacial polymerization between *m*-phenylenediamine (MPD) and trimesoyl chloride (TMC), as shown in Figure 6.18. The double-skinned TFC membrane exhibits a mild fouling property in wastewater. Clearly, the secondary skin effectively alleviates the internal concentration polarization and fouling caused by wastewater.

However, the rough secondary skin induces a certain amount of foulant deposition on the surface. Furthermore, the carboxylate groups of the thin film increase the foulant–surface interactions. To solve this problem, chemical cleaning is necessary to recover performance of the double-skinned TFC membranes, as shown in Figure 6.19.

Hu and coworkers (2016) designed another double-skinned PRO membrane for fouling control. They simultaneously fabricated a dense graphene oxide (GO) barrier on top of the membrane support as well as a GO barrier

Aminosilane

Polyimide

SCHEME 6.4
Structure of 3-aminopropyltrimethoxysilane and the possible chemical reaction between the polyimide membrane and aminosilane.

Reprinted with permission from Elsevier (Zhang et al., 2016a).

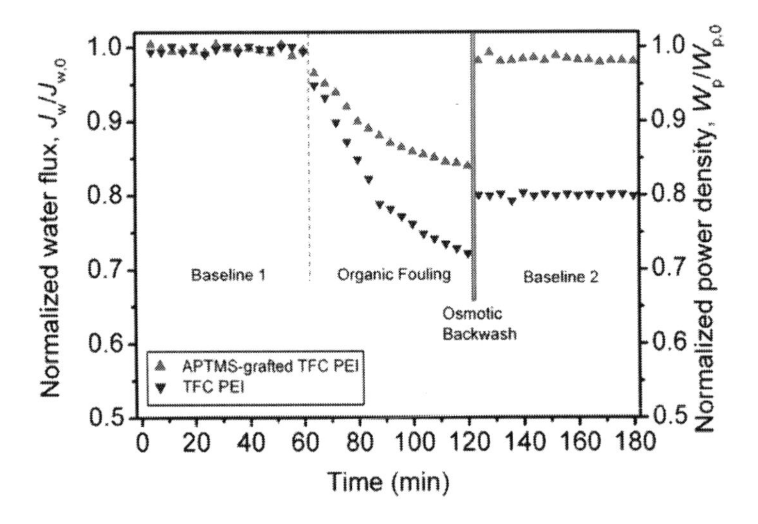

FIGURE 6.16
Fouling performance of the unmodified and modified membranes in 200 mg/L humic acid, 3 mM $CaCl_2$ and 10 mM NaCl; draw solution, 1.0 M NaCl.

Reprinted with permission from Elsevier (Zhang et al., 2016a).

on its back to repel foulants from the feed (Figure 6.20). GO is a novel single-layer sheet possessing abundant oxygenated groups and good anti-fouling properties. Hu and coworkers synthesized the double barriers by LbL assembly in an alternate depositing of anionic GO nanosheets and cationic

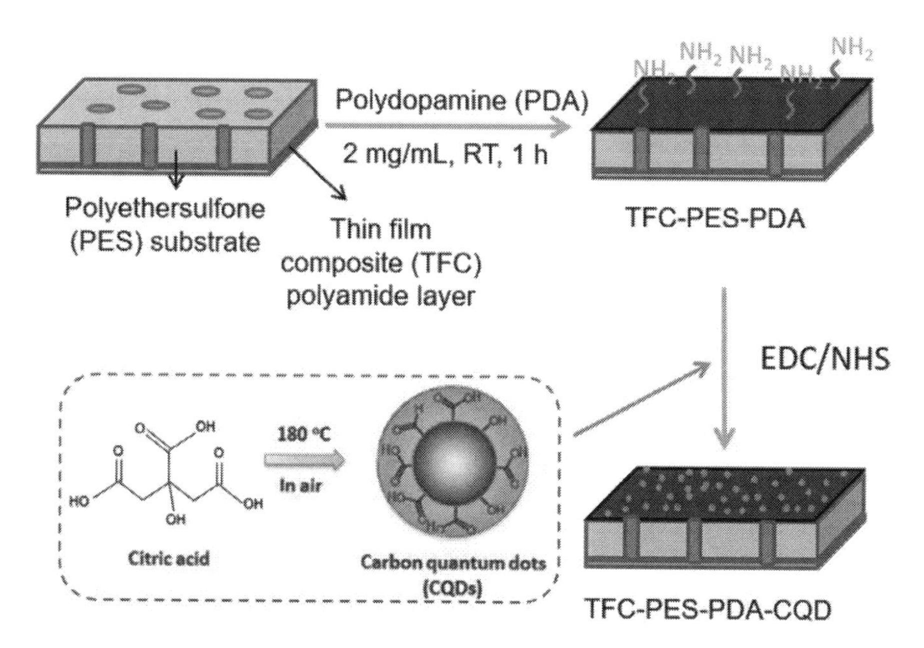

FIGURE 6.17
Fabrication route of CQD-modified PRO membranes.

Reprinted with permission from Elsevier (Zhao et al., 2017).

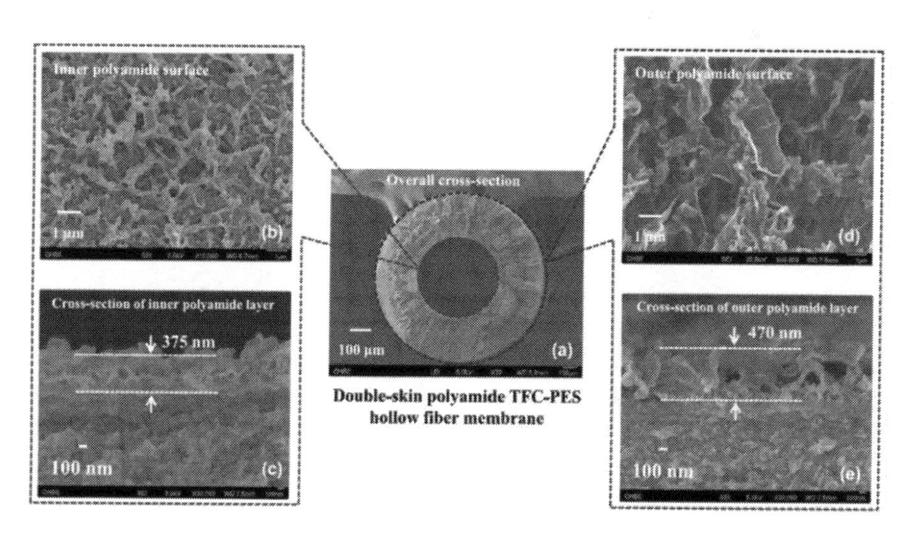

FIGURE 6.18
FESEM images of the double-skinned thin-film composite hollow fiber membrane.

Reprinted with permission from Elsevier (Han et al., 2017).

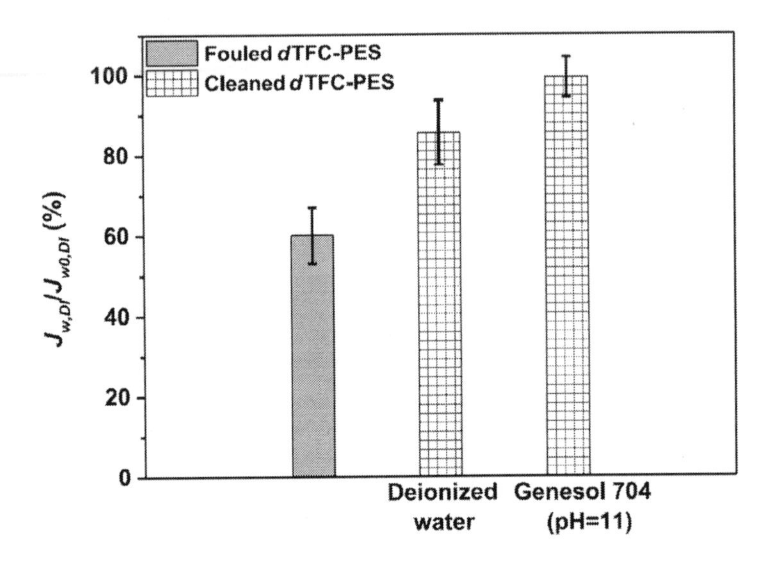

FIGURE 6.19
Normalized initial water fluxes of the fouled and cleaned double-skinned membranes. The membranes were cleaned by flushing with freshwater for 6 h or with chemical Genesol 704 (pH=11.0) for 0.5 h on the fouled surface.

Reprinted with permission from Elsevier (Han et al., 2017).

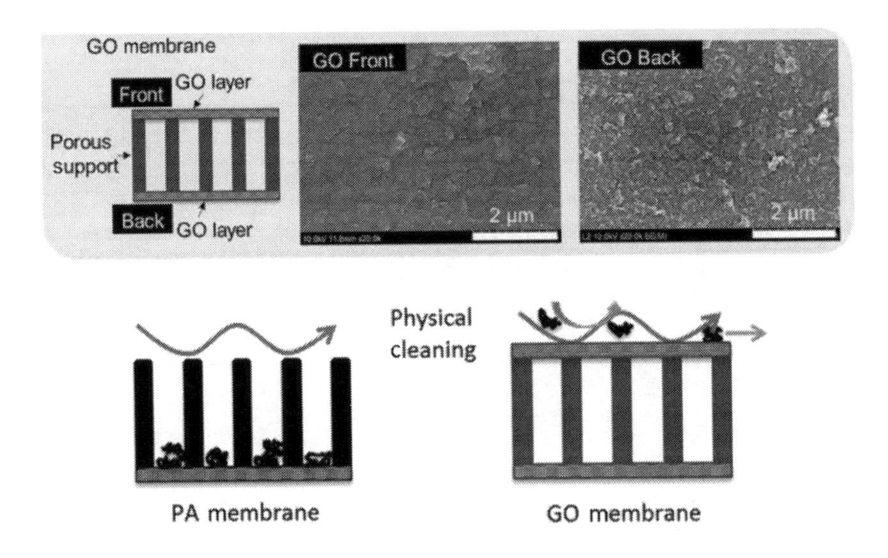

FIGURE 6.20
Structure and surface morphology of a double-skinned GO membrane and illustration of physical cleaning of its PA and GO membranes.

Adapted with permission from Hu et al. (2016). Copyright (2016) American Chemical Society.

TABLE 6.1

PRO performance of different antifouling membranes

Membrane	Surface Modification Method	Feed Solution	Test Duration	Flux Reduction After 1st run	Cleaning Agent	Flux Recovery	Reference
HPG-COOH	HPG grafting	WWRe	3 h	20~30%	DI	~92%	Li et al. (2016)
HPG-SO_4H	HPG grafting	WWRe	10 h	60~70%	DI	~100%	Zhang et al. (2018)
PES-P(MPC-*co*-AEMA)	Zwitterionic grafting	WWRe	~11 h	~30%	DI	98%	Han et al. (2018)
PES-PDA-MPC	Zwitterionic grafting	WWRe	3 h	~30%	DI	~90%	Zhao et al. (2016)
PES-*g*-P(DMAPS-*co*-DMEL), PES-*g*-P(MPC-*co*-DMEL)	Zwitterionic grafting	WWRe	3 h	20~30%	DI	>96%	Cai et al. (2016)
*d*TFC-PES	Double-skinned	WWRe	–	29%	DI or Genesol 704	80%(DI), or 98% (Genesol)	Han et al. (2017)
HPG-SH	HPG grafting	200 ppm BSA	9 h	11%	DI	~92%	Li et al. (2014)
PAH/PAA	Polyelectrolytes deposition	200 ppm BSA	5 h	16%	–	–	Li et al. (2017b)
M 0.25/6	Zwitterionic grafting	200 ppm BSA	9 h	15%	DI	~91%	Le et al. (2017)
GO	Double-skinned	200 ppm BSA, 0.5 mM Ca^{2+}	16 ~ 20 h	~20%	DI	~80%	Hu et al. (2016)

(*Continued*)

TABLE 6.1 (Cont.)

Membrane	Surface Modification Method	Feed Solution	Test Duration	Flux Reduction After 1st run	Cleaning Agent	Flux Recovery	Reference
CQD	CQDs modification	1000 ppm Alginate	3 h	18%	DI	94%	Zhao et al. (2017)
PAH/PAA	Polyelectrolytes deposition	1000 ppm Alginate	5 h	~0%	–	–	Li et al. (2017b)
GO	Double-skinned	200 ppm Alginate, 0.5 mM Ca^{2+}	16 ~ 20 h	~30%	DI	>80%	Hu et al. (2016)
PES-PDA-PVA	PVA grafting	200 ppm Alginate, 1.5 mM Ca^{2+}	~11 h	~20–25%	DI	~90%	Zhang et al. (2016b)
APTMS-TFC	Aminosilane grafting	3 mM Ca^{2+}, 10 mM Na^{+}, 200 ppm humic acid	1 h	~16%	DI	~98%	Zhang et al. (2016a)

polyelectrolyte PAH. Compared with a normal polyamide membrane, the double-skinned GO membrane has better antifouling behavior in the PRO testing dosing organic foulants. The back GO layer sufficiently prevents the foulants from permeating into the support. Physical washing can readily flush off the foulants and recover the flux (Figure 6.20).

6.5 Conclusions

Fabrication of antifouling PRO membranes is one of the effective methods for fouling mitigation. Even though antifouling PRO membranes have very different structures and preparation approaches, they possess similar functions, for example high hydrophilicity and low adhesion to foulants. Table 6.1 summarizes the PRO performance of antifouling membranes. Clearly, the effectiveness of antifouling modification varies with different types of feeds, test duration, and modification methods. The diverse structure and properties furnish antifouling PRO membranes with broad application areas. Surface grafting, nanoparticle deposition, secondary skin layer formation, and other newly emerging approaches make the vitality of antifouling membranes versatile and robust. Therefore, the progress of antifouling PRO membranes can not only advance the development of pressure retarded osmosis process as well as other membrane subjects, but also motivate wider application realms. One may expect more and more applications of antifouling PRO membranes in the future.

Acknowledgments

The financial supports from the National Natural Science Foundation of China (grant nos 51703167 and 51773156), the Science and Technology Program of Shenzhen (grant nos JCYJ20170303170340122 and JCYJ20170303170403122), and the Natural Science Foundation of Jiangsu Province (grant numbers BK20160384 and BK20170412) are acknowledged.

References

Cai, T., Li, M., Neoh, K.G., Kang, E.T., 2013. Surface-functionalizable membranes of polycaprolactone-click-hyperbranched polyglycerol copolymers from combined atom transfer radical polymerization, ring-opening polymerization and click chemistry. *J. Mater. Chem. B* 1, 1304–1315.

Cai, T., Li, X., Wan, C.F., Chung, T.S., 2016. Zwitterionic polymers grafted poly-(ethersulfone) hollow fiber membranes and their antifouling behaviors for osmotic power generation. *J. Membr. Sci.* 497, 142–152.

Chang, Y., Chang, W.J., Shih, Y.J., Wei, T.C., Hsiue, G.H., 2011. Zwitterionic sulfonbetaine-grafted poly(vinylidene fluoride) membrane with improved blood compatibility via atmospheric plasma-induced surface copolymerization. *ACS Appl. Mater. Interfaces* 3, 1228–1237.

Chen, S.C., Amy, G.L., Chung, T.S., 2016. Membrane fouling and antifouling strategies using RO retentate from a municipal water recycling plant as the feed for osmotic power generation. *Water Res.* 88, 144–155.

Chen, S.C., Fu, X.Z., Chung, T.S., 2014. Fouling behaviors of polybenzimidazole (PBI)-polyhedral oligomeric silsesquioxane (POSS)/polyacrylonitrile (PAN) hollow fiber membranes for engineering osmosis processes. *Desalination* 335, 17–26.

Chen, S.C., Wan, C.F., Chung, T.S., 2015. Enhanced fouling by inorganic and organic foulants on pressure retarded osmosis (PRO) hollow fiber membranes under high pressures. *J. Membr. Sci.* 479, 190–203.

Chen, X., Lawrence, J., Parelkar, S., Emrick, T., 2013. Novel zwitterionic copolymers with dihydrolipoic acid: synthesis and preparation of nonfouling nanorods. *Macromolecules* 46, 119–127.

Chung, T.S., Li, X., Ong, R.C., Ge, Q.C., Wang, H.L., Han, G., 2012. Emerging forward osmosis (FO) technologies and challenges ahead for clean water and clean energy applications. *Curr. Opin. Chem. Eng.* 1, 246–257.

Díez, B., Amariei, G., Rosal, R., 2018. Electrospun composite membranes for fouling and biofouling control. *Ind. Eng. Chem. Res.* 57, 14561–14570.

Duong, P.H.H., Chung, T.S., Shawn, W., Lana, I., 2014. Highly permeable double-skinned forward osmosis membranes for antifouling in the emulsified oil-water separation process. *Environ. Sci. Technol.* 48, 4537–4545.

Fang, L.-F., Cheng, L., Jeon, S., Wang, S.-Y., Takahashi, T., Matsuyama, H., 2018. Effect of the supporting layer structures on antifouling properties of forward osmosis membranes in AL-DS mode. *J. Membr. Sci.* 552, 265–273.

Ghanbari, M., Emadzadeh, D., Lau, W.J., Matsuura, T., Davoody, M., Ismail, A.F., 2015. Super hydrophilic TiO2/HNT nanocomposites as a new approach for fabrication of high performance thin film nanocomposite membranes for FO application. *Desalination* 371, 104–114.

Han, G., Cheng, Z.L., Chung, T.S., 2017. Thin-film composite (TFC) hollow fiber membrane with double-polyamide active layers for internal concentration polarization and fouling mitigation in osmotic processes. *J. Membr. Sci.* 523, 97–504.

Han, G., Liu, J.T., Lu, K.J., Chung, T.S., 2018. Advanced antifouling membranes for osmotic power generation from wastewater via pressure retarded osmosis (PRO). *Environ. Sci. Technol.* 52, 6686–6694.

Han, G., Zhang, S., Li, X., Chung, T.S., 2015. Progress in pressure retarded osmosis (PRO) membranes for osmotic power generation. *Progr. Polym. Sci.* 51, 1–27.

Hansson, S., Trouillet, V., Tischer, T., Goldmann, A.S., Carlmark, A., Barner-Kowollik, C., Malmstrom, E., 2013. Grafting efficiency of synthetic polymers onto biomaterials: a comparative study of grafting-from versus grafting-to. *Biomacromolecules* 14, 64–74.

Herzberger, J., Niederer, K., Pohlit, H., Seiwert, J., Worm, M., Wurm, F.R., Frey, H., 2016. Polymerization of ethylene oxide, propylene oxide, and other alkylene oxides: synthesis, novel polymer architectures, and bioconjugation. *Chem. Rev. C.* 116, 2170–2243.

Hu, M., Zheng, S.X., Mi, B.X., 2016. Organic Fouling of graphene oxide membranes and its implications for membrane fouling control in engineered osmosis. *Environ. Sci. Technol.* 50, 685–693.

Jiang, S., Cao, Z., 2010. Ultralow-fouling, functionalizable, and hydrolysable zwitterionic materials and their derivatives for biological applications. *Adv. Mater.* 22, 920–932.

Kainthan, R.K., Muliawan, E.B., Hatzikiriakos, S.G., Brooks, D.E., 2006. Synthesis, characterization, and viscoelastic properties of high molecular weight hyperbranched polyglycerols. *Macromolecules* 39, 7708–7717.

Khandare, J., Calderón, M., Dagia, N.M., Haag, R., 2012. Multifunctional dendritic polymers in nanomedicine: opportunities and challenges. *Chem. Soc. Rev.* 41, 2824–2848.

Kim, D.I., Kim, J.W., Shon, H.K., Hong, S., 2015. Pressure retarded osmosis (PRO) for integrating seawater desalination and wastewater reclamation: energy consumption and fouling. *J. Membr. Sci.* 483, 34–41.

Le, N. L., Quilitzsch, M., Cheng, H., Hong, P. Y., Ulbricht, M., Nunes, S. P., Chung, T.S., 2017. Hollow fiber membrane lumen modified by polyzwitterionic grafting. *J. Membr. Sci.* 522, 1–11.

Li, G., Cheng, G., Xue, H., Chen, S., Zhang, F., Jiang, S., 2008. Ultralow fouling zwitterionic polymers with a biomimetic adhesive group. *Biomaterials* 29, 4592–4597.

Li, X., Cai, T., Amy, G.L., Chung, T.S., 2017a. Cleaning strategies and membrane flux recovery on antifouling membranes for pressure retarded osmosis. *J. Membr. Sci.* 522, 116–123.

Li, X., Cai, T., Chen, C.Y., Chung, T.S., 2016. Negatively charged hyperbranched polyglycerol grafted membranes for osmotic power generation from municipal wastewater. *Water Res.* 89, 50–58.

Li, X., Cai, T., Chung, T.S., 2014. Antifouling behavior of hyperbranched polyglycerol-grafted poly(ether sulfone) hollow fiber membranes for osmotic power generation. *Environ. Sci. Technol.* 48, 9898–9907.

Li, Y., Qi, S., Wang, Y., Setiawan, L., Wang, R., 2017b. Modification of thin film composite hollow fiber membranes for osmotic energy generation with low organic fouling tendency. *Desalination* 424, 131–139.

Liu, Y., Ai, K., Lu, L., 2014. Polydopamine and its derivative materials: synthesis and promising applications in energy, environmental, and biomedical fields. *Chem. Rev.* 114, 5057–5115.

Liu, Y., Mi, B., 2012. Combined fouling of forward osmosis membranes: synergistic foulant interaction and direct observation of fouling layer formation. *J. Membr. Sci.* 407–408, 136–144.

Logan, B.E., Elimelech, M., 2012. Membrane-based processes for sustainable power generation using water. *Nature* 488, 313–319.

Lowe, A.B., McCormick, C.L., 2002. Synthesis and solution properties of zwitterionic polymers. *Chem. Rev.* 102, 4177–4189.

Nguyen, A., Zou, L., Priest, C., 2014. Evaluating the antifouling effects of silver nanoparticles regenerated by TiO_2 on forward osmosis membrane. *J. Membr. Sci.* 454, 264–271.

Paulus, F., Steinhilber, D., Welker, P., Mangoldt, D., Licha, K., Depner, H., Sigrist, S., Haag, R., 2014. Structure related transport properties and cellular uptake of hyperbranched polyglycerol sulfates with hydrophobic cores. *Polym. Chem.* 5, 5020–5028.

She, Q., Wong, Y.K.W., Zhao, S., Tang, C.Y., 2013. Organic fouling in pressure retarded osmosis: experiments, mechanisms and implications. *J. Membr. Sci.* 428, 181–189.

She, Q., Zhang, L., Wang, R., Krantz, W.B., Fane, A.G., 2017. Pressure-retarded osmosis with wastewater concentrate feed: Fouling process considerations. *J. Membr. Sci.* 542, 233–244.

Sofia, S.J., Premnath, V., Merrill, E.W., 1998. Poly(ethylene oxide) grafted to silicon surfaces: grafting density and protein adsorption. *Macromolecules* 31, 5059–5070.

Sunder, A., Hanselmann, R., Frey, H., Mulhaupt, R., 1999. Controlled synthesis of hyperbranched polyglycerols by ring-opening multibranching polymerization. *Macromolecules* 32, 4240–4246.

Thelin, W.R., Sivertsen, E., Holt, T., Brekke, G., 2013. Natural organic matter fouling in pressure retarded osmosis. *J. Membr. Sci.* 438, 46–56.

Touati, K., Tadeo, F., 2016. Green energy generation by pressure retarded osmosis: state of the art and technical advancement – review. *Int. J. Green Energy* 14, 337–360.

Wan, C.F., Chung, T.S., 2015. Osmotic power generation by pressure retarded osmosis using seawater brine as the draw solution and wastewater brine as the feed. *J. Membr. Sci.* 479, 148–158.

Wang, K.Y., Ong, R.C., Chung, T.S., 2010. Double-skinned forward osmosis membranes for reducing internal concentration polarization within the porous sublayer. *Ind. Eng. Chem. Res.* 49, 4824–4831.

Wilms, D., Stiriba, S.E., Frey, H., 2010. Hyperbranched polyglycerols: from the controlled synthesis of biocompatible polyether polyols to multipurpose applications. *Acc. Chem. Res.* 43, 129–141.

Yan, D.Y., Hou, J., Zhu, X.Y., Kosman, J.J., Wu, H.S., 2000. A new approach to control crystallinity of resulting polymers: self-condensing ring opening polymerization. *Macromol. Rapid Commun.* 21, 557–561.

Yang, R., Xu, J., Ozaydin-Ince, G., Wong, S.Y., Gleason, K.K., 2011. Surface-tethered zwitterionic ultrathin antifouling coatings on reverse osmosis membranes by initiated chemical vapor deposition. *Chem. Mater.* 23, 1263–1272.

Yang, W.J., Tao, X., Zhao, T., Weng, L., Kang, E.-T., Wang, L., 2015. Antifouling and antibacterial hydrogel coatings with self-healing properties based on a dynamic disulfide exchange reaction. *Polym. Chem.* 6, 7027–7035.

Yeh, P.-Y.J., Kainthan, R.K., Zou, Y., Chiao, M., Kizhakkedathu, J.N., 2008. Self-assembled monothiol-terminated hyperbranched polyglycerols on a gold surface: a comparative study on the structure, morphology, and protein adsorption characteristics with linear poly(ethylene glycol)s. *Langmuir* 24, 4907–4916.

Yip, N.Y., Elimelech, M., 2013. Influence of natural organic matter fouling and osmotic backwash on pressure retarded osmosis energy production from natural salinity gradients. *Environ. Sci. Technol.* 47, 12607–12616.

Yu, H.-Y., Kang, Y., Liu, Y., Mi, B., 2014. Grafting polyzwitterions onto polyamide by click chemistry and nucleophilic substitution on nitrogen: a novel approach to enhance membrane fouling resistance. *J. Membr. Sci.* 449, 50–57.

Zhang, L., She, Q., Wang, R., Wongchitphimon, S., Chen, Y., Fane, A.G., 2016a. Unique roles of aminosilane in developing antifouling thin film composite (TFC) membranes for pressure retarded osmosis (PRO). *Desalination* 389, 119–128.

Zhang, S., Wang, K.Y., Chung, T.S., Chen, H., Jean, Y.C., Amy, G., 2010. Well-constructed cellulose acetate membranes for forward osmosis: minimized internal concentration polarization with an ultra-thin selective layer. *J. Membr. Sci.* 360, 522–535.

Zhang, S., Zhang, Y., Chung, T.S., 2016b. Facile preparation of antifouling hollow fiber membranes for sustainable osmotic power generation. *ACS Sustain. Chem. Eng.* 4, 1154–1160.

Zhang, Y., Li, J.L., Cai, T., Cheng, Z.L., Li, X., Chung, T.S., 2018. Sulfonated hyper-branched polyglycerol grafted membranes with antifouling properties for sustainable osmotic power generation using municipal wastewater. *J. Membr. Sci.* 563, 521–530.

Zhao, D.L., Das, S., Chung, T.S., 2017. Carbon quantum dots grafted antifouling membranes for osmotic power generation via pressure-retarded osmosis process. *Environ. Sci. Technol.* 51, 14016–14023.

Zhao, D.L., Qiu, G., Li, X., Wan, C.F., Lu, K., Chung, T.S., 2016. Zwitterions coated hollow fiber membranes with enhanced antifouling properties for osmotic power generation from municipal wastewater. *Water Res.* 104, 389–396.

7

PRO Pretreatment

Tianshi Yang
Department of Chemical and Biomolecular Engineering
National University of Singapore
Singapore

Chakravarthy S. Gudipati
Separation Technologies Applied Research and Translation (START) – NTUitive
Cleantech One, Singapore

Tai-Shung Chung
Department of Chemical and Biomolecular Engineering
National University of Singapore
Singapore

CONTENTS

7.1 Introduction

The concerns of global warming and highly fluctuated oil prices have stimulated the research on alternative renewable and sustainable energies. Among various alternatives, osmotic energy is a promising, green and sustainable one that can be harnessed worldwide. Pressure retarded osmosis (PRO) is one of the technologies to harvest osmotic energy through an osmotically-driven membrane process (Han et al. 2015, Loeb and Norman 1975, Thorsen and Holt 2009). Various solution pairs have been studied and used as feed pairs in PRO processes. The most common one in early PRO studies was the feed pair of seawater and river water (Chu and Majumdar 2012, Chung et al. 2015, Kim et al. 2013b, Loeb 2002, O'Toole et al. 2016, Sharqawy et al. 2011, Thorsen and Holt 2009), but using this pair for PRO was not economically feasible because it produced a low energy density due to its small salinity difference (Bui et al. 2011, Chung et al. 2015, Han et al. 2015, Skilhagen et al. 2008). The feed pair consisting of concentrated seawater brine (denoted as SWBr) from seawater reverse osmosis (SWRO) plants and discharged wastewater retentate (denoted as WWRe) from municipal wastewater plants has drawn global attention (Chung et al. 2012, Kim et al. 2013a, Logan and Elimelech 2012, Saito et al. 2012). In addition to producing a higher energy output, it also lowers the energy consumption for SWRO, mitigates the environmental issues of brine disposal, reduces the water production cost, and produces a greener SWRO process (Chung et al. 2015).

Although SWBr and WWRe have a high salinity difference that can potentially produce a high power density, severe membrane fouling has been observed especially by WWRe into the porous layer of PRO membranes during operations, which causes extra operational costs associated with membrane cleaning and fouling control strategies (Chen et al. 2014, Han et al. 2016, Li et al. 2017, She et al. 2013, Yang et al. 2019, Zhao et al. 2010). Among various fouling control strategies, pretreatment of the feed and draw solutions is one of the most straightforward ways that could effectively mitigate foulants in the feed pair (Chung et al. 2015, Kim et al. 2013a). In this chapter, we will introduce several commonly used pretreatment methods, namely membrane filtrations, chemical additives, and coagulations, and their efficacies in detail.

7.2 Pretreatment Using Membrane Filtrations

Membrane modules are used to filter the feed and draw solutions prior to sending the solution into the PRO unit so that the major foulants or scalants inside the feed pair could be removed in advance. Depending on the type and size of the foulants or scalants, different types of membrane could be used to remove the target foulants. In this section, we will introduce some self-developed and commercially available membranes as well as membrane modules as pretreatment devices for PRO and its pilot plants.

7.2.1 Pretreatment Using Self-Developed Membranes and Modules

Three kinds of polyethersulfone (PES) hollow fiber membrane were developed and fabricated into 1-inch modules in Yang et al.'s work (Yang et al. 2019). Based on their pore sizes, the hollow fiber membranes were classified as ultrafiltration (UF), nanofiltration (NF), and low-pressure reverse osmosis (LP-RO) membranes. The UF membranes were prepared via a dry-jet wet spinning process using the conditions reported in their previous work (Wan and Chung 2015). The detailed spinning conditions are summarized in Table 7.1. Then, the UF hollow fiber membranes were fabricated into 1-inch modules using the same module fabrication method described elsewhere (Wan et al. 2017). To prepare NF and LP-RO membranes, two types of polyamide selective layer were formed on the inner surface (i.e., the lumen side) of PES hollow fiber substrates by means of interfacial polymerization. The LP-RO membrane had a dense polyamide selective layer formed via interfacial polymerization between an *m*-phenylenediamine (MPD) aqueous solution and a trimesoyl chloride (TMC) hexane solution. Table 7.1 summarizes the detailed parameters of interfacial polymerization (Wan and Chung 2015, Zhang et al. 2014). The resultant thin-film composite (TFC)-PES membrane modules were used for LP-RO operations. The NF membrane had a loose polyamide selective layer made by interfacial polymerization between a piperazine (PIP) aqueous

TABLE 7.1

Summary of dope formulation, monomers in thin-film polymerizations, hollow fiber, and module specifications (Yang et al. 2019)

Dope composition of the hollow fiber support (wt%)	20-22/37-40/37-40/2-6 (PES/NMP/PEG/water)
Chemicals used to form a dense polyamide layer (wt%)	2/0.1 (MPD/SDS in water) 0.15 (TMC in hexane)
Chemicals used to form a loose polyamide layer (wt%)	2 (PIP in water) 0.15 (TMC in hexane)
Fiber outer diameter (μm)	1025
Fiber inner diameter (μm)	575
Fiber number in lab-scale modules and packing density (%)	3/2.5
Fiber number in 1-inch modules and packing density (%)	200/40

solution and a TMC hexane solution. The resultant TFC-PES membrane modules were used for NF pre-treatment.

The characterization results of UF, NF, and LP-RO membranes are summarized in Table 7.2. The mean pore size of the UF membrane was 3.847 nm. A tight UF membrane with this pore size could remove most of large organic particles, but it would not be effective to remove inorganic salts because salts have much smaller sizes. The PWP of this UF membrane was 576.6 LMH/bar, which was much higher than those of NF and LP-RO membranes. The NF membrane with a loose polyamide selective layer had a mean pore size of 0.422 nm and a PWP of 2.78 LMH/bar. Its high rejections toward divalent salts indicated its ability for scalant removal, such as Ca^{2+} and SO_4^{2-}. Since TFC membranes with a polyamide selective layer often exhibited negative charges at a neutral pH (He et al. 2016), they showed higher rejections to negative divalent ions based on the Donnan effect. This explained why both NF and LP-RO membranes had slightly higher rejection to $MgSO_4$ than $MgCl_2$. Besides, the LP-RO membrane showed a high rejection to NaCl even though both Na^+ and Cl^- diameters are much smaller than Mg^{2+} and SO_4^{2-}, respectively. Thus, the LP-RO membrane had a small mean pore size that could not be measured using the conventional solute rejection method (Aimar et al. 1990, Han and Chung 2014, Singh et al. 1998). This implied that the LP-RO membrane would produce the best quality solution for PRO operations compared with UF and NF pretreatments. However, the pretreatment by the LP-RO membrane might not be cost effective because its PWP was only 1.51 LMH/bar which was the lowest among the three membranes.

TABLE 7.2

Characteristics of the UF, NF, and LP-RO membranes (Yang et al. 2019)

	UF	NF	LP-RO
PWP (LMH/bar)	576.6[a]	2.78[b]	1.51[b]
NaCl rejection	–	51.4%[c]	96.1%[c]
$MgCl_2$ rejection	–	88.2%[c]	96.2%[c]
$MgSO_4$ rejection	–	90.3%[c]	97.3%[c]
Na_2SO_4 rejection	–	90.1%[c]	97.7%[c]
Pore size (nm)	3.847	0.422	–
MWCO	18743	263.5	–

[a] Tested at 2.5 bar using DI water.

[b] Tested at 10 bar using DI water.

[c] Tested at 10 bar using 1000 ppm respective salt solution.

The original WWRe was collected from a local municipal wastewater reclamation plant and used as the feed solution for the UF, NF, and LP-RO pretreatment processes. The pretreatment processes were carried out in batches, during which WWRe was pumped into the lumen side of modules at a constant flow rate of 1 L/min and circulated back to the feed tank. The UF pretreatment was carried out at a constant pressure of 2.5 bar, while the NF and LP-RO pretreatments were conducted at a constant pressure of 10 bar. The filtrate was collected through the shell side of modules until the total amount of filtrate reached 70 wt% of the feed solution. The filtrates collected from each pre-treatment process were denoted as UF-, NF- and LP-RO-filtrate, respectively. Their compositions are summarized in Table 7.3.

A synthetic SWBr solution of 0.8 M NaCl was chosen as the draw solution for the following PRO tests. Four solutions, including UF-filtrate, NF-filtrate, LP-RO-filtrate, and the original WWRe were examined as the feed solutions in PRO tests. In order to study the effectiveness of each pretreatment process, two groups of bench-scale PRO tests were carried out: (1) under the PRO mode where active layer faced the draw solution (denoted as AL-DS mode thereafter) at ΔP = 0 bar for 24 h, and (2) under the AL-DS mode at ΔP = 15 bar for 6 h. Long operating hours were employed because we aimed to study the long-term PRO performance stability in this work. The detailed setup, configuration, and operation procedures of the bench-scale PRO had been reported in previous studies

TABLE 7.3

Characteristics of filtrates and percentages of ion and particle removal (Yang et al. 2019)

	UF-filtrate[a]	NF-filtrate[b]	LP-RO-filtrate[b]
pH	7.4	7.1	7.4
Conductivity (μS/cm)	1557	1076	69.7
Turbidity (NTU)	0.15	0.45	0.65
TDS (ppm)	6.8%	35.7%	96.8%
SO_4^{2-} (ppm)	0%	82%	100%
PO_4^{3-} (ppm)	0.7%	65.5%	100%
Ca^{2+} (ppm)	1.5%	67.2%	99.1%
Mg^{2+} (ppm)	1.3%	73.6%	100%
Silica (ppm)	2.4%	15.4%	98.8%
TOC (ppm)	13.2%	76.6%	94.0%

[a] UF pretreatment was conducted at 2.5 bar.

[b] NF and LP-RO pretreatments were conducted at 10 bar.

(Sukitpaneenit and Chung 2012). The draw solution (i.e., the synthetic SWBr solution) was pumped into the lumen of the lab-scale PRO modules while the feed solution (i.e., the original WWRe, and UF-, NF-, or LP-RO-filtrate) was flowing along the shell side in a counter-current mode. The flow rate of the draw solution was kept at 0.1 L/min while that of the feed solution was at 0.2 L/min. The temperatures of both solutions were maintained at room temperature.

Figure 7.1 shows the water flux patterns in PRO tests under 0 bar using the original WWRe, and UF-, NF- and LP-RO-filtrates as the feed solutions, respectively. Each test lasted more than 24 h in order to reveal the long-term stability of PRO performance as well as the effectiveness of each pretreatment method. The baseline was set by using the original WWRe as the feed solution without any pretreatment. Its initial water flux was around 13 LMH and started to drop from the very beginning of the test. As time went by, the decline of water flux became slower and the water flux finally stabilized at around 5 LMH after 10 h, which was only 38.5% of the initial flux. Since there was no hydraulic pressure in the lumen of PRO hollow fibers, the permeate water across the membrane from the shell side to the lumen was driven by the osmotic pressure difference. The permeate

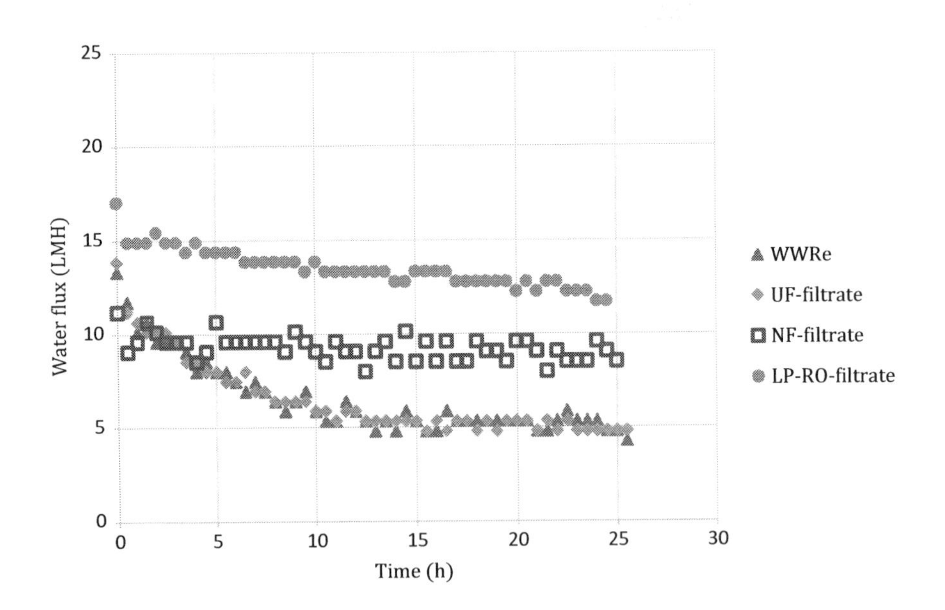

FIGURE 7.1
Flux patterns of PRO tests at a trans-membrane pressure of 0 bar using synthetic SWBr as the draw solution and four different feed solutions (Yang et al. 2019).

water also brought the ions, silica, and organic compounds in WWRe into the lumen. When they reached the interface between the support layer and the polyamide selective layer, the permeate water continued to diffuse through the polyamide layer, but the ions, silica, and small organic compounds were blocked by the polyamide layer and accumulated underneath it. Thus, the local ion concentrations in the interface region became higher and higher. This phenomenon was also known as the internal concentration polarization (ICP) (Chen et al. 2016, Wang et al. 2010, Xiong et al. 2016). As the ion concentrations increased to certain levels, various inorganic scaling occurred inside the porous support, which would further block membrane pores and reduce water flux. By analyzing the composition of WWRe and the saturation status of various salts (Table 7.3), Ca^{2+} and PO_4^{3-} were found to form hydroxyapatite inside the porous layer and block the passage of water permeance across the PRO membranes (Chen et al. 2016).

The water flux pattern showed exactly the same trend in the PRO test using the UF-filtrate as the PRO feed solution. Since this tight UF membrane had a pore size too big to reject various ions in WWRe, it was not effective to pretreat WWRe and maintain the PRO performance. Although UF pretreatment could work in other cases if the foulant size was bigger, it did not work for this WWRe.

Compared to the UF-filtrate, the PRO test using the NF-filtrate as the feed solution produced a more stable water flux pattern. The initial water flux was 11 LMH which was slightly lower than the previous two cases. However, during the entire 25-h test, there was no sign of significant water flux decline. The final water flux at the end of the test was 8.5 LMH which meant a water flux drop of only 22.7% after 1 day of operation. The PRO test using the LP-RO-filtrate as the feed solution also produced a stable water flux pattern since its water quality was better than the NF-filtrate. The initial water flux was 16.5 LMH and was the highest flux in all tests. The final water flux after the 25-h test was 12.5 LMH which was equivalent to 75.8% of the initial water flux. Clearly, the pretreatments by NF and LP-RO membranes were effective to maintain the PRO performance because their pore sizes were small enough to remove Ca^{2+} and PO_4^{3-} from WWRe. As a result, less fouling occurred in the PRO tests as discussed.

Figure 7.2 shows the water flux patterns of PRO tests at 15 bar using the original WWRe and UF-, NF- and LP-RO-filtrates as feed solutions, respectively. Similar to Figure 7.1, the water flux showed a big decrease when using the original WWRe and UF-filtrate as the feed solutions, indicating that UF was not effective to pre-treat WWRe. By incorporating NF- and LP-RO-filtrates as feeds in PRO tests at 15 bar, their corresponding initial water fluxes were 19.7 LMH and 21.6 LMH, respectively. Both initial fluxes were higher than their initial fluxes tested at 0 bar. In addition, both flux patterns were more stable, which further proved that both NF and LP-RO pretreatments were effective to mitigate fouling caused by WWRe. The

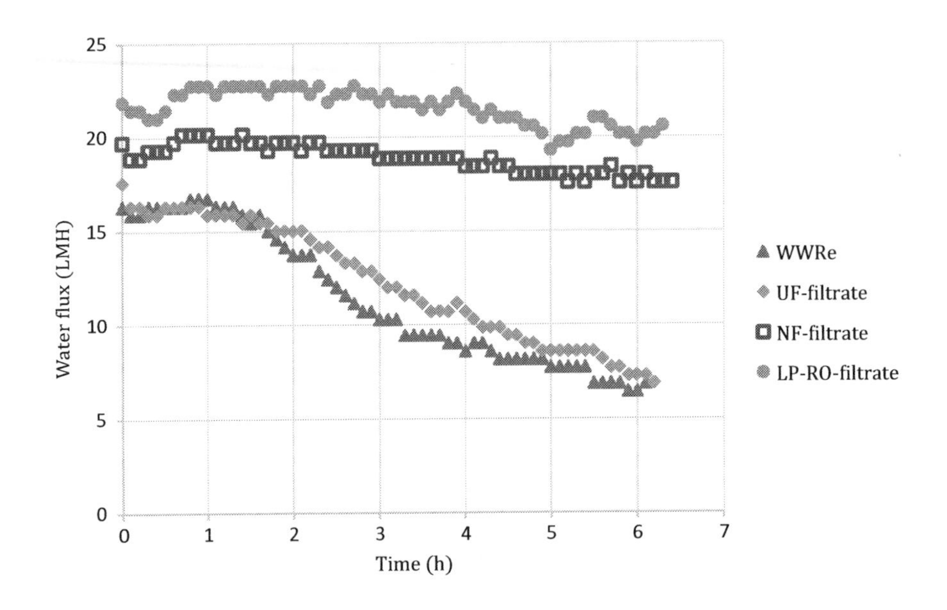

FIGURE 7.2
Flux patterns of PRO tests at a trans-membrane pressure of 15 bar using synthetic SWBr as the draw solution and four different feed solutions (Yang et al. 2019).

final water fluxes after 6-h PRO tests at 15 bar were 17.6 LMH and 20.1 LMH, respectively, when using NF- and LP-RO-filtrates as feeds.

Similarly, Chen et al. also developed their lab-made NF membranes and membrane modules to pretreat WWRe as the feed solution for PRO (Chen et al. 2017). A polyelectrolyte deposited small NF (S-NF) module was fabricated based on the method described elsewhere (Liu et al. 2015) together with a relatively large 2-inch NF (L-NF) module with an improved selectivity. Table 7.4 summarizes their characteristics and includes a commercially available NF270 membrane for comparison. The glucose rejection was used to compare their indicative pore sizes. Generally, a higher glucose rejection of L-NF suggested that it had a smaller pore size than S-NF. However, the former had a lower pure water permeability (PWP) of 10 LMH/bar than the latter that had ~15 LMH/bar.

In addition, these NF membranes possess different charge properties. The isoelectric points of these membranes were measured to identify where the membrane was considered uncharged and at which pH values. Below the points, the NF membranes were generally positively charged and vice versa (Cheng et al. 2008, Childress and Elimelech 2000, Li and Chung 2008). According to Table 7.4, both S-NF and L-NF were positively

TABLE 7.4

Characteristics of membrane NF270, S-NF, and L-NF (Chen et al. 2017)

	NF270	S-NF	L-NF
PWP[a] (LMH/bar)	15.3	15.5	10
Glucose rejection[b]	84.0%	88.5%	93.2%
$MgCl_2$ rejection[c]	42.9%	98.2%	99.2%
Isoelectric point[d] (pH)	4	8	9

[a] Tested at 1 bar using DI water.
[b] Tested at 1 bar using 200 ppm glucose (180 Da).
[c] Tested at 2 bar using 1000 ppm $MgCl_2$, respectively.
[d] Tested with 1000 ppm NaCl at various pH.

charged at neutral pH, and thus they had higher rejections to Ca^{2+} and Mg^{2+} ions, as shown in Table 7.4. In addition, since membranes with a smaller pore size would show greater effects of dielectric exclusion and steric hindrance (Labban et al. 2017), the L-NF/P membrane could have higher ion rejections and produce a better quality feed water for PRO.

All NF pretreatment experiments were operated at 2 bar, and the water that permeated from the S-NF and L-NF modules were denoted as S-NF/P and L-NF/P, respectively. The polyetherimide (PEI) PRO hollow fiber membranes were used for PRO tests. NaCl was used to prepare the draw solution while DI water, WWRe, S-NF/P, and L-NF/P were then used as feed solutions to evaluate the PRO performance and to understand the effect of the NF pretreatment on the PRO fouling mitigation.

The PRO fouling test results using DI water and WWRe are shown in Figure 7.3. The intrinsic PRO water flux profile using DI water as the feed solution (Figure 7.3(a)) indicated that the PRO water flux was stable under each pressure if fouling was not occurring. When there was a step change in the applied pressure (either went up or went down), the water flux also changed accordingly. In contrast, a fast flux decline was observed during the first several minutes of the none pressurized PRO operation using WWRe as the feed solution, as shown in Figure 7.3(b). After that, even though the applied pressure changed, there were no corresponding changes in the water flux profile, which indicated that the PRO membrane was fouled severely. The water flux was almost stabilized at ~9 LMH at 16 bar, producing a power density of 4.4 W/m^2. This value was much smaller than the 23.6 W/m^2 obtained using DI water as the feed.

The PRO water flux profiles with different NF-treated feed solutions are presented in Figure 7.4. It was observed that the water flux significantly increased after NF pretreatment compared with the case of untreated WWRe

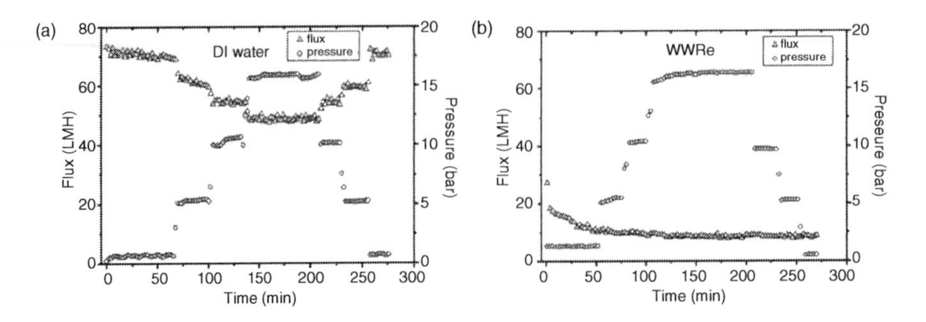

FIGURE 7.3
PRO water flux as a function of applied pressure in the draw solution in PRO tests using 1 M NaCl solution as the draw solution and (a) DI water and (b) WWRe as the feed solutions (Chen et al. 2017).

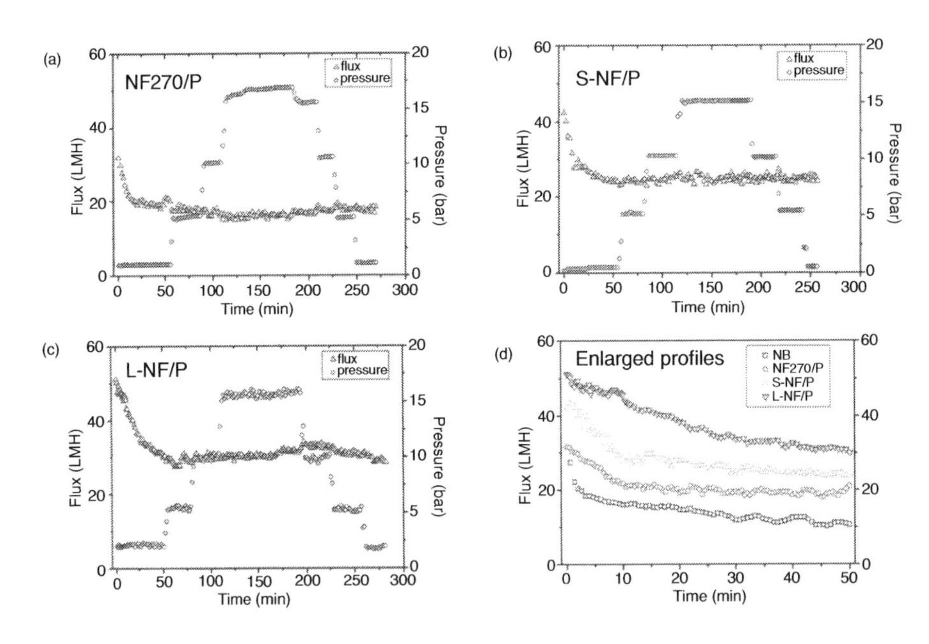

FIGURE 7.4
PRO flux profiles of PRO membranes using (a) NF270/P, (b) S-NF/P, and (c) L-NF/P as feed and 1 M NaCl solution as draw and (d) enlarged water flux profiles of (a–c) in the first 50 min of operation. NB stands for Newater brine which is the same as WWRe (Chen et al. 2017).

(Figure 7.3(b)), which suggested that NF pretreatment greatly improved the quality of the PRO feed solution. The stabilized water fluxes at 16 bar were about 24 LMH and 30.5 LMH for S-NF/P and L-NF/P as the feed, respectively. Compared with 9 LMH using WWRe as the feed, the water flux increased at least 166% by the NF pretreatment. However, similar to the case of WWRe fouled membrane, the step changes of the applied pressure had no significant effect on the water flux, indicating that fouling on the PRO membrane still occurred but in a mild way. By analyzing the composition of the original WWRe and solutions after NF pretreatments, it was found that S-NF pretreatment rejected 86% of calcium, 93% of sulfate, and 61% phosphate while L-NF pretreatment rejected 97% of calcium, 99% of sulfate, and 72% phosphate. Due to the removal of the majority of those ions, the PRO water flux using the NF pretreated feed solution improved a lot. However, silicon concentration in NF permeates remained unchanged. The increased water flux triggered higher internal silica concentration in the porous substrate layer and more severe silica scaling. Consequently, silica scaling was the most significant for the PRO membrane using L-NF permeate as the feed solution due to the high water flux (>30 LMH). It also revealed that an even higher level of cleanness of PRO feed was required when PRO process was operated at higher water flux.

7.2.2 Pretreatment Using Commercial Membranes and Modules

DOW filmtec™ NF270 membrane was used to pretreat the original WWRe for performance comparison in Chen et al.'s work as mentioned earlier (Chen et al. 2017). The pretreatment experiment was also operated at 2 bar. The characteristics of NF270 membrane are shown in Table 7.4. A slightly lower glucose and ion rejections of NF270 compared with S-NF and L-NF membranes indicated that NF270 membranes had bigger pores. An isoelectric point of 4 meant this membrane was negatively charged under neutral environment. Thus, NF270 membranes could have better rejection to negatively charged ions.

The PRO water flux profile of the following PRO test using NF270 pretreated solution as the feed is shown in Figure 7.4(a). Compared with the fouling behavior using the original WWRe, NF270 pretreatment improved the PRO water flux from 9 to 18 LMH at the applied pressure of 16 bar on the draw solution, which accounted for 100% increment. However, due to the larger pore size and less rejection ability than S-NF and L-NF, NF270 membrane was not able to remove as many ions as the other two NF membranes, and it could not remove silica either. Consequently, a similar PRO water flux profile was observed where the water flux was stable but showed no response to step changes of the applied pressure. In other words, fouling on the PRO membrane was mitigated to some extent by the NF pretreatment, but this NF pretreatment could not fully prevent fouling from occurring in this case. In order to achieve a complete fouling mitigation, high rejection abilities of both ions and silica were required for membranes used for PRO pretreatment.

Wan and Chung (2015) also used commercial membranes to pretreat WWRe for PRO processes. NADIR® UH050 UF membranes (Microdyn-Nadir, GmbH, Germany) and CSM® NE2540-70 NF membranes (Woongjin Chemical, Co. Ltd., Korea) were selected, and properties of these UF and NF membranes are presented in Table 7.5. The UF membrane had a much larger MWCO value than the NF membrane, indicating that it had bigger pores. Thus, this UF membrane could not be used to remove ions from WWRe. The NF membrane, on the contrary, had a nearly perfect rejection to $MgSO_4$, showing that it was supposed to remove most of multivalent ions. In this work, both UF and NF membranes were installed on a dead-end permeation cell, where WWRe was filled and pressurized. Pressures were 1 bar and 5 bar for UF and NF processes, respectively. Filtrates from each pretreatment process were collected and used as feed solutions in the following PRO tests.

The PRO test was similar to Yang et al.'s work (Yang et al. 2019) where the TFC-PES hollow fiber membrane was used as the PRO membrane. Both the 0.81 M NaCl solution and the real seawater brine (SWBr) were used as the draw solution. The original WWRe, the WWRe UF filtrate, and the WWRe NF filtrate were used as the feed solution. The PRO testing results are shown in Figures 7.5 and 7.6.

Figure 7.5(a) and (b) shows a comparison of the water flux and power density of the PRO process using a 0.81 M NaCl solution as the draw solution. The UF pretreatment reduced the TOC content from 44.05 to 32.23 ppm, but ion concentrations remained almost the same. As a result, only a marginal improvement was observed in the pressure range from 0 bar to 10 bar. Beyond 10 bar, the improvement became significant. Since there were negligible changes in ion concentrations, it was very likely that the improvement was resulted from the removal of organic foulants by the

TABLE 7.5

Characteristics of the UH050 flat-sheet UF and NE2540-70 flat-sheet NF membranes (Wan and Chung 2015)

	UH050	NE2540-70
Material	Polyethersulfone	Polyamide
PWP (E-12 m^3/m^2sPa)	231.6	12.0
NaCl rejection	NA	40–70%
$MgSO_4$ rejection	NA	99.5%
MWCO (Da)	50,000	200

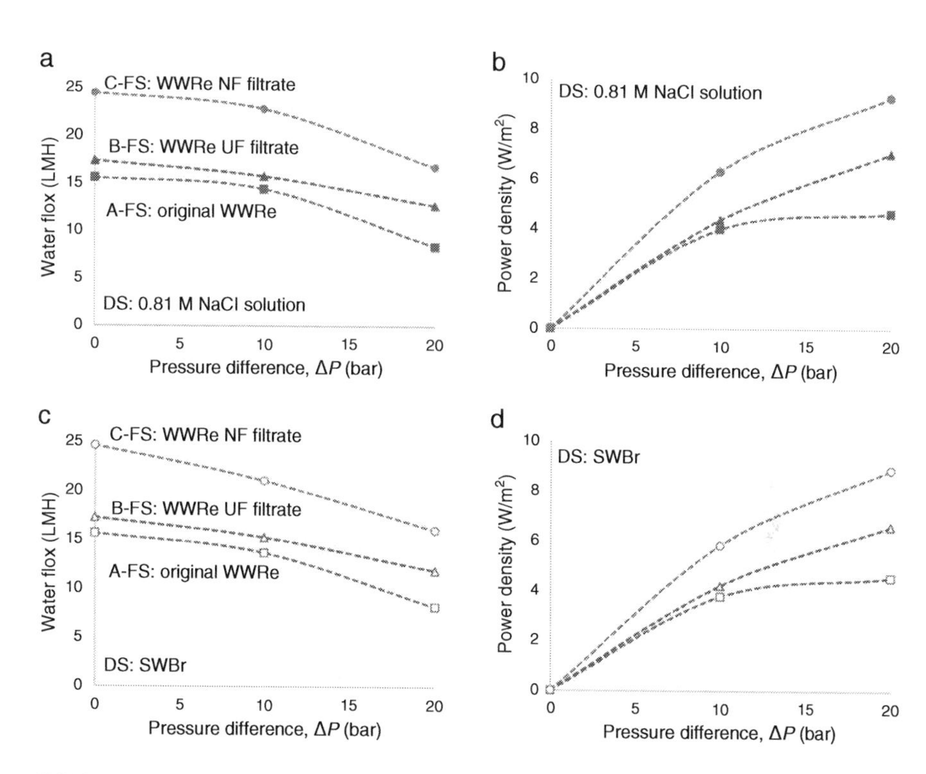

FIGURE 7.5
(a) Water flux and (b) power density as functions of transmembrane pressure using a 0.81 M NaCl solution as the draw solution (DS) and (A) original WWRe, (B) WWRe UF filtrate, and (C) WWRe NF filtrate as feed solutions (FS). (c) Water flux and (d) power density as functions of transmembrane pressure using SWBr as the DS and (A) original WWRe, (B) WWRe UF filtrate, and (C) WWRe NF filtrate as the FS (Wan and Chung 2015).

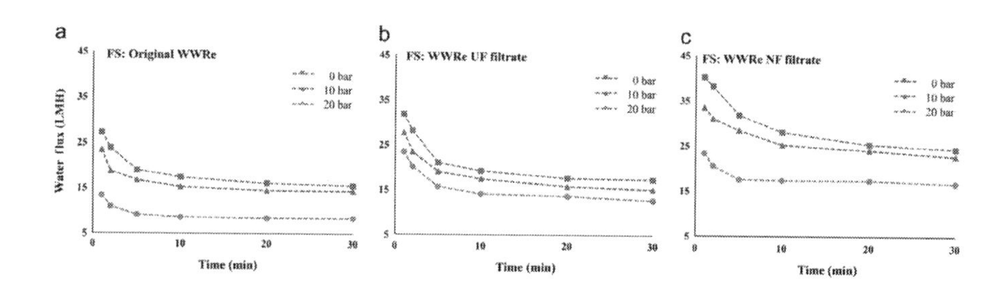

FIGURE 7.6
Water flux as a function of time using a 0.81 M NaCl solution as the draw solution and (a) original WWRe, (b) WWRe UF filtrate and (c) WWRe NF filtrate as feed solutions at different pressures (Wan and Chung 2015).

UF pretreatment. The NF treated WWRe further improved the PRO performance. This was because the NF membrane not only effectively removed almost half of the TOC but also reduced 66% and 43% of the sulfate and calcium ions, respectively. The performance enhancement might result from the reduced organic and gypsum fouling.

The SWBr had a more complicated composition than the 0.81 M NaCl solution. Thus, the reverse salt flux from SWBr could interact with organic foulants in the feed solution and enhance fouling (Chen et al. 2014, She et al. 2013). In order to find out the fouling effect from the SWBr, the PRO experiments were repeated using the SWBr as the draw solution. The results shown in Figure 7.5(c) and (d) proved that fouling by the SWBr from the lumen side was negligible because the PRO water flux profile using SWBr as the draw solution was almost identical to the water flux profile using 0.81 M NaCl solution as the draw solution. It could be concluded that the pretreatment of SWBr was not necessary for PRO processes if TFC-PES membranes were used as the PRO membrane. The highest power density reached 6.6 W/m^2 and 8.9 W/m^2 when using SWBr as the draw solution, and UF- and NF-treated WWRe as feed solutions, respectively.

Figure 7.6 shows the water flux for the first 30 min of PRO experiments using WWRe, and UF- and NF-treated WWRe as feed solutions, respectively. In all cases, the water flux dropped sharply in the first 5 min, slowed down gradually, and continued to drop at a much slower rate over a long period. Although the stable water flux obtained using the WWRe NF filtrate was greater than that obtained using the WWRe UF filtrate, which was in turn greater than that obtained using the original WWRe, the sharp drop in the water flux at the beginning still indicated that there might be fouling or scaling occurring.

7.2.3 Pretreatment Using Membranes and Modules in PRO Prototype Plants

In Japan, a PRO prototype plant was built near Fukuoka SWRO plant and operated using SWBr as the draw solution and wastewater from a local wastewater treatment plant as the feed solution (Sakai et al. 2016). In order to achieve a stable permeate flow, the wastewater was pretreated with the on-site UF and the LP-RO method. Good water quality was obtained by the LP-RO pretreatment, so there was no more fouling inside the membrane and the PRO performance was maintained for more than a year. However, the LP-RO pretreatment required a large pump power. In some cases, the energy consumption even exceeded the energy generation. Thus, applying the LP-RO pretreatment on site for PRO processes was not economically feasible.

The UF pretreatment, which consumed much less energy than LP-RO pretreatment, became an alternative option. Figure 7.7 shows the daily variation of the PRO permeate flow when only the UF pretreatment was applied on site. Within 7 days, the permeate flow decreased by approximately 30%,

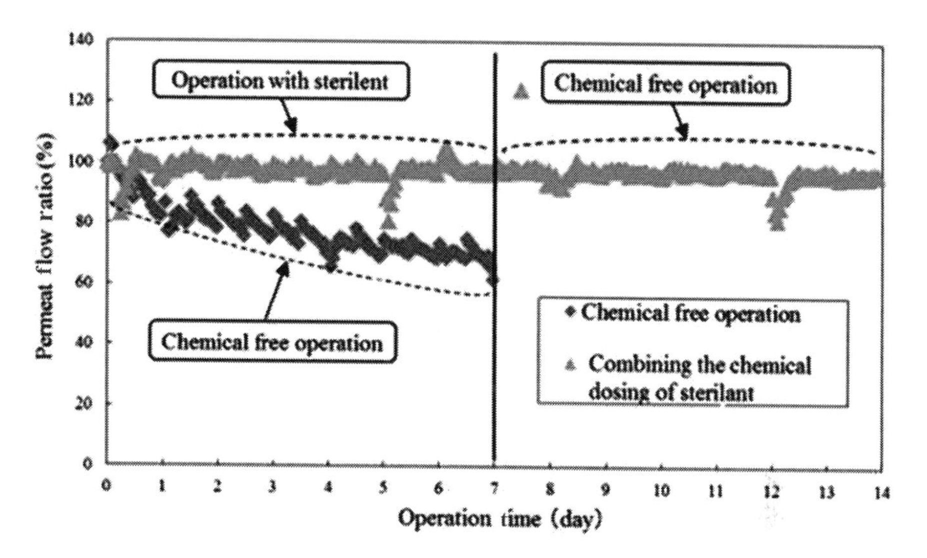

FIGURE 7.7
Daily change of the permeate flow, when only using the UF membrane treatment (Sakai et al. 2016).

which indicated that fouling still occurred inside PRO membrane modules. In this case, the wastewater contained some organic molecules that could not be removed by the UF pretreatment, such as fulvic acid and humic acid. Thus, chemical sterilant was needed to prevent organic fouling on PRO membranes. By employing both the on-site UF pretreatment and chemical sterilant, the PRO permeate flow became stable for as long as 14 days, as shown in Figure 7.7. This result proved that PRO could operate without fouling, even if LP-RO pretreatment was not employed in this case.

7.3 Pretreatment Using Chemical Additives

Besides the pretreatment using membrane filtration methods to remove major foulants from PRO feed solutions, chemical additives could also be added into PRO feed solutions in order to prevent membrane fouling. The purpose of adding chemicals is to alter the chemical environment of PRO feed solutions so that foulants will not be able to stick on PRO membrane surfaces or accumulate inside membranes easily. In this section, we will introduce two commonly used methods using chemical additives to pretreat PRO feed solutions.

7.3.1 Pretreatment with pH Adjustment

PH value affects the electrostatic interaction of foulant molecules and the scaling process of inorganic salts. By adjusting pH, the inorganic salt scaling would be either promoted or weakened (Antony et al. 2011, Yu et al. 2010). Both Chen et al. (2016) and Han et al. (2016) have found that the inorganic scaling dominated in the fouling behavior when the original WWRe was used as the feed solution for PRO processes. Thus, by adjusting pH of WWRe, the inorganic scaling is supposed to be controlled and the PRO membrane fouling can be mitigated.

TFC-PES hollow fiber membranes were used for PRO tests in Han et al.'s work and both DI water and the original WWRe were used as the PRO feed solution independently. Figure 7.8 shows the normalized PRO water flux patterns under various pressures. As indicated in this figure, the normalized water flux dropped sharply in the first 60 min, meaning fouling induced by the WWRe occurred very quickly. Over the entire 12-h PRO test, the water flux dropped to only 60% of its initial value when the applied pressure on the draw solution was 0 bar. Compared with the drop of 10% in water flux when using DI water as the feed solution, this fouling was severe.

In order to mitigate the inorganic fouling within the porous substrate layer, hydrochloric acid (HCl, 37%) was added into the WWRe to lower its pH value. As the original WWRe possessed a pH of 7.4 and a relatively high concentration of calcium and phosphate ions (Wan and Chung 2015), scaling of calcium phosphate and hydroxyapatite was demonstrated to be significant (Chen et al. 2016). By lowering the pH value of WWRe to 2.0, the solubility of calcium salts increased and thus

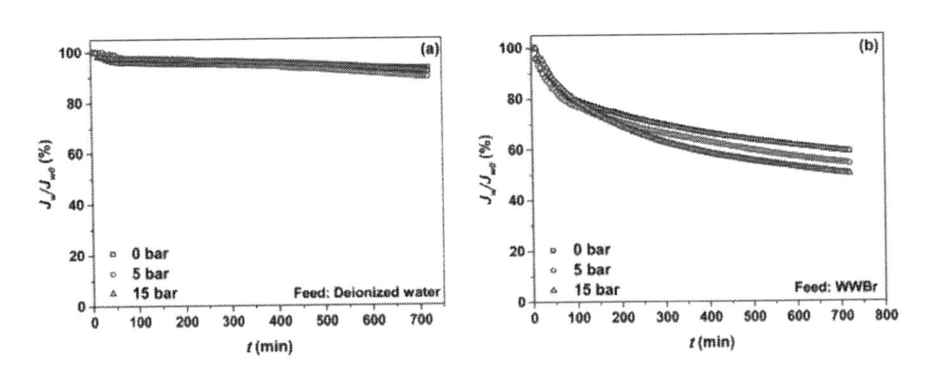

FIGURE 7.8
Normalized water flux, J_w/J_{w0}, of the TFC-PES membrane at various ΔP as a function of operation time during PRO tests using (a) deionized water and (b) WWBr as the feed, respectively. Draw solution: 2 M NaCl; initial feed volume: 0.5 L. WWBr stands for wastewater brine which is same as WWRe (Han et al. 2016).

the scaling was controlled. As displayed in Figure 7.9, the flux decline when using the HCl treated WWRe was significantly lower than that using the original WWRe. Even at a high permeate volume of 400 ml or a feed recovery of 80%, the flux drop was reduced from 75% to 40%, suggesting that the inorganic scaling was effectively controlled by the pH adjustment.

Similar work has been done by Chen et al. (2016). TFC-PES hollow fiber membranes were used for PRO tests. The baseline PRO power density profile using the original RO retentate (which is the same as WWRe) as the feed solution is shown in Figure 7.10(a). Since PRO power density was directly calculated from water flux, the water flux profile should have the same pattern as the power density profile. Two rounds of PRO test produced the same power density pattern. The initial power density was as high as 17.1 W/m^2, but the inorganic scaling in the membrane porous substrate layer led to an abrupt drop of power density. At t = 1000 min,

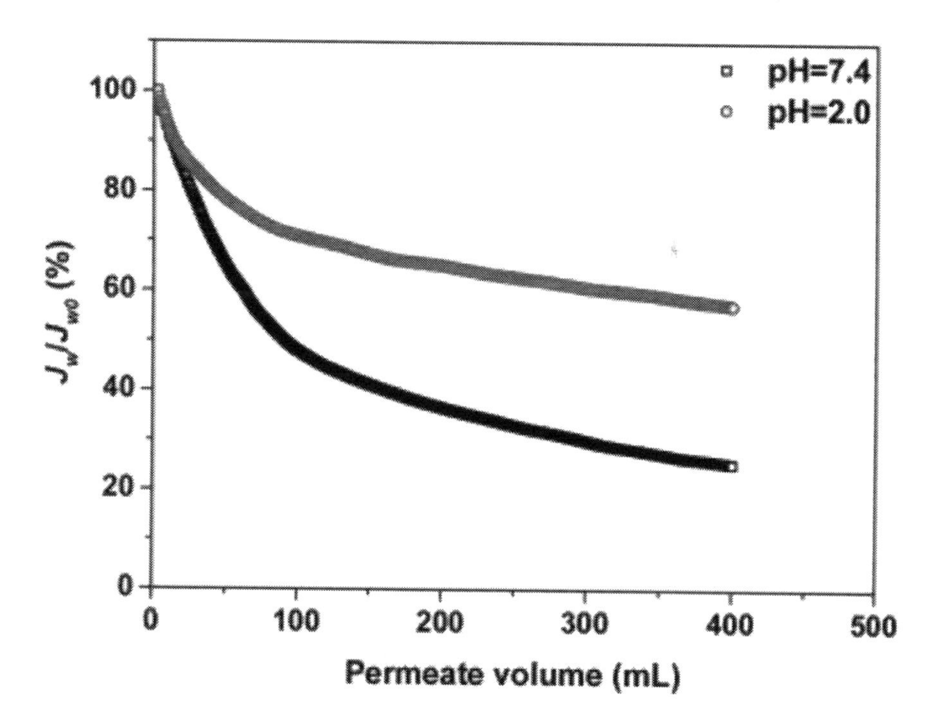

FIGURE 7.9
Normalized water flux, J_w/J_{w0}, of the TFC-PES membrane as a function of permeate volume during PRO tests at ΔP = 0 bar using the original WWRe (pH = 7.4) and HCl pretreated WWRe (pH = 2.0) as feed solutions. Draw solution: 2M NaCl; initial feed volume: 0.5 L; initial water flux: averaged flux within the first 5 min (Han et al. 2016).

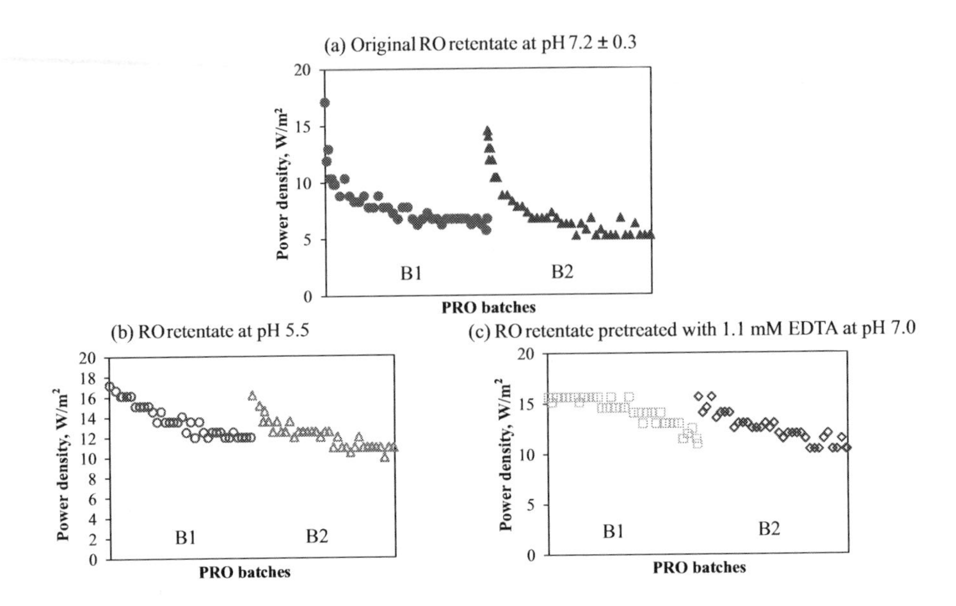

FIGURE 7.10
Power density using original and modified RO retentate as feeds. The original or modified RO retentate was circulating in the shell side with a crossflow rate of 11.6 cm/s at 0 bar. The draw solution, 1 M NaCl, was circulating in the lumen side with a crossflow rate of 2.2 m/s at 20 bar. B1 used fresh modules. B2 used the cleaned modules by cleaning B1 on the shell side with DI water flushing under a high crossflow velocity of 23.3 cm/s (Re = 2497) for 30 min. Each batch was tested for 1000 min (Chen et al. 2016).

the power density was 5.7 W/m^2, which was only 33.3% of the initial power density.

With the adjustment of pH to 5.5 (Figure 7.10b), a gradual decline of power density was observed, meaning the scaling of inorganic salts was inhibited to a certain extent. However, due to the existence of silica in the treated WWRe, PRO membrane might still suffer from silica fouling, which could explain why the power density still dropped using the treated WWRe as the feed solution. Compared with the previous case of using WWRe as the feed, the current case had a similar initial power density of 17.2 W/m^2. The final power density using the pH-adjusted WWRe as the feed solution was much higher than that in the previous case, around 12 W/m^2 or 70.0% of the initial power density.

7.3.2 Pretreatment with Anti-Scalant

Adding anti-scalant into the solution has been reported to be effective in inorganic scaling inhibition (Al-Mutairi et al. 2009, Yang et al. 2010b).

Chen et al. (2016) found that by adding ethylenediaminetetraacetic acid (EDTA) into the WWRe, the concentration of free calcium ions would be reduced. As depicted in Figure 7.11, the concentration of free calcium ions was reduced to 1.3 mg/L when 1.1 mM EDTA was added into the WWRe. At this concentration of free calcium ions, the calculated saturation index indicated that hydroxyapatite was at under-saturated status.

When DI water was used as the feed (B*), the PRO flux reduction is not due to fouling but due to the dilution of draw and the effect of reverse salt flux as shown in Figure 7.12(a). Using the WWRe pretreated with 1.1 mM EDTA as the feed (B1), the same flux trend was observed as B*, indicating that the inorganic scaling was inhibited due to the addition of EDTA. Figure 7.12(b) shows that the flux at t = 200 min remained the same as the initial flux of 28.2 LMH. After DI water flushing, the initial flux of the second PRO test (B2) was 28.1 LMH. At t = 200 min, the normalized flux of B2 was 0.900 (Figure 7.12c).

Han et al. (2016) added two different anti-scalants into the WWRe to reduce the concentration of free calcium ions. EDTA and 1-hydroxyethane 1,1-diphosphonic acid (HEDP, 60% aqueous solution) were added drop-wise into the WWRe feed under stirring with a final dosage of 400 ppm independently. It was worth noting that the addition of anti-scalants would not change the pH of WWRe, and pH values stayed almost the same as the original WWRe (i.e., 7.2–7.4 vs. 7.4).

Figure 7.13 shows their normalized PRO water fluxes as a function of permeate volume. Effective anti-scaling performances were achieved by

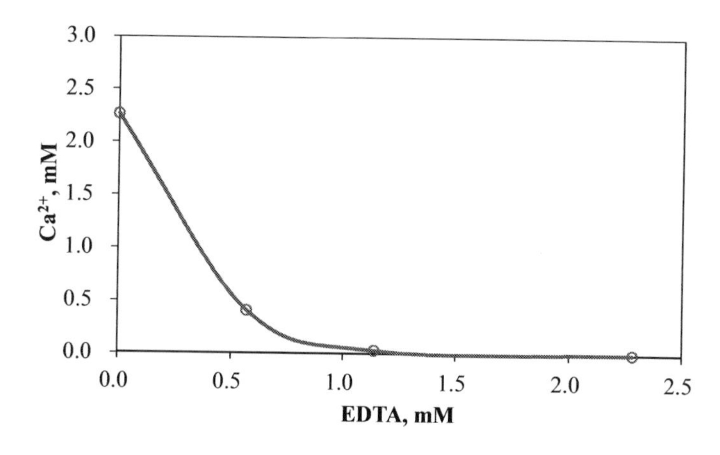

FIGURE 7.11
The concentration of free calcium ions in RO retentate versus the EDTA concentration added. The EDTA solution had a pH value of 7.0 (Chen et al. 2016).

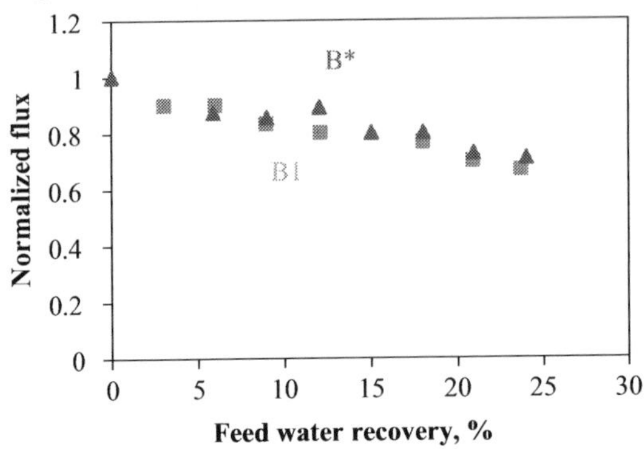

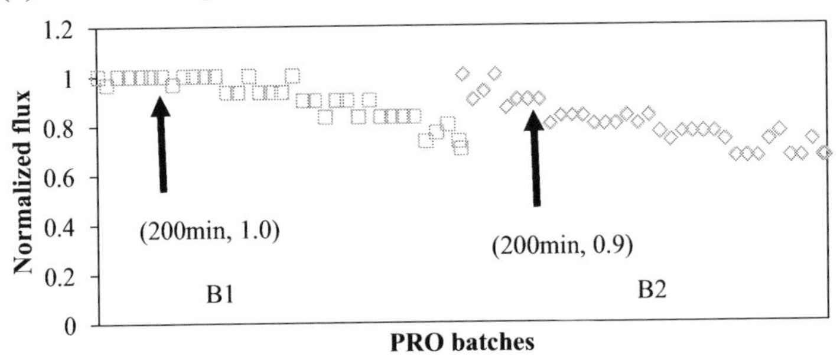

FIGURE 7.12
Flux patterns of RO retentate pretreated with 1.1 mM EDTA (B1–B2) and DI water (B*). The feed, modified RO retentate or DI water, was circulating in the shell side with a crossflow rate of 11.6 cm/s at 0 bar. The draw, 1 M NaCl, was circulating in the lumen side with a crossflow rate of 2.2 m/s at 20 bar. After physical cleaning, concentrated feed and diluted draw solutions were replaced by fresh ones and the PRO process was continued to the next batch. Each batch was tested for 1000 min (Chen et al. 2016).

both anti-scalants. The water flux decline at a feed recovery of 80% were reduced from 75% to 43% and 50%, respectively, when using the EDTA- and HEDP-treated WWRe as the PRO feed solution. Since EDTA and HEDP were powerful chelating agents with certain functional groups, they could form stable complexes with free ions such as Ca^{2+} and effectively suppress the inorganic scaling (Kim et al. 2015, Qin et al. 2009). In summary, to sustain PRO operations, proper pretreatments of the feed wastewater by anti-scalants could effectively mitigate the inorganic fouling and facilitate a high water flux.

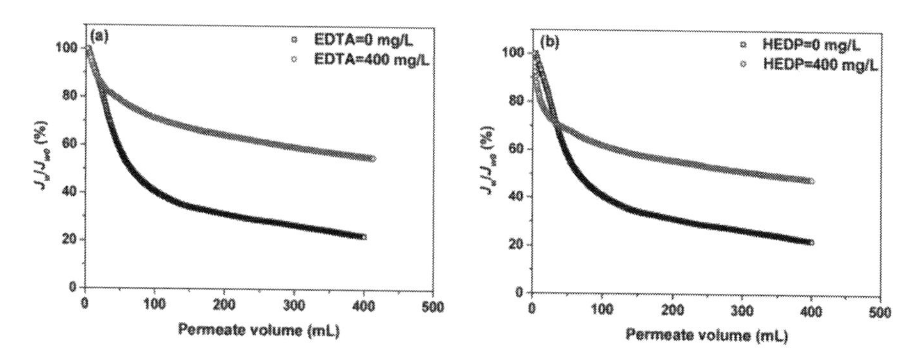

FIGURE 7.13
Normalized water flux, J_w/J_{w0}, of the TFC-PES membrane as a function of permeate volume during PRO tests at $\Delta P = 0$ bar using the original WWRe and (a) EDTA-pretreated WWRe and (b) HEDP-pretreated WWRe as feed solutions. Draw solution: 2 M NaCl; initial feed volume: 0.5 L; initial water flux: averaged flux within the first 5 min (Han et al. 2016).

7.4 Pretreatment Using Coagulations

Coagulation is another effective method to remove ions from the WWRe (Katz and Dosoretz 2008, Yang et al. 2010a). Compared with anti-scalants, coagulants are much cheaper. Wan et al. employed aluminum chloride ($AlCl_3$) and sodium aluminate ($NaAlO_2$) as the acidic and caustic coagulants, respectively (Wan et al. 2019). At pH of 7.2, both aluminum ions and $NaAlO_2$ hydrolyzed into amorphous $Al(OH)_3$, which absorbed phosphate ions from the WWRe and formed $Al(OH)_{3-3x}(PO_4)_x$ flocculants (Katz and Dosoretz 2008). In order to achieve an initial aluminum to phosphorous (Al:P) molar ratio up to 12:1 in the solutions, various dosages of coagulants were tested. The addition of coagulants was followed by continuous mixing for 120 min at 250 rpm and settlement for another 120 min. The upper clear solutions and the untreated WWRe were filtered by 8 μm filtration papers before PRO tests.

Figure 7.14(a) shows the concentration changes of phosphate and aluminum as $AlCl_3$ was added into the WWRe. The phosphate concentration in the WWRe was in excess for $AlCl_3$ until a 6-time $AlCl_3$ dosage was added. Above the 6-time dosage, the unreacted aluminum became excess while phosphate became undetectable in the WWRe. Thus, a 6-time $AlCl_3$ dosage was chosen as the optimal dosage. Similar trends of aluminum and phosphate concentration changes were observed with the $NaAlO_2$ pretreatment in Figure 7.14(b). Different from $AlCl_3$ dosing, $NaAlO_2$ pretreatment could not completely remove the phosphate. Therefore, the optimal $NaAlO_2$ dosage was determined as 8-time, beyond which a further increase in the $NaAlO_2$ dosage only

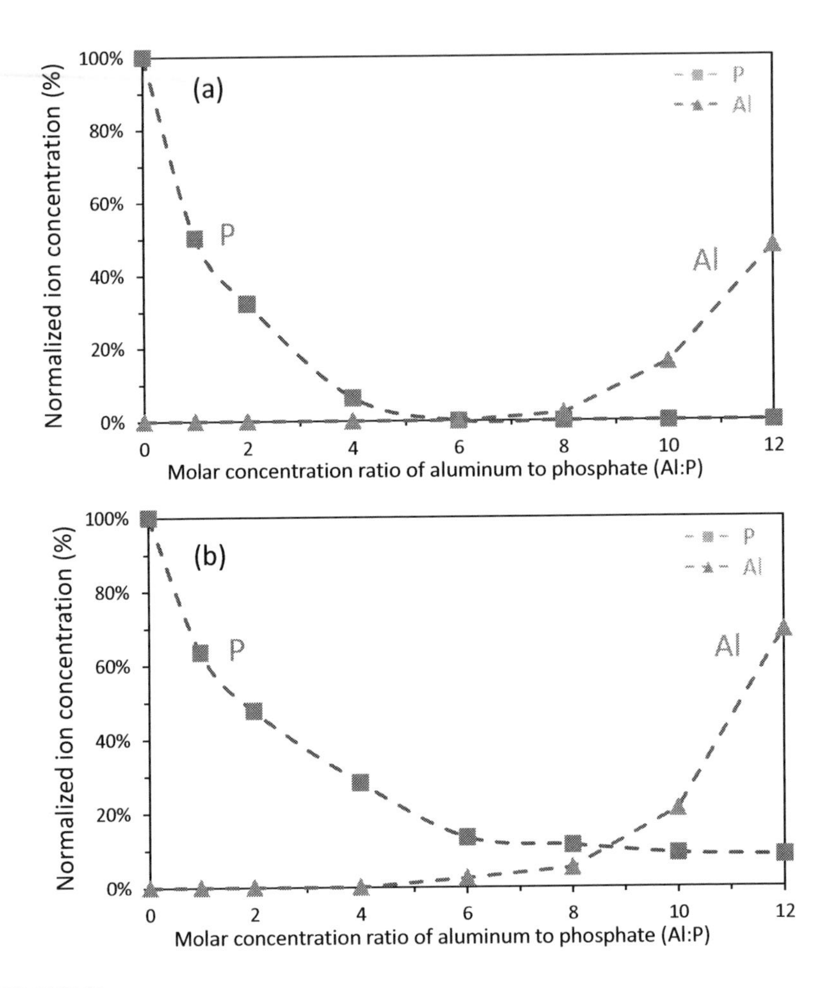

FIGURE 7.14
Aluminum and phosphate concentrations at different dosages of (a) $AlCl_3$ and (b) $NaAlO_2$ (Wan et al. 2019).

resulted in a diminishing reduction of phosphate concentration but a significant increase in aluminum concentration. At the optimal dosages of $AlCl_3$ and $NaAlO_2$, the respective pH values of the pretreated WWRe were 6 and 8.9, right within the discharge limits of 6–9 required in Singapore (Public Utility Board 2018).

Figure 7.15 summarizes the evolution of PRO water flux versus time at 20 bar using different feed solutions to achieve a 250 ml permeation volume. The highest initial water flux was achieved using DI water as the feed, which was 36.4 LMH. It took 697 min to obtain a total permeation volume of

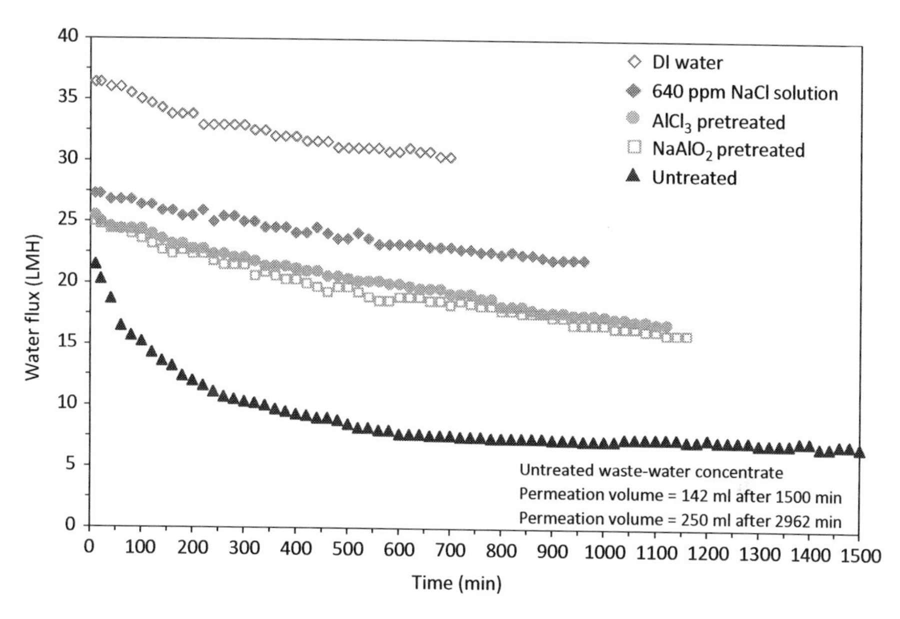

FIGURE 7.15
Water flux as a function of time using 1000 ml of DI water, 640 ppm NaCl solution, and $AlCl_3$- and $NaAlO_2$-pretreated and untreated wastewater concentrates as the respective feed solutions in PRO at 20 bar to achieve a total permeation volume of 250 ml (Wan et al. 2019).

250 ml, which was also the shortest among all the feed solutions. The final water flux and power density using DI water as the feed solution were 30.4 LMH and 16.9 W/m^2, respectively. A 640 ppm NaCl solution was used to simulate the total salinity of the real WWRe. It reached a total permeation volume of 250 ml within 942 min during which the water flux dropped from 27.3 LMH to 22 LMH. The relatively lower water flux indicated that the internal concentration polarization caused by the 640 ppm salt in the feed solution significantly reduced the effective driving force across the TFC-PES membrane (She et al. 2012, Xiong et al. 2016). When the untreated WWRe was employed as the feed solution, the initial water flux was 21.5 LMH and quickly decreased to 9.7 LMH within 360 min. In 2962 min, the water flux further decreased to 6.6 LMH at a 25% recovery of the feed solution. For a clear presentation, the time axis was stopped at 1500 min when only 142 ml of the untreated WWRe passing through the membrane was achieved. Compared with the original WWRe, PRO tests using $AlCl_3$ and $NaAlO_2$ pretreated WWRes showed significant improvements with an initial water flux of 25.5 LMH and 24.8 LMH, respectively. Both PRO performances exhibited similar trends. At a total permeation volume of 250 ml, the former still produced a water flux of 16.8 LMH and a power density of

9.3 W/m^2, while the latter generated a water flux of 15.9 LMH and a power density of 8.8 W/m^2. In summary, pretreatment by coagulations using $AlCl_3$ and $NaAlO_2$ were effective to remove foulants from PRO feed solutions and sustain high PRO performance.

7.5 Summary

Pretreatment of PRO feed streams is essential in order to maintain high water flux and power density and prolong PRO membrane life, especially when the quality of the feed stream is low. This chapter only introduces the most commonly used pretreatment methods. Depending on the specific situation, various pretreatment methods could be applied. Pretreatment using UF membranes can only remove large organic compounds from PRO feed solutions. NF membranes are effective in removing ions but cannot reject silica. Pretreatment using LP-RO membranes can produce the best quality of PRO feed solutions. In addition, using pH adjustment and anti-scalants can prohibit inorganic salt scaling, while coagulation followed by simple filtration is effective in removing ions from PRO feed solutions. However, apart from the PRO performance after the pretreatment, we need to consider the cost of pretreatment and the energy input as well. For example, LP-RO could produce the cleanest permeate, but it is not economically applicable due to high energy input demand. Thus, a trade-off between PRO performance after the pretreatment and the cost of the pretreatment is inevitable. Nevertheless, an optimum pretreatment method could always be found after thorough research and consideration.

Acknowledgments

This research grant is supported by the National Research Foundation, Prime Minister's Office, Republic of Singapore, under its Environmental & Water Technologies Strategic Research Programme and administered by the Environment & Water Industry Programme Office (EWI) of the PUB. This research work was funded by the project entitled "Development of 8 inch Novel High Efficiency Pressure Retarded Osmosis (PRO) Membrane Modules towards Potential Pilot Testing and Field Validation" (USS-IF-2018-1) and NUS grant number of R-279-000-555-592. Special thanks to Dr. Chun Feng Wan for his kind help.

References

Aimar, P., Meireles, M., Sanchez, V., 1990. A contribution to the translation of retention curves into pore size distributions for sieving membranes. *J. Membr. Sci.* 54(3), 321–338.

Al-Mutairi, N.N., Aleem, F.A., Al-Ahmad, M.I., 2009. Effect of antiscalants for inhibition of calcium sulfate deposition in thermal desalination systems. *Desalination Water Treat.* 10(1–3), 39–46.

Antony, A., Low, J.H., Gray, S., Childress, A.E., Le-Clech, P., Leslie, G, 2011. Scale formation and control in high pressure membrane water treatment systems: A review. *J. Membr. Sci.* 383(1), 1–16.

Bui, N.-N., Lind, M.L., Hoek, E.M.V., McCutcheon, J.R., 2011. Electrospun nanofiber supported thin film composite membranes for engineered osmosis. *J. Membr. Sci.* 385–386, 10–19.

Chen, S.C., Amy, G.L., Chung, T.S., 2016. Membrane fouling and anti-fouling strategies using RO retentate from a municipal water recycling plant as the feed for osmotic power generation. *Water Res.* 88, 144–155.

Chen, S.C., Fu, X.Z., Chung, T.S., 2014. Fouling behaviors of polybenzimidazole (PBI)–polyhedral oligomeric silsesquioxane (POSS)/polyacrylonitrile (PAN) hollow fiber membranes for engineering osmosis processes. *Desalination* 335(1), 17–26.

Chen, Y., Liu, C., Setiawan, L., Wang, Y.-N., Hu, X., Wang, R., 2017. Enhancing pressure retarded osmosis performance with low-pressure nanofiltration pretreatment: Membrane fouling analysis and mitigation. *J. Membr. Sci.* 543, 114–122.

Cheng, J., Li, Y., Chung, T.S., Chen, S.-B., Krantz, W.B., 2008. High-performance protein separation by ion exchange membrane partitioned free-flow isoelectric focusing system. *Chem. Eng. Sci.* 63(8), 2241–2251.

Childress, A.E., Elimelech, M., 2000. Relating nanofiltration membrane performance to membrane charge (electrokinetic) characteristics. *Environ. Sci. Technol.* 34(17), 3710–3716.

Chu, S., Majumdar, A., 2012. Opportunities and challenges for a sustainable energy future. *Nature* 488, 294.

Chung, T.S., Luo, L., Wan, C.F., Cui, Y., Amy, G., 2015. What is next for forward osmosis (FO) and pressure retarded osmosis (PRO). *Sep. Purif. Technol.* 156, 856–860.

Chung, T.S., Zhang, S., Wang, K.Y., Su, J., Ling, M.M., 2012. Forward osmosis processes: Yesterday, today and tomorrow. *Desalination* 287, 78–81.

Han, G., Chung, T.S., 2014. Robust and high performance pressure retarded osmosis hollow fiber membranes for osmotic power generation. *AIChE J.* 60(3), 1107–1119.

Han, G., Zhang, S., Li, X., Chung, T.S., 2015. Progress in pressure retarded osmosis (PRO) membranes for osmotic power generation. *Prog. Polym. Sci.* 51, 1–27.

Han, G., Zhou, J., Wan, C., Yang, T., Chung, T.S., 2016. Investigations of inorganic and organic fouling behaviors, antifouling and cleaning strategies for pressure retarded osmosis (PRO) membrane using seawater desalination brine and wastewater. *Water Res.* 103, 264–275.

He, Y., Tang, Y.P., Chung, T.S., 2016. Concurrent removal of selenium and arsenic from water using polyhedral oligomeric silsesquioxane (POSS)–polyamide thin-film nanocomposite nanofiltration membranes. *Ind. Eng. Chem. Res.* 55(50), 12929–12938.

Katz, I., Dosoretz, C.G., 2008. Desalination of domestic wastewater effluents: phosphate removal as pretreatment. *Desalination* 222(1), 230–242.

Kim, D.I., Kim, J., Shon, H.K., Hong, S., 2015. Pressure retarded osmosis (PRO) for integrating seawater desalination and wastewater reclamation: Energy consumption and fouling. *J. Membr. Sci.* 483, 34–41.

Kim, J., Park, M., Snyder, S.A., Kim, J.H., 2013a. Reverse osmosis (RO) and pressure retarded osmosis (PRO) hybrid processes: Model-based scenario study. *Desalination* 322, 121–130.

Kim, Y.C., Kim, Y., Oh, D., Lee, K.H., 2013b. Experimental investigation of a spiral-wound pressure-retarded osmosis membrane module for osmotic power generation. *Environ. Sci. Technol.* 47(6), 2966–2973.

Labban, O., Liu, C., Chong, T.H., Lienhard V.J.H., 2017. Fundamentals of low-pressure nanofiltration: Membrane characterization, modeling, and understanding the multi-ionic interactions in water softening. *J. Membr. Sci.* 521, 18–32.

Li, X., Cai, T., Amy, G.L., Chung, T.S., 2017. Cleaning strategies and membrane flux recovery on anti-fouling membranes for pressure retarded osmosis. *J. Membr. Sci.* 522, 116–123.

Li, Y., Chung, T.-S., 2008. Exploration of highly sulfonated polyethersulfone (SPES) as a membrane material with the aid of dual-layer hollow fiber fabrication technology for protein separation. *J. Membr. Sci.* 309(1), 45–55.

Liu, C., Shi, L., Wang, R., 2015. Crosslinked layer-by-layer polyelectrolyte nanofiltration hollow fiber membrane for low-pressure water softening with the presence of SO_4^{2-} in feed water. *J. Membr. Sci.* 486, 169–176.

Loeb, S., 2002. Large-scale power production by pressure-retarded osmosis, using river water and sea water passing through spiral modules. *Desalination* 143(2), 115–122.

Loeb, S., Norman, R.S., 1975. Osmotic power plants. *Science* 189(4203), 654–655.

Logan, B.E., Elimelech, M., 2012. Membrane-based processes for sustainable power generation using water. *Nature* 488, 313.

O'Toole, G., Jones, L., Coutinho, C., Hayes, C., Napoles, M., Achilli, A., 2016. River-to-sea pressure retarded osmosis: Resource utilization in a full-scale facility. *Desalination* 389, 39–51.

Public Utility Board, Requirements for discharge of trade effluent into the public sewers. Retrieved on 24 April 2018 from www.pub.gov.sg/Documents/requirements_UW.pdf.

Qin, J.J., Wai, M.N., Oo, M.H., Kekre, K.A., Seah, H., 2009. Impact of anti-scalant on fouling of reverse osmosis membranes in reclamation of secondary effluent. *Water Sci. Technol.* 60(11), 2767–2774.

Saito, K., Irie, M., Zaitsu, S., Sakai, H., Hayashi, H., Tanioka, A., 2012. Power generation with salinity gradient by pressure retarded osmosis using concentrated brine from SWRO system and treated sewage as pure water. *Desalination Water Treat.* 41(1–3), 114–121.

Sakai, H., Ueyama, T., Irie, M., Matsuyama, K., Tanioka, A., Saito, K., Kumano, A., 2016. Energy recovery by PRO in sea water desalination plant. *Desalination* 389, 52–57.

Sharqawy, M.H., Zubair, S.M., Lienhard, J.H., 2011. Second law analysis of reverse osmosis desalination plants: An alternative design using pressure retarded osmosis. *Energy* 36(11), 6617–6626.

She, Q., Jin, X., Tang, C.Y., 2012. Osmotic power production from salinity gradient resource by pressure retarded osmosis: Effects of operating conditions and reverse solute diffusion. *J. Membr. Sci.* 401–402, 262–273.

She, Q., Wong, Y.K.W., Zhao, S., Tang, C.Y., 2013. Organic fouling in pressure retarded osmosis: Experiments, mechanisms and implications. *J. Membr. Sci.* 428, 181–189.

Singh, S., Khulbe, K.C., Matsuura, T., Ramamurthy, P., 1998. Membrane characterization by solute transport and atomic force microscopy. *J. Membr. Sci.* 142(1), 111–127.

Skilhagen, S.E., Dugstad, J.E., Aaberg, R.J., 2008. Osmotic power – Power production based on the osmotic pressure difference between waters with varying salt gradients. *Desalination* 220(1), 476–482.

Sukitpaneenit, P., Chung, T.S., 2012. High performance thin-film composite forward osmosis hollow fiber membranes with macrovoid-free and highly porous structure for sustainable water production. *Environ. Sci. Technol.* 46(13), 7358–7365.

Thorsen, T., Holt, T., 2009. The potential for power production from salinity gradients by pressure retarded osmosis. *J. Membr. Sci.* 335(1), 103–110.

Wan, C.F., Chung, T.S., 2015. Osmotic power generation by pressure retarded osmosis using seawater brine as the draw solution and wastewater retentate as the feed. *J. Membr. Sci.* 479, 148–158.

Wan, C.F., Jin, S., Chung, T.S., 2019. Mitigation of inorganic fouling on pressure retarded osmosis (PRO) membranes by coagulation pretreatment of the wastewater concentrate feed. *J. Membr. Sci.* 572, 658–667.

Wan, C.F., Li, B., Yang, T., Chung, T.S., 2017. Design and fabrication of inner-selective thin-film composite (TFC) hollow fiber modules for pressure retarded osmosis (PRO). *Sep. Purif. Technol.* 172, 32–42.

Wang, K.Y., Ong, R.C., Chung, T.S., 2010. Double-skinned forward osmosis membranes for reducing internal concentration polarization within the porous sublayer. *Ind. Eng. Chem. Res.* 49(10), 4824–4831.

Xiong, J.Y., Cheng, Z.L., Wan, C.F., Chen, S.C., Chung, T.S., 2016. Analysis of flux reduction behaviors of PRO hollow fiber membranes: Experiments, mechanisms, and implications. *J. Membr. Sci.* 505, 1–14.

Yang, K., Li, Z., Zhang, H., Qian, J., Chen, G., 2010a. Municipal wastewater phosphorus removal by coagulation. *Environ. Technol.* 31(6), 601–609.

Yang, Q., Liu, Y., Li, Y., 2010b. Humic acid fouling mitigation by antiscalant in reverse osmosis system. *Environ. Sci. Technol.* 44(13), 5153–5158.

Yang, T., Wan, C.F., Xiong, J.Y., Chung, T.S., 2019. Pre-treatment of wastewater retentate to mitigate fouling on the pressure retarded osmosis (PRO) process. *Sep. Purif. Technol.* 215, 390–397.

Yu, Y., Lee, S., Hong, S., 2010. Effect of solution chemistry on organic fouling of reverse osmosis membranes in seawater desalination. *J. Membr. Sci.* 351(1), 205–213.

Zhang, S., Sukitpaneenit, P., Chung, T.S., 2014. Design of robust hollow fiber membranes with high power density for osmotic energy production. *Chem. Eng. J.* 241, 457–465.

Zhao, Y., Song, L., Ong, S.L., 2010. Fouling behavior and foulant characteristics of reverse osmosis membranes for treated secondary effluent reclamation. *J. Membr. Sci.* 349(1), 65–74.

8

Design and Fabrication of Thin-Film Composite Hollow Fiber Modules for Pressure Retarded Osmosis

Chun Feng Wan and Tianshi Yang

Department of Chemical and Biomolecular Engineering
National University of Singapore
Singapore

CONTENTS

8.1 Introduction

As the heart of pressure retarded osmosis (PRO) processes, PRO membranes have received great attention and significant progress since the testing of commercial RO membranes for PRO applications by Loeb and Mehta in the 1970s (Loeb 1976, Loeb et al. 1976, Mehta and Loeb 1979, Son et al. 2016, Zhou et al. 2014). In the last decades, thin-film composite (TFC) membranes gained increasing considerations because of their higher water fluxes and higher power densities (Chou et al. 2013, Cui et al. 2014, Gerstandt et al. 2008, Song et al. 2013, Yip et al. 2011). Most TFC membranes are formed in situ onto the surface of microporous substrates via interfacial polymerization of aromatic diamine such as piperazine (PIP) (Wang and Li 2010) and *m*-phenylenediamine (MPD) (Veríssimo et al. 2005) with acid chloride monomers such as trimesoyl chloride (TMC) (Liu et al. 2007) and isophthaloyl chloride (IPC) (Korikov et al. 2006). Compared to TFC flat-sheet membranes, TFC hollow fiber membranes have a higher surface area per volume, a self-mechanical support, and no shadow effects due to the usage of spacers (Han et al. 2015, Peng et al. 2012). In spite of the advantages of TFC hollow fiber membranes, there are no commercially available TFC hollow fiber membrane modules possibly because of the difficulties to fabricate them cost effectively.

Recently, a robust inner-selective TFC polyethersulfone (PES) hollow fiber membrane that exhibited a power density of 27 W/m^2 using 1 M NaCl solution and DI water as the feed pair at 20 bar has been developed (Wan and Chung 2015, Zhang et al. 2014). In previous studies, interfacial polymerization was conducted on mini-modules comprising 3–10 pieces of hollow fiber membranes (Sarp et al. 2016, Straub et al. 2016). In order to modularize installation and test membranes on a pilot system, this novel TFC hollow fiber membrane must be assembled into an useful and operational unit – a module (Mat et al. 2014, Sarp et al. 2016, Straub et al. 2016). Design and fabrication of hollow fiber modules is a complicated process that involves different disciplines and requires a thorough understanding of the intended application. Module engineering plays an important role in optimizing membrane performance for PRO in terms of increasing membrane area per module, optimizing flow pattern, and enhancing mechanical properties of the modules (Mat et al. 2014, Sivertsen et al. 2013). Despite the importance of module design and fabrication, the literature in this area is sparse relative to that devoted to membrane materials. Additionally, significant knowledge is held as trade secrets because of its high market value. The majority of the public knowledge is taught in patents and a limited number of studies and reviews are available in the literature (Chong et al. 2013, Li et al. 2004, Teoh et al. 2008, Wan et al. 2017, Yang et al. 2011).

In this chapter, we will investigate the science and engineering to fabricate TFC hollow fiber modules by assembling hollow fibers into semi-pilot-scale modules and subsequently conducting interfacial polymerization to form a thin-film polyamide layer on top of the inner surface of hollow fibers. These knowledge and skills are important to fabricate TFC hollow fiber modules for PRO, reverse osmosis, nanofiltration, and other applications, bridging the gaps between membrane fabrication and module fabrication.

8.2 Experimental

8.2.1 Materials

Radel® A polyethersulfone (PES, Solvay Advanced Polymer), *N*-methyl-2-pyrrolidone (NMP, > 99.5%, Merck), polyethylene glycol 400 (PEG, MW = 400 g/mol, Sigma-Aldrich), and deionized (DI) water were used as the polymer, solvent, and nonsolvent additives, respectively, to prepare the polymer dope of hollow fiber substrates. A 50/50 wt% mixture of glycerol (Industrial grade, Aik Moh Pains & Chemicals, Singapore) and DI water was employed for the post-treatment of as-spun hollow fiber substrates. *m*-phenylenediamine (MPD, >99%, Sigma-Aldrich), sodium dodecyl sulphate (SDS, >97%, Fluka), and DI water were acquired to prepare the MPD solution; trimesoyl chloride (TMC, >98%, Tokyo Chemical Industry, Japan) and hexane (>99.9%, Fisher Chemicals) were utilized to prepare the TMC solution for the interfacial polymerization. Epoxy (KSbond, Kuo Seng Enterprise, Taiwan) was purchased to cast the tube-sheets of hollow fiber modules.

8.2.2 Fabrication of TFC-PES Hollow Fiber Membranes

The PES hollow fiber substrates were prepared by a dry jet wet spinning process (Han and Chung 2014). The polymer dope composition and spinning conditions have been reported previously (Zhang and Chung 2013, Zhang et al. 2014). Minor modifications were made to further enhance the mechanical properties and PRO performance. The modified spinning parameters are presented in Table 8.1. The inner and outer diameters of the hollow fiber substrates were 575 μm and 1023 μm, respectively. The as-spun hollow fiber substrates were collected from the spinning line and soaked in water for 2 days to remove the residual solvent and nonsolvent additives. The PES hollow fiber substrates were then posted in a 50/50 wt% glycerol/water solution for another 2 days. After drying, three pieces of the PES substrates were assembled into a mini-module. The TFC-PES hollow fiber membranes were then formed via interfacial polymerization of MPD and TMC on the inner surfaces of hollow fiber membranes (Han and Chung 2014, Zhang et al. 2014).

TABLE 8.1
Dope composition and spinning conditions of the PES hollow fiber substrates

Spinning Parameters	PES Membrane Substrate
Bore fluid	Deionized Water
Dope composition (wt%)	20-22/36-39/36-39/1-4 (PES/PEG/NMP/Water)
Outer channel	NMP
External coagulant	Water
Outer flow rate (ml/min)	0.03–0.1
Inner dope flow rate (ml/min)	2.0–3.0
Bore fluid flow rate (ml/min)	0.8–1.2
Air gap length (cm)	1–3
Take-up speed (m/min)	1–5
Dual layer spinneret dimension (mm)	Outer OD = 1.6 mm, Inner OD = 1.3 mm, ID = 1.14 mm

8.2.3 Fabrication of TFC-PES Hollow Fiber Modules

Figure 8.1 shows a general configuration of an inner-selective PRO hollow fiber module with the draw solution flowing along the lumen side and the feed solution flowing along the shell side. The housing has two ports for the introduction and exit of the feed solution, respectively. The draw solution enters the port from one end cap, flows though the fiber lumens, and exits from the other end cap. Tubesheets are formed at the two ends of the housing with openings to the lumens of hollow fiber membranes. The tubesheets work as physical barriers to separate the feed solution from the draw solution and prevent mixing. Therefore, the only path for the feed solution to meet the draw solution is by passing through the walls of hollow fiber membranes.

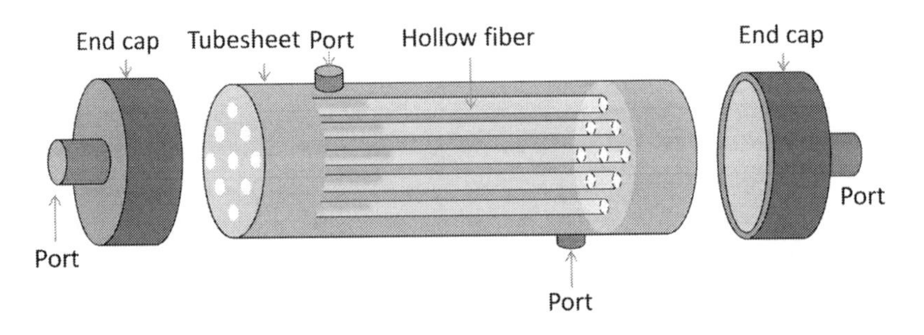

FIGURE 8.1
Schematic drawing of a PRO hollow fiber module.

8.2.3.1 Fabrication of PES Hollow Fiber Modules

The step-by-step fabrication of a hollow fiber module is presented in Figure 8.2. The PES hollow fiber substrates after the glycerol post-treatment were collected into a bundle. To meet a target packing density (ϕ) that is defined as the volumetric fraction of the module occupied by the hollow fibers, the number of hollow fibers (n) in each bundle could be calculated as follows:

$$n = \frac{\phi D^2}{d^2} \tag{1}$$

where D is the inner diameter of the module housing and d is the outer diameter of the hollow fibers.

Hollow fiber modules with 30% and 50% packing densities were prepared. The simplest way to prepare a hollow fiber bundle is to collect the hollow fibers and arrange them in parallel randomly to form a bundle. However, when the packing density is low, the hollow fibers may not distribute uniformly and fill up the housing (Wang et al. 2003). To overcome these shortcomings, structured hollow fiber bundles were prepared by laying them down in parallel at a uniform spacing to form a hollow fiber mat and then rolling up the mat to form a structured bundle (Niermeyer 1997, Philip et al. 1997). This method is relatively complicated but yields a more uniformly packed bundle even at a low packing density. At a high packing density, the hollow fibers can support one

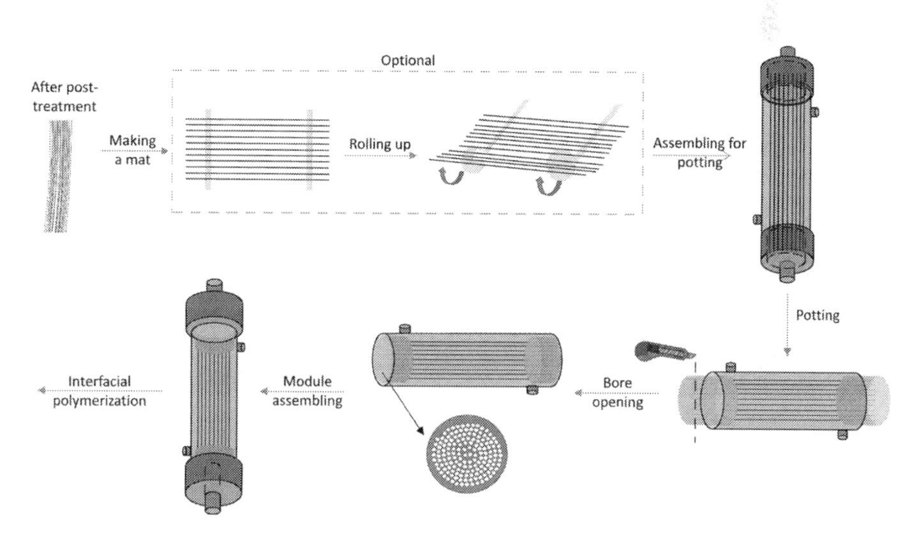

FIGURE 8.2
Schematic drawings of the procedures to fabricate hollow fiber modules.

another and distribute more uniformly in a random packing bundle (Wang et al. 2003).

The bundle was then cut to a desired length and inserted to the module housing. The lumens of hollow fibers were temporarily sealed with glue or cement to prevent intrusion of epoxy during the tubesheet formation that was conducted by potting the hollow fibers into a mold filled with epoxy. Two potting techniques were widely used, namely gravitationally potting (Azran and Dagan 2001) and centrifugal potting (Maxwell et al. 1967). In gravitationally potting, as shown in Figure 8.3(a), the module was held vertically and the epoxy was introduced from the bottom of the mold. The epoxy gradually penetrated into the space among hollow fibers and settled down under the gravitational and capillary forces. To produce a tubesheet with a high packing density, centrifugal potting was often used to ensure that the space among hollow fibers was filled up with epoxy. During centrifugal potting, the module and the epoxy container were fixed horizontally along the central axis of the centrifuge machine as shown in Figure 8.3(b). Two flow channels connected the epoxy container and the two molds. Under the centrifugal force, the epoxy was sent to fill up the molds and form the tubesheets. However, the drawbacks of centrifugal potting are (1) electricity-consuming and (2) productivity limited by the capacity of the centrifuge machine. After the epoxy was solidified, the molds were removed and the additional epoxy was cut off to re-expose the lumens of hollow fibers.

8.2.3.2 Interfacial Polymerization on the Inner Surface of Hollow Fiber Modules

Interfacial polymerization was carried out on the inner surface of each hollow fiber to form a polyamide selective layer. First, the MPD solution

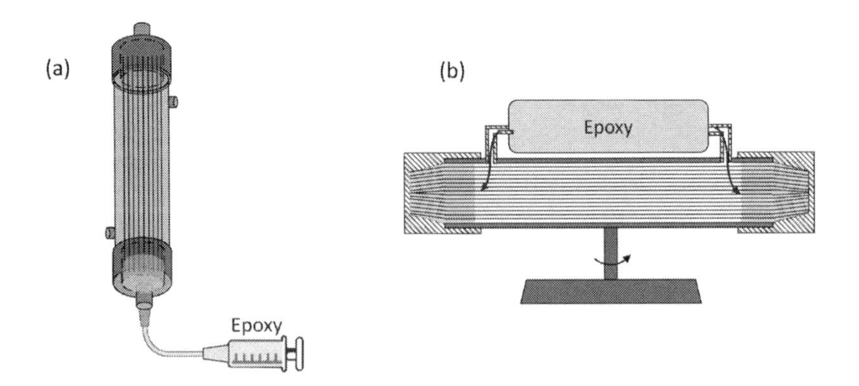

FIGURE 8.3
Formation of the tubesheet by (a) gravitational potting and (b) centrifugal potting.

was circulated in the lumen at a flowrate of 0.5 ml/min per hollow fiber for a certain duration using the setup sketched in Figure 8.4(a). A duration of 3 min was employed to fabricate coupons of hollow fibers (Han and Chung 2014, Zhang et al. 2014), but the duration was increased to 5 min and 7 min to ensure all the hollow fibers were effectively rinsed with the MPD solution. The excessive MPD solution was then removed by purging it out with compressed nitrogen for 5 min using the setup sketched in Figure 8.4(b). Second, the TMC solution was circulated in the lumens at a flowrate of 0.5 ml/min per hollow fiber for a certain duration, ranging from 5 min when fabricating three pieces of TFC-PES hollow fibers to 7 min and 9 min, to ensure a polyamide layer was successfully formed on each hollow fiber. The excessive TMC solution was then removed by purging it with compressed nitrogen for 2 min.

8.2.4 Detection and Repair of Hollow Fiber Membranes

8.2.4.1 Detection of Broken TFC-PES Hollow Fibers

The hollow fibers could be damaged during handling and high pressure testing. Such broken fibers must be identified and plugged to restore the module performance. In Figure 8.5(a), the module housing was pressurized with compressed nitrogen at 10 bar. One end of the cap was plugged and the other end remained open. If there were any gas leakages from the

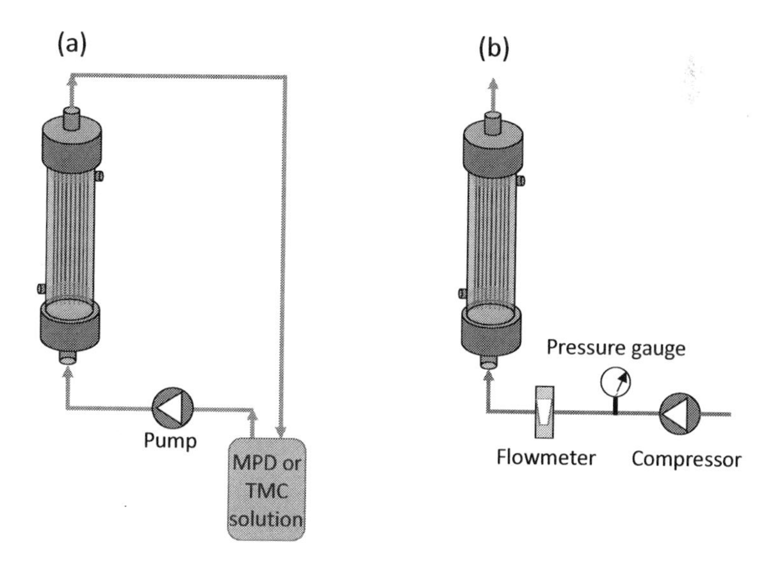

FIGURE 8.4
Procedures for interfacial polymerization of hollow fiber modules: (a) circulation of MPD and TMC solutions and (b) purging.

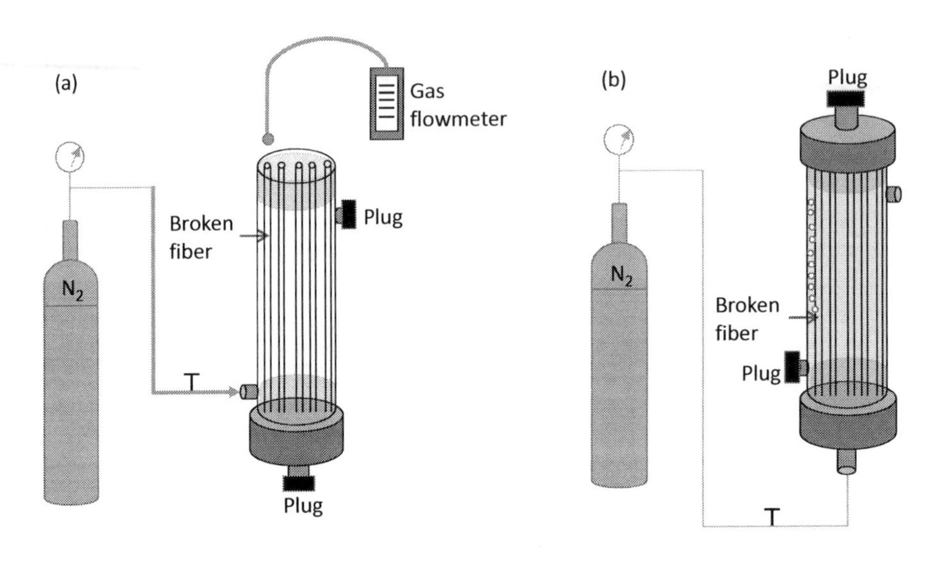

FIGURE 8.5
(a) A leakage test on the tubesheet to identify the broken fiber and (b) a leakage test on the hollow fiber bundle to identify the spot of leakage.

shell side to the lumens, they should be detected by a gas detector at the open end of the module. By moving the gas detector across the tubesheet, the broken fibers could be located.

In order to further identify the most vulnerable portion of hollow fibers along the axis, bubbling tests were carried out by filling the housing with water and compressing the hollow fibers from inside out with nitrogen. As shown in Figure 8.5(b), at the spot of leakage, continuous bubbling was observed.

8.2.4.2 Repair of Hollow Fiber Modules

The broken fibers were subsequently repaired after identification. One must use epoxy that had been partially solidified before injection to prevent the epoxy from flowing out. As shown in Figure 8.6(a), a syringe pump was used to push the viscous epoxy through a small nozzle into the defective fiber. Another method to plug the broken fibers was by means of filler insertion, as shown in Figure 8.6(b). The epoxy-coated filler had to be inserted at least 3 cm deep into the broken hollow fiber to prevent its leakage and provide adequate mechanical supports against the pressure difference across the tubesheet. The epoxy was then fully solidified before tests. During repair, it was important to apply the epoxy only to the leakage area without plugging other fibers in order to minimize the loss of effective areas.

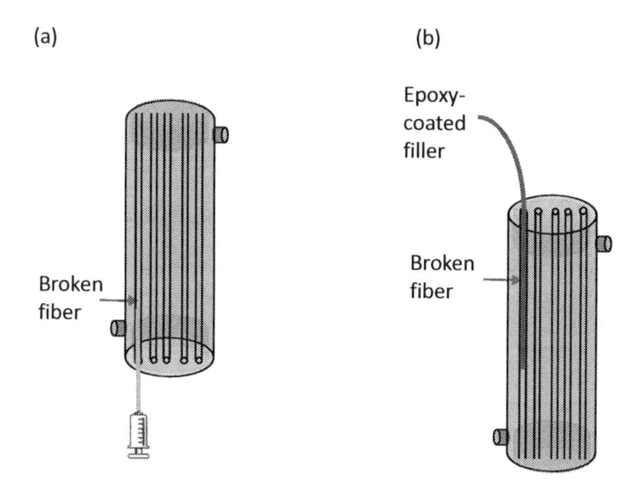

FIGURE 8.6
Repair of hollow fiber modules by (a) injection of epoxy and (b) insertion of epoxy-coated fillers.

8.2.5 Measurements of Pure Water Permeability and Salt Permeability of the TFC-PES Membrane Modules

The pure water permeability (A) and the salt permeability (B) of TFC-PES hollow fiber modules were measured under the RO mode using the PRO setup shown in Figure 8.7. The lumen side of the modules was pressurized with DI water and the permeate was collected from the shell side.

A (L/m^2/h/bar, LMH/bar) was calculated as follows:

$$A = \frac{\Delta V}{M \Delta t \Delta P} \tag{2}$$

where ΔV is the volumetric change of the permeate collected over a period of Δt(hr) during the test, M (m^2) is the effective permeation area of the module, and ΔP (bar) is the transmembrane pressure.

To measure the membrane rejection (R), DI water in the above test was replaced with a 1000 ppm NaCl solution. The salt concentrations of the permeate collected from the shell and feed sides were measured to calculate R using the following equation:

$$R = \left(1 - \frac{C_p}{C_f}\right) \times 100\% \tag{3}$$

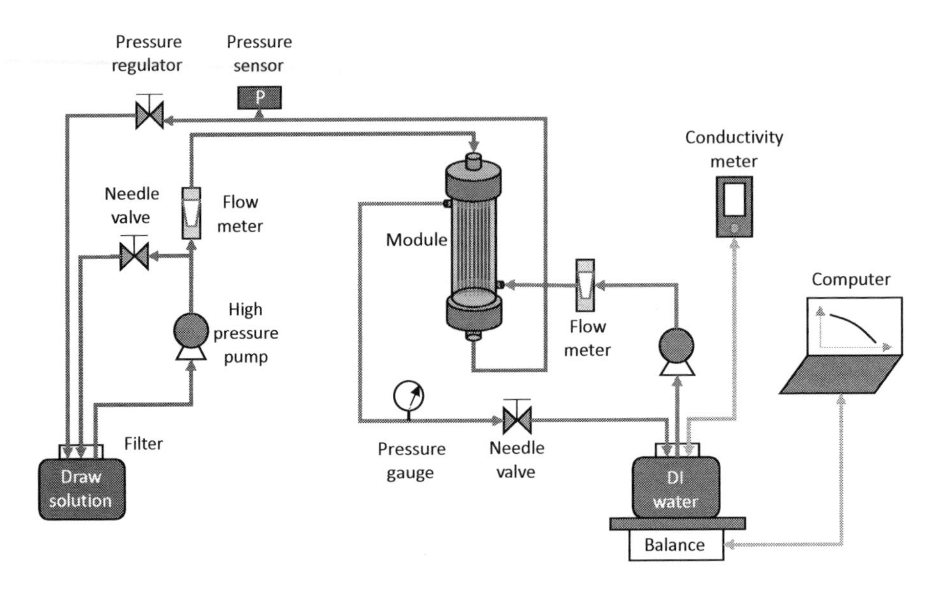

FIGURE 8.7
Schematic drawing of the PRO setup to test hollow fiber modules.

where C_f and C_p refer to the concentrations of the feed and permeate, respectively. Salt permeability B (L/m^2/h, LMH) was then calculated using the following equation (Wan and Chung 2015, Zhang et al. 2014):

$$B = \frac{1 - R}{R}(\Delta P - \Delta\pi)A \tag{4}$$

8.2.6 PRO Performance Tests

Figure 8.7 shows the setup to test the PRO performance of a hollow fiber module. A 1 M NaCl solution was employed as the draw solution flowing through the module's lumen (i.e., facing the inner selective skin), while DI water as the feed flowing along the shell side (i.e., facing the outer PES substrates). The modules were pre-pressurized at 20 bar for 30 min prior to the tests. The flowrate of the draw solution was maintained at 6.67 ml/min per hollow fiber. Since the burst pressure of the TFC hollow fiber membrane was 22–23 bar, the maximal pressure tested was 20 bar for safe and stable operations. The pressure of the draw solution was increased from 0 bar to 20 bar at an increment of 5 bar, while the feed was maintained at atmospheric pressure. The flowrate ratio of the draw solution to the feed was kept at 1:1 throughout all the tests. All the tests were carried out at room temperature (25±1°C). The weight and conductivity of the feed solution were recorded for the calculations of water flux, reverse salt flux, and power density.

The water permeation flux J_w (L/m^2/h, LMH) was calculated from the volumetric change (ΔV_f) of the feed solution:

$$J_W = \frac{\Delta V_f}{M\Delta t} \quad (5)$$

The theoretical osmotic power density, W (W/m^2), was calculated as the product of water flux and the pressure of the draw solution:

$$W = \frac{J_w P}{36} \quad (6)$$

where P (bar) is the hydraulic pressure of the draw solution and the denominator 36 is a factor for unit conversion.

The reverse salt flux from the draw solution to the feed solution, J_s (g/m^2/hr, gMH), was calculated from the following equation:

$$J_s = \frac{\Delta(C_f V_f)}{M\Delta t} \quad (7)$$

where C_f (g/L) and V_f (L) are the salt concentration and volume of the feed solution, respectively.

8.3 Results and Discussions

8.3.1 Morphology of the TFC-PES Hollow Fibers

Figure 8.8(a) shows the cross section of the membrane. The inner and outer diameters of the hollow fiber membrane are 575 μm and 1023 μm, respectively. The membrane consists of two layers of uniform macrovoids as presented in Figure 8.8(b), caused by water penetration from both inner and outer surfaces (Peng et al. 2008, Widjojo and Chung 2006). According to our previous studies, uniform macrovoids could not only enhance the mechanical strength of the membranes but also provide open channels for the transmembrane water flux (Cheng et al. 2016, Zhang et al. 2014). Figure 8.8(c) confirms the successful formation of the thin-film polyamide selective layer on the inner surface of hollow fibers. The selective layer had the nodular "ridge-and-valley" polyamide morphology. The outer surface of the TFC-PES hollow fiber, as shown in Figure 8.8(d), was fully porous owing to the

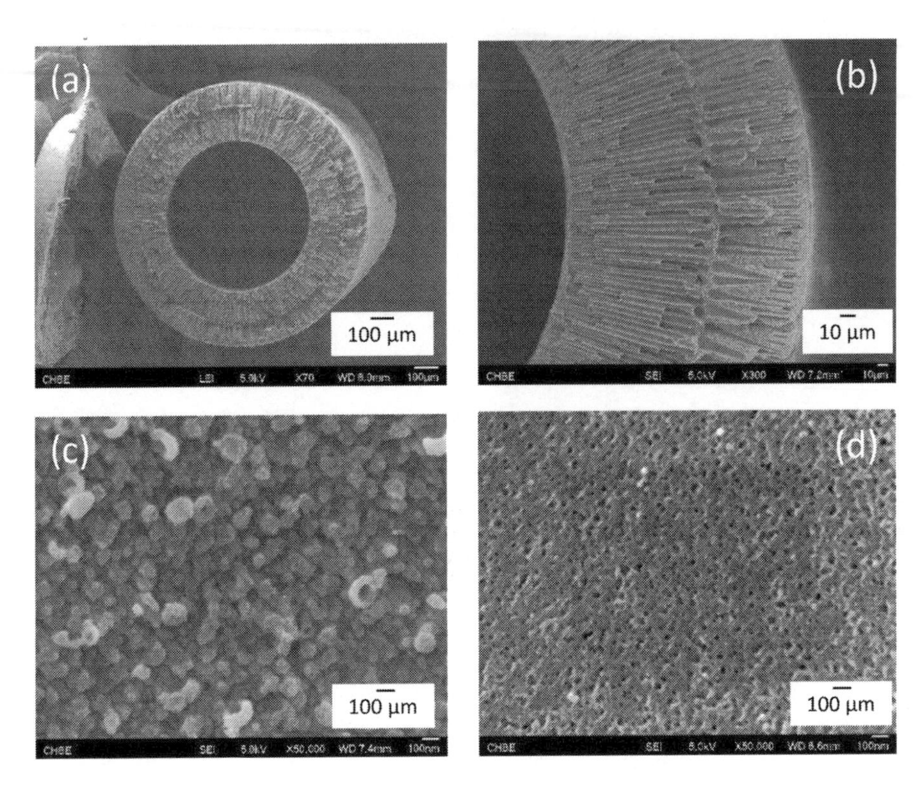

FIGURE 8.8
The morphologies of (a) cross section (X70), (b) enlarged cross section (X300) (c) inner surface and (X50000), (d) outer surface (X50000) of the TFC-PES hollow fiber.

delayed demixing induced by the NMP solvent in the outer channel of the dual layer spinneret (Zhang et al. 2014). Detailed characterizations of the TFC-PES hollow fiber membrane were presented in our previous study (Zhang et al. 2014). The morphologies in Figure 8.8 confirmed that the same TFC-PES membranes are duplicated on a semi-pilot-scale module.

8.3.2 Structured and Random Packing of Hollow Fibers

At a packing density of 30% with 180 pieces of hollow fibers, the hollow fibers could not disperse uniformly in the housing if they were randomly collected into a bundle. As shown in Figure 8.9(a), the hollow fibers were densely packed in the central and right areas. However, there was a gap between the bundle and the housing on the left side, which would cause channeling and bypassing in the shell. Bypassing would reduce the flow going into the bundle, worsen the concentration polarization, and compromise the PRO performance (Bao and Lipscomb 2002a, 2002b, 2002c,

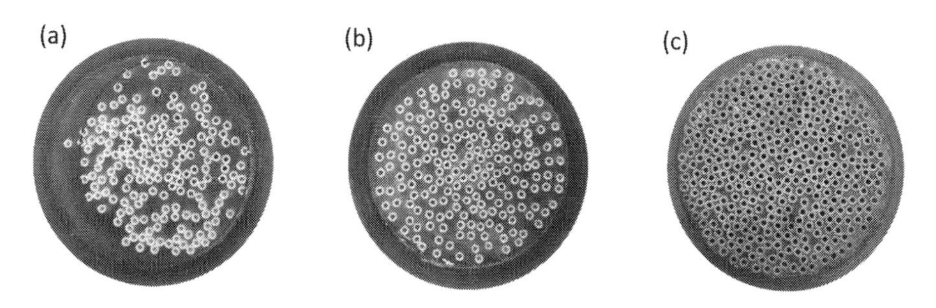

FIGURE 8.9
Cross sections of (a) a tubesheet of 30% random packing, (b) a tubesheet of 30% structured packing, and (c) a tubesheet of 50% random packing.

Merdaw et al. 2011a). These issues were overcome by arranging the fibers in an orderly structure as presented in Figure 8.9(b). The diameter of the bundle could be controlled by changing the spacing among hollow fibers (Niermeyer 1997, Philip et al. 1997) to fill up the housing and therefore minimize bypassing. Another way to improve the uniformity of packing was to increase the packing density of the hollow fiber bundle (Wang et al. 2003, Wu and Chen 2000). At an increased packing density of 50% with 300 pieces of hollow fibers, hollow fibers effectively filled up the housing and left little room for fibers to move around. Although some areas had a denser packing and some areas had a looser packing, the total uniformity of the bundle was significantly improved when comparing Figure 8.9(c) to 8.9(a). However, a higher packing density may introduce more dead zones in the shell that may enhance concentration polarization and also make the module more prone to fouling (Sivertsen et al. 2012, 2013). The optimal packing density of TFC-PES hollow fiber modules for PRO will be investigated under different flow and fouling conditions in the future (Chen et al. 2015, 2016, Xiao et al. 2012). In this study, hollow fiber modules with 30% structured packing and 50% random packing were chosen for further investigation.

8.3.3 Effects of MPD and TMC Circulation Durations on Interfacial Polymerization

To maximize module performance, the effects of MPD and TMC circulation durations during interfacial polymerization on permeability properties of the polyamide layer were studied. The MPD solution was circulated through the module lumen for 3 min, 5 min, or 7 min while the circulation time of the TMC solution was maintained at 5 min. Table 8.2 summarizes the A and B values of the resultant modules from each experimental condition. Both A and B values decreased with an increase in MPD circulation duration because MPD may penetrate deeper into the substrate at a longer circulation time, possibly leading to a thicker polyamide layer (Ghosh and Hoek 2009, Li et al. 2012).

TABLE 8.2
Pure water permeability (A) and salt permeability (B) at different circulation times of the MPD solution

MPD Time (min)	A (LMH/bar)	B (LMH)
3	2.5	0.9
5	2.3	0.6
7	1.9	0.4

To determine the optimal circulation time, the A and B values were input into a PRO model developed in our previous studies (Wan and Chung 2015, 2016). The simulated PRO water flux and reverse salt flux as a function of pressure were plotted in Figure 8.10. At 3 min, even though the A value was higher, the water flux was the lowest because of the significantly higher reverse salt flux and the resulting enhanced concentration polarization (Achilli et al. 2009, 2014, Merdaw et al. 2011b, She et al. 2012). When the MPD circulation time was increased from 3 min to 5 min, the water flux was increased by 4.2% despite a 4% decrease in the A value. This flux enhancement resulted from a 39.2% decrease in reverse salt flux and less concentration polarization in the PES substrates. A further increase in the MPD circulation time to 7 min only resulted in slight drops in both water flux and reverse salt flux. Therefore, the optimal duration to circulate the MPD solution was 5 min, which yielded the highest water flux and a moderately low reverse salt flux.

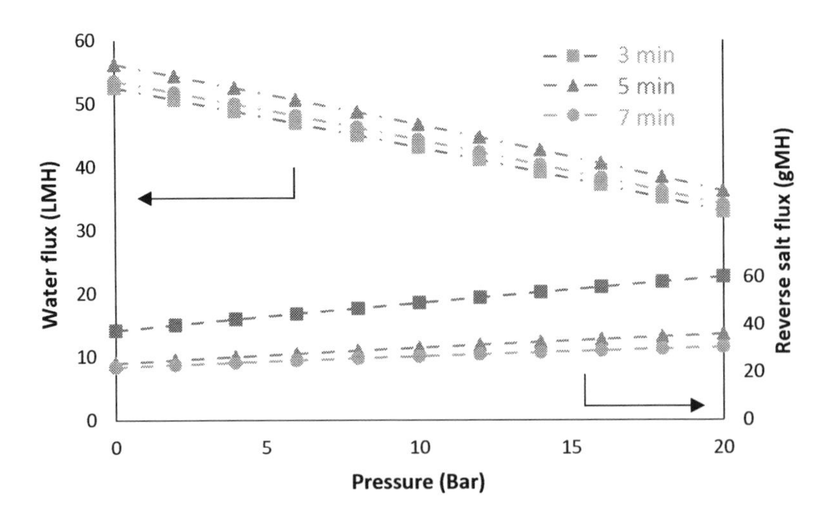

FIGURE 8.10
Modelled PRO performance at different circulation times of the MPD solution.

Similarly, the optimal circulation time of the TMC solution was determined by maintaining the MPD circulation time at 5 min. Table 8.3 tabulates the A and B values, while Figure 8.11 shows the calculated water flux and reverse salt flux as a function of pressure. They did not vary much when the TMC circulation time increased from 5 to 9 min. This was because after 5 min, the polyamide layer was thick enough to prevent contact of the reactants and stop the polyamide layer from growing (Liu et al. 2012). Therefore, the optimal TMC circulation time was 5 min.

8.3.4 PRO Performance of Hollow Fiber Modules

Three types of modules with packing densities of 2%, 30%, and 50% were prepared. They contained 3, 180, and 300 pieces of hollow fibers, respectively. Figure 8.12 displays the evolution of their water fluxes, power

TABLE 8.3

Pure water permeability (A) and salt permeability (B) at different circulation times of the TMC solution

TMC Time (min)	A (LMH/bar)	B (LMH)
5	2.3	0.5
7	2.3	0.5
9	2.2	0.5

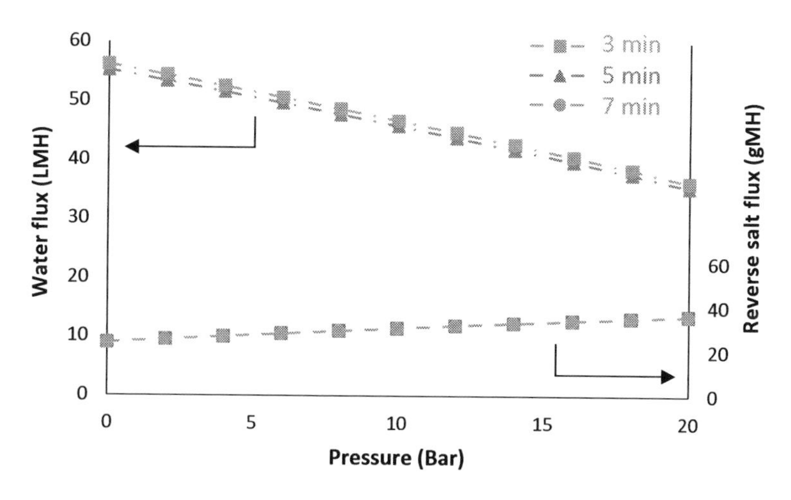

FIGURE 8.11
Modelled PRO performances at different circulation times of the TMC solution.

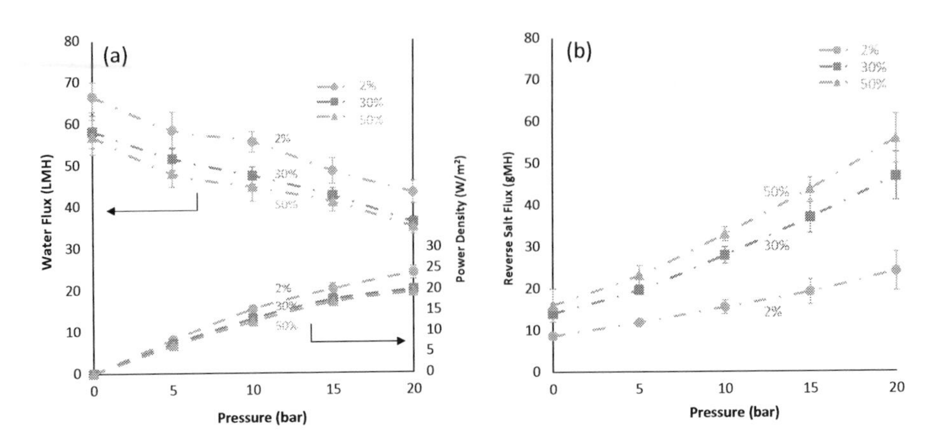

FIGURE 8.12
(a) Water flux and power density and (b) reverse salt flux of modules with 2%, 30%, and 50% packing densities under PRO tests.

densities, and reverse salt fluxes as a function of pressure from 0 bar to 20 bar. In all three cases, the water flux dropped at higher pressures as the driving force, defined as the difference between the osmotic pressure difference and hydraulic pressure difference, diminishes. The power densities increased continuously with increasing pressure and peak at 20 bar. Theoretically, the maximal power density could be achieved at one half of the osmotic pressure difference, which was 24 bar when the feed pair of 1M NaCl solution and DI water was used.

The 2% packing density mini-module containing three hollow fibers yielded the highest water flux of 66.5 LMH at 0 bar and 43.4 LMH at 20 bar. In contrast, the water flux was reduced by 13.5% on average when the packing density was increased by 30%, and it further declined by another 4.5% when the packing density was increased by 50%. As a consequence, the peak power density at 20 bar dropped to 20.0 W/m^2 and 19.4 W/m^2 at 30% and 50% packing, respectively. The reasons of reductions in water flux and power density were (1) less effective interfacial polymerization for more hollow fibers, (2) less efficient mass transfer on the shell side with an increase in packing density, and (3) loss of membrane area due to epoxy wicking.

First, it was much more challenging to ensure a uniform flow distribution of the MPD and TMC solutions during interfacial polymerization in large modules than small modules (Park and Chang 1986). This led to variations in the formation of polyamide selective layers. This phenomenon was also indicated by the significant higher reverse salt flux at higher packing densities as shown in Figure 8.12(b). The reverse salt flux increased with pressure and became more than doubled when the packing density was increased by 30% and 50%.

Moreover, the mass transport on the shell side became less effective at higher packing densities (Bao and Lipscomb 2002a, 2002b, 2002c). As a result, the salt accumulated in the substrates could not be effectively washed out, causing severer concentration polarization. The combined effects of the increased reverse salt flux and the less effective mass transfer on the shell side resulted in reductions of water flux and power density at higher packing densities.

Autopsy of the hollow fiber modules shows that wicking of epoxy reduced the effective membrane areas. After injecting epoxy into the potting mold, the epoxy climbed up to 8 mm on each end of the hollow fiber bundle due to the capillary actions. To measure the true effective length of the module, we dissembled the module and took out the hollow fiber bundle. The bundle was then cut off by 1 mm slices until the space among hollow fibers was no more filled with epoxy. The true effective length of the module was 140 ± 3 mm, which was 9.7% short than the apparent length of 155 mm used in the calculations of water flux and power density. The loss of effective membrane area is presented in Figure 8.13, contributing to reductions in water flux and power density in the semi-pilot-scale hollow fiber modules as shown in Figure 8.12(a). Therefore, the water fluxes and power densities of the 30% and 50% packing modules would be improved by 9.7% if wicking of epoxy could be eliminated. It is worth noting that wicking of epoxy can be less restrictive if the modules are enlarged to commercial scales, as the percentage loss of membrane areas becomes negligible.

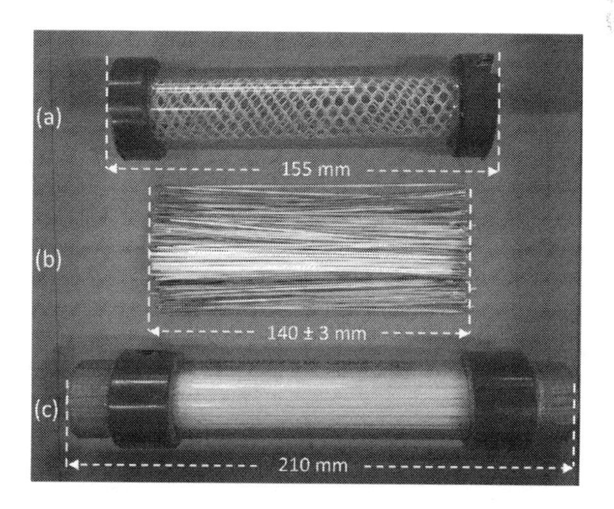

FIGURE 8.13
(a) Apparent effective length (used in calculations), (b) true effective length, and (c) module length.

8.3.5 Distribution of Broken Hollow Fibers

Some hollow fibers may be damaged during handling and testing of modules. The methods shown in Figure 8.5(a) and (b) were employed to find out which hollow fiber was broken and which portion along the hollow fiber was broken, respectively. Figure 8.14 shows the radial distribution of broken membranes in the hollow fiber bundle. The x-axis is the relative location of the bundle with 0% being the center and 100% being the outermost edge of the bundle. Ninety percent of the broken fibers were on the outer region of hollow fiber bundles. The broken membranes were probably caused by damages during handling and the damaged areas became the weak points of the membranes. They may burst under high pressure tests.

The axial distribution of the broken spots is shown in Figure 8.15. For module without the protective net, 70% of the damages occurred in the middle portion of the hollow fiber bundle. This may arise from the fact that the hollow fiber bundle was axially compressed and radially expanded during insertion into the housing. It not only formed an expanded belly region in the middle of the bundle but also was being pushed against the housing, and thus it got damaged due to abrasion. This also explained why most broken fibers were on the outermost region of the bundle. This problem was solved by wrapping the bundle with a protective net during insertion into the housing. The net needed to be porous and flexible to prevent introducing additional resistance to the flow or hurt the hollow fiber membranes. Figure 8.16(a) and (b) shows the modules with and without the protective net, respectively. This method could prevent the bundle from expanding and protecting the fibers from abrasion.

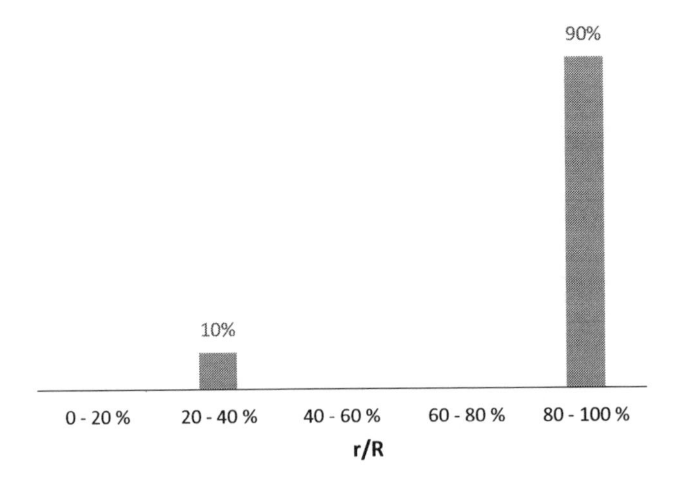

FIGURE 8.14
The radical distribution of broken hollow fibers in the module.

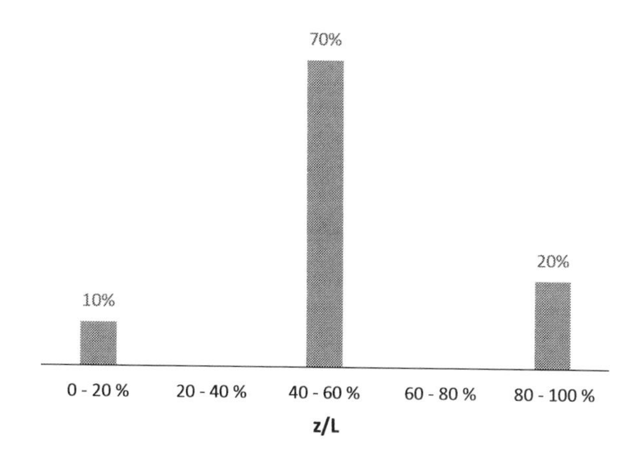

FIGURE 8.15
The axial distribution of leakages in the module.

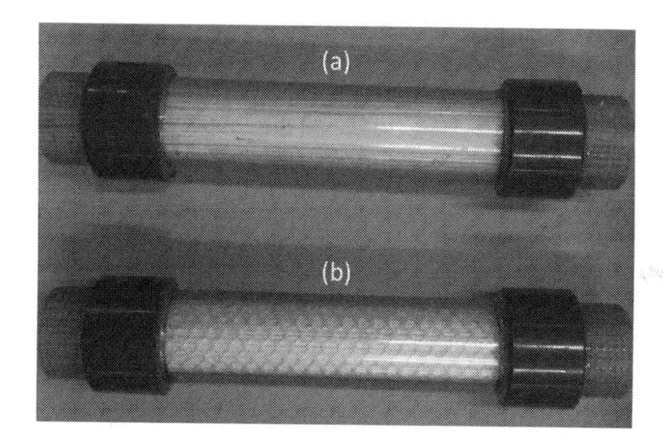

FIGURE 8.16
Hollow fiber modules at 50% packing (a) without and (b) with the protective net.

Another 30% of damages were observed on the two ends of the bundle. Because the two ends contacting with the tubesheets were partially coated with epoxy due to epoxy wicking, this portion of the fibers became brittle and fragile. They could be easily damaged by the impact force of water and fluctuating pressure and flowrate.

8.3.6 Modules Repaired by Epoxy-Coated Fillers and Injection of Epoxy

The damaged hollow fiber modules were then repaired by using epoxy-coated fillers and injection of epoxy. Figure 8.17(a) and (b) shows the

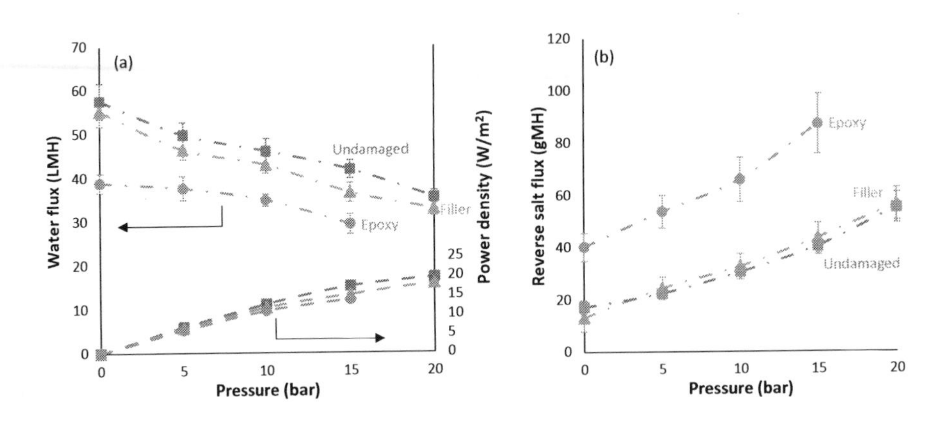

FIGURE 8.17
(a) Water flux and power density and (b) reverse salt flux of 30% packing hollow fiber modules (undamaged and damaged but being repaired by epoxy or fillers).

comparisons of their PRO performances with undamaged ones. Modules repaired with fillers had a peak power density of 18.4 W/m^2, 9% lower than 20.0 W/m^2 of the undamaged modules, while those repaired with injection of epoxy only had a peak power density of 14.0 W/m^2 at 15 bar. In summary, modules repaired with fillers had a 12.3% higher power density and a 55.6% lower reverse salt flux than those repaired with epoxy injection. This indicates that the injection of epoxy still left some small unsealed defects on the membranes. Moreover, the mechanical strength of the modules repaired with epoxy injection was weaker and could only sustain a pressure of 15 bar. Clearly, injection of epoxy was less effective because the injected epoxy may flow out from the fibers through the small broken area and contact other fibers. The fibers contacted with epoxy would become brittle and break under vibration and impact. Therefore, insertion of the epoxy-coated fillers was the more effective method. This repair method was also less tedious because it only required to identify the broken hollow fibers from one end of the modules instead of both ends. The fillers need to have good adhesion with epoxy and a slightly bigger dimension than the inner diameter of the hollow fibers to ensure an effective sealing.

8.4 Conclusions

For the first time, the protocols to fabricate semi-pilot-scale hollow fiber modules, conduct interfacial polymerization, and identify and repair leakages have been developed. TFC-PES hollow fiber modules with 30%

structured packing and 50% random packing were successfully prepared. When the packing of hollow fibers is loose, it is important to arrange the hollow fibers in an orderly structure to achieve a uniform distribution of hollow fibers. The optimal packing density of the TFC-PES hollow fiber modules would be determined by different flow, fouling conditions, and economic factors. The respective peak power densities of a 30% packing module and a 50% packing module at 20 bar using 1 M NaCl solution and DI water as feed pair were 20.0 W/m^2 and 19.4 W/m^2, respectively. Even though they were lower than the value of 24.1 W/m^2 when only three coupons of hollow fiber membranes were tested, the power densities achieved are the highest among scalable PRO modules globally under the same testing condition. Broken hollow fiber membranes can be effectively plugged with an epoxy-coated filler to restore the PRO performance. The performance of PRO modules can be further improved by (1) optimizing the interfacial polymerization conditions, (2) promoting the mixing on the shell side, and (3) preventing epoxy from wicking onto the hollow fiber membranes. Other engineering issues, such as enhancement of mechanical strength and durability of modules, need to be addressed before making commercial scale TFC-PES hollow fiber modules.

Acknowledgments

This research grant is supported by the Singapore National Research Foundation under its Environment and Water Res. Programme and administered by Public Utility Board (PUB), under the following projects (1) "Membrane Development for Osmotic Power Generation, Part 1. Materials Development and Membrane Fabrication" (1102-IRIS-11-01) and NUS grant no. R-279-000-381-279 (2) "Membrane Development for Osmotic Power Generation, Part 2, Module Fabrication and System Integration" (1102-IRIS-11-01) and NUS grant no. R-279-000-382-279, and (3) "Development of 8 inch Novel High Efficiency Pressure Retarded Osmosis (PRO) Membrane Modules towards Potential Pilot Testing and Field Validation" (USS-IF-2018-1) and NUS grant number of R-279-000-555-592. Special thanks would be given to Dr. Chakravarthy S. Gudipati and Mr. Junyou Zhang for their kind help.

References

Achilli, A., Cath, T.Y., Childress, A.E., 2009. Power generation with pressure retarded osmosis: an experimental and theoretical investigation. *J. Membr. Sci.* 343, 42–52.

Achilli, A., Prante, J.L., Hancock, N.T., Maxwell, E.B., Childress, A.E., 2014. Experimental results from RO-PRO: a next generation system for low-energy desalination. *Environ. Sci. Technol.* 48, 6437–6443.

Azran, A., Dagan, G., 2001. Method for potting or casting inorganic hollow fiber membranes into tube sheets. US Patent 6,270,714 B1.

Bao, L., Lipscomb, G.G., 2002a. Effect of random fiber packing on the performance of shell-fed hollow-fiber gas separation modules. *Desalination* 146, 243–248.

Bao, L., Lipscomb, G.G., 2002b. Mass transfer in axial flows through randomly packed fiber bundles with constant wall concentration. *J. Membr. Sci.* 204, 207–220.

Bao, L., Lipscomb, G.G., 2002c. Well-developed mass transfer in axial flows through randomly packed fiber bundles with constant wall flux. *Chem. Eng. Sci.* 57, 125–132.

Chen, S.C., Amy, G.L., Chung, T.S., 2016. Membrane fouling and anti-fouling strategies using RO retentate from a municipal water recycling plant as the feed for osmotic power generation. *Water Res.* 88, 144–155.

Chen, S.C., Wan, C.F., Chung, T.S., 2015. Enhanced fouling by inorganic and organic foulants on pressure retarded osmosis (PRO) hollow fiber membranes under high pressures. *J. Membr. Sci.* 479, 190–203.

Cheng, Z.L., Li, X., Liu, Y.D., Chung, T.S., 2016. Robust outer-selective thin-film composite polyethersulfone hollow fiber membranes with low reverse salt flux for renewable salinity-gradient energy generation. *J. Membr. Sci.* 506, 119–129.

Chong, K.L., Peng, N., Yin, H., Lipscomb, G.G., Chung, T.S., 2013. Food sustainability by designing and modelling a membrane controlled atmosphere storage system. *J. Food Eng.* 114, 361–374.

Chou, S., Wang, R., Fane, A.G., 2013. Robust and high performance hollow fiber membranes for energy harvesting from salinity gradients by pressure retarded osmosis. *J. Membr. Sci.* 448, 44–54.

Cui, Y., Liu, X.Y., Chung, T.S., 2014. Enhanced osmotic energy generation from salinity gradients by modifying thin film composite membranes. *Chem. Eng. J.* 242, 195–203.

Gerstandt, K., Peinemann, K.V., Skilhagen, S.E., Thorsen, T., Holt, T., 2008. Membrane processes in energy supply for an osmotic power plant. *Desalination* 224, 64–70.

Ghosh, A.K., Hoek, E.M.V., 2009. Impacts of support membrane structure and chemistry on polyamide–polysulfone interfacial composite membranes. *J. Membr. Sci.* 336, 140–148.

Han, G., Chung, T.S., 2014. Robust and high performance pressure retarded osmosis hollow fiber membranes for osmotic power generation. *AIChE J.* 60, 1107–1119.

Han, G., Zhang, S., Li, X., Chung, T.S., 2015. Progress in pressure retarded osmosis (PRO) membranes for osmotic power generation. *Prog. Polym. Sci.* 51, 1–27.

Korikov, A.P., Kosaraju, P.B., Sirkar, K.K., 2006. Interfacially polymerized hydrophilic microporous thin film composite membranes on porous polypropylene hollow fibers and flat films. *J. Membr. Sci.* 279, 588–600.

Li, D., Wang, R., Chung, T.S., 2004. Fabrication of lab-scale hollow fiber membrane modules with high packing density. *Sep. Sci. Technol.* 40, 15–30.

Li, X., Wang, K.Y., Helmer, B., Chung, T.S., 2012. Thin-film composite membranes and formation mechanism of thin-film layers on hydrophilic cellulose acetate propionate substrates for forward osmosis processes. *Ind. Eng. Chem. Res.* 51, 10039–10050.

Liu, J., Xu, Z., Li, X., Zhang, Y., Zhou, Y., Wang, Z., Wang, X., 2007. An improved process to prepare high separation performance PA/PVDF hollow fiber composite nanofiltration membranes. *Sep. Sci. Technol.* 58, 53–60.

Liu, M., Zheng, Y., Shuai, S., Zhou, Q., Yu, S., Gao, C., 2012. Thin-film composite membrane formed by interfacial polymerization of polyvinylamine (PVAm) and trimesoyl chloride (TMC) for nanofiltration. *Desalination* 288, 98–107.

Loeb, S., 1976. Production of energy from concentrated brines by pressure-retarded osmosis: I. Preliminary technical and economic correlations. *J. Membr. Sci.* 1, 49–63.

Loeb, S., Hessen, F.V., Shahaf, D., 1976. Production of energy from concentrated brines by pressure-retarded osmosis: II. Experimental results and projected energy costs. *J. Membr. Sci.* 1, 249–269.

Mat, N.C., Lou, Y., Lipscomb, G.G., 2014. Hollow fiber membrane modules. *Curr. Opin. Chem. Eng.* 4, 18–24.

Maxwell, J.M., Moore, W.E., Rego, R.D., 1967. Fluid separation process and apparatus. US Patent 3,339,341.

Mehta, G.D., Loeb, S., 1979. Performance of permasep B-9 and B-10 membranes in various osmotic refions and at high osmotic pressures. *J. Membr. Sci.* 4, 335–349.

Merdaw, A.A., Sharif, A.O., Derwish, G.A.W., 2011a. Mass transfer in pressure-driven membrane separation processes, Part I. *Chem. Eng. J.* 168, 215–228.

Merdaw, A.A., Sharif, A.O., Derwish, G.A.W., 2011b. Mass transfer in pressure-driven membrane separation processes, Part II. *Chem. Eng. J.* 168, 229–240.

Niermeyer, J.K., 1997. Thermoplastic hollow fiber membrane module and method of manufacture. US Patent 5,695,702.

Park, J.K., Chang, H.N., 1986. Flow distribution in the fiber lumen side of a hollow-fiber module. *AICHE J.* 32, 1937–1947.

Peng, N., Chung, T.S., Wang, K.Y., 2008. Macrovoid evolution and critical factors to form macrovoid-free hollow fiber membranes. *J. Membr. Sci.* 318, 363–372.

Peng, N., Widjojo, N., Sukitpaneenit, P., Teoh, M.M., Lipscomb, G.G., Chung, T.S., Lai, J.-Y., 2012. Evolution of polymeric hollow fibers as sustainable technologies: past, present, and future. *Prog. Polym. Sci.* 37, 1401–1424.

Philip, E.A., Schletz, J.C., Jensvold, J.A., Tegrotenhuis, W.E., Allen, W., Coan, F.L., Skala, K.L., Clark, D.O., Wait, V.W.J., 1997. Loom processing of hollow fiber membranes. US Patent 5,598,874.

Sarp, S., Li, Z., Saththasivam, J., 2016. Pressure retarded osmosis (PRO): past experiences, current developments, and future prospects. *Desalination* 389, 2–14.

She, Q., Jin, X., Tang, C.Y., 2012. Osmotic power production from salinity gradient resource by pressure retarded osmosis: effects of operating conditions and reverse solute diffusion. *J. Membr. Sci.* 401–402, 262–273.

Sivertsen, E., Holt, T., Thelin, W., Brekke, G., 2012. Modelling mass transport in hollow fibre membranes used for pressure retarded osmosis. *J. Membr. Sci.* 417–418, 69–79.

Sivertsen, E., Holt, T., Thelin, W., Brekke, G., 2013. Pressure retarded osmosis efficiency for different hollow fibre membrane module flow configurations. *Desalination* 312, 107–123.

Son, M., Park, H., Liu, L., Choi, H., Kim, J.H., Choi, H., 2016. Thin-film nanocomposite membrane with CNT positioning in support layer for energy harvesting from saline water. *Chem. Eng. J.* 284, 68–77.

Song, X., Liu, Z., Sun, D.D., 2013. Energy recovery from concentrated seawater brine by thin-film nanofiber composite pressure retarded osmosis membranes with high power density. *Energy Environ. Sci.* 6, 1199–1210.

Straub, A.P., Deshmukh, A., Elimelech, M., 2016. Pressure-retarded osmosis for power generation from salinity gradients: is it viable? *Energy Environ. Sci.* 9, 31–48.

Teoh, M.M., Bonyadi, S., Chung, T.S., 2008. Investigation of different hollow fiber module designs for flux enhancement in the membrane distillation process. *J. Membr. Sci.* 311, 371–379.

Veríssimo, S., Peinemann, K.V., Bordado, J., 2005. Thin-film composite hollow fiber membranes: an optimized manufacturing method. *J. Membr. Sci.* 264, 48–55.

Wan, C.F., Chung, T.S., 2015. Osmotic power generation by pressure retarded osmosis using seawater brine as the draw solution and wastewater brine as the feed. *J. Membr. Sci.* 479, 148–158.

Wan, C.F., Chung, T.S., 2016. Energy recovery by pressure retarded osmosis (PRO) in SWRO–PRO integrated processes. *Appl. Energy* 162, 687–698.

Wan, C.F., Yang, T., Lipscomb, G.G., Stookey, D.J., Chung, T.S., 2017. Design and fabrication of hollow fiber membrane modules. *J. Membr. Sci.* 538, 96–107.

Wang, W., Li, G., 2010. One-step fabrication of high selective hollow fiber nanofiltration membrane module. *Fiber. Polym.* 11, 1041–1048.

Wang, Y., Chen, F., Wang, Y., Luo, G., Dai, Y., 2003. Effect of random packing on shell-side flow and mass transfer in hollow fiber module described by normal distribution function. *J. Membr. Sci.* 216, 81–93.

Widjojo, N., Chung, T.S., 2006. Thickness and air gap dependence of macrovoids evolution in phase-inversion asymmetric hollow fiber membranes. *Ind. Eng. Chem. Res.* 45, 7618–7626.

Wu, J., Chen, V., 2000. Shell-side mass transfer performance of randomly packed hollow fiber modules. *J. Membr. Sci.* 172, 59–74.

Xiao, D., Li, W., Chou, S., Wang, R., Tang, C.Y., 2012. A modeling investigation on optimizing the design of forward osmosis hollow fiber modules. *J. Membr. Sci.* 392–393, 76–87.

Yang, X., Wang, R., Fane, A.G., 2011. Novel designs for improving the performance of hollow fiber membrane distillation modules. *J. Membr. Sci.* 384, 52–62.

Yip, N.Y., Tiraferri, A., Phillip, W.A., Schiffman, J.D., Hoover, L.A., Kim, Y.C., Elimelech, M., 2011. Thin-film composite pressure retarded osmosis membranes for sustainable power generation from salinity gradients. *Environ. Sci. Technol.* 45, 4360–4369.

Zhang, S., Chung, T.S., 2013. Minimizing the instant and accumulative effects of salt permeability to sustain ultrahigh osmotic power density. *Environ. Sci. Technol.* 47, 10085–10092.

Zhang, S., Sukitpaneenit, P., Chung, T.S., 2014. Design of robust hollow fiber membranes with high power density for osmotic energy production. *Chem. Eng. J.* 241, 457–465.

Zhou, Z., Lee, J.Y., Chung, T.S., 2014. Thin film composite forward-osmosis membranes with enhanced internal osmotic pressure for internal concentration polarization reduction. *Chem. Eng. J.* 249, 236–245.

9

Design and Optimization of Seawater Reverse Osmosis and Pressure Retarded Osmosis-Integrated Pilot Processes

Chun Feng Wan

Department of Chemical and Biomolecular Engineering
National University of Singapore
Singapore

CONTENTS

Nomenclature and Units

A pure water permeability (PWP) ($m^3/(m^2sPa)$)
B salt permeability ($m^3/(m^2s)$)
C concentration (g/m^3)
D diffusivity (m^2/s)
d_i inner diameter (m)
f fraction
i Van't Hoff factor
J_s reverse salt flux ($g/(m^2s)$)
J_w water flux ($m^3/(m^2s)$)
k mass transfer coefficient ($m^3/(m^2s)$)
M membrane area (m^2)
P pressure (Pa)
R universal gas constant (J/(mol K))
S structure parameter (m)
t residence time (s)
T temperature (K)
V flow rate (m^3/s)
W work (J)

Greek Symbols

α price of water ($/$m^3$ of freshwater)
β price of electricity ($/J)
γ price ratio
ε porosity
η efficiency
τ tortuosity
λ wall thickness (m)
π osmotic pressure (Pa)

Abbreviations

DF dilutive factor
PD power density
PR pressure ratio
Re Reynold number

Rec recovery
Sc Schmidt number
Sh Sherwood number
SEC specific energy consumption (J/m^3)
SER specific energy recovery (J/m^3)

Subscripts

a process in Figure 9.1 (a)
b process in Figure 9.1 (b)
E energy recovery device
D draw solution
F feed solution
P pump
PRO pressure retarded osmosis
ERD1 energy recovery device 1
ERD2 energy recovery device 2
SW seawater
SWRO seawater reverse osmosis

Superscripts

0 initial state
f final state
PRO pressure retarded osmosis
ERD energy recovery device

9.1 Introduction

Although the current seawater reverse osmosis (SWRO) process is highly energy-efficient, it still consumes a large amount of energy to pressurize and pump water (Greenlee et al. 2009, Liyanaarachchi et al. 2013, Peñate and García-Rodríguez 2012). Moreover, disposal of the SWRO concentrated brine may disturb the marine ecosystem (Greenlee et al. 2009, Liyanaarachchi et al. 2013, Peñate and García-Rodríguez 2012). Development of high-efficiency pumps and high-efficiency energy recovery devices (ERDs) has significantly reduced the specific energy consumption (SEC) of SWRO toward its thermodynamic minimum (Greenlee et al. 2009, Liyanaarachchi et al. 2013, Peñate and García-Rodríguez 2012). To further reduce the SEC of SWRO, pressure-retarded osmosis (PRO) can be integrated with SWRO and recover the osmotic energy from seawater brine to compensate the energy consumption of SWRO. Compared with the conventional seawater-freshwater PRO process, the SWRO–PRO integrated process offers a number of advantages (Chung et al. 2012): (1) a higher power density is possible due to the increased difference in osmotic pressure; (2) as the seawater brine is pretreated in the SWRO system, it will cause less fouling in the PRO system; (3) even though the pretreatment of the feed solution to the PRO system is still required, the overall pretreatment units can be significantly downsized if the brine is diluted to the seawater level in PRO and recycled to SWRO. More importantly, to take full advantages of the synergic SWRO–PRO process, a strategic collocation of the SWRO plant and low salinity water sources is required during urban planning (Sim et al. 2013).

In this chapter, conceptual and detailed process designs of a closed-loop and an open-loop SWRO–PRO integrated processes are presented for simulation and optimization. The closed-loop SWRO–PRO may significantly reduce the seawater pretreatment cost and recover more osmotic energy from the brine, while the open-loop SWRO–PRO may offer greater flexibility for process design. In addition, mathematical models that describe both the transport phenomena on a module level and the energy flow on a system level are developed to evaluate the performances of SWRO–PRO processes. These two models are closely related by the flow rate and pressure of the seawater/brine stream, which can be used to optimize the entire SWRO–PRO system.

9.2 Theory

9.2.1 Process Descriptions

Figure 9.1(a) presents a general process of the integrated SWRO+ERD+PRO. Since it usually takes a huge membrane area to transport the same amount of water in PRO as in SWRO, the permeate volume in PRO is usually less than that in SWRO (i.e., $\Delta V_{PRO} < \Delta V_{SWRO}$) in the actual application. Therefore, the pretreated seawater under this condition is split into two streams before ERD2. The fraction of pretreated seawater pressurized by ERD2 is assumed to be f, and then the fraction of the pretreated seawater directly pressurized to the operating pressure of SWRO by high-pressure pump HP1 will be $(1 - f)$. In ERD2, the pretreated seawater is pressurized to a medium-high pressure by the diluted draw solution from PRO. Subsequently, the

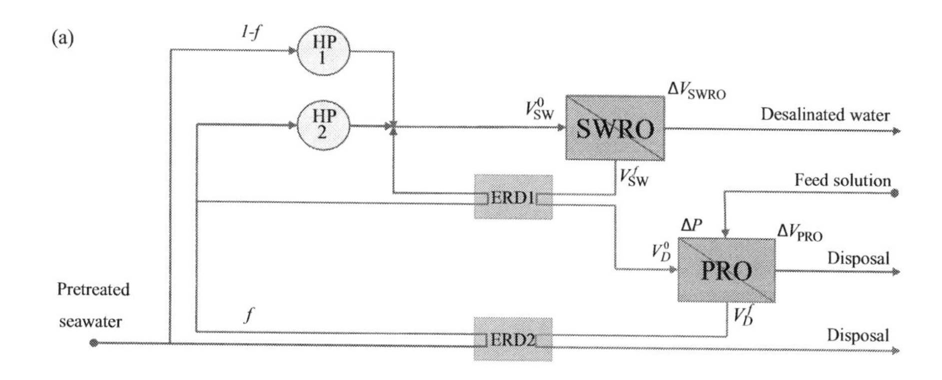

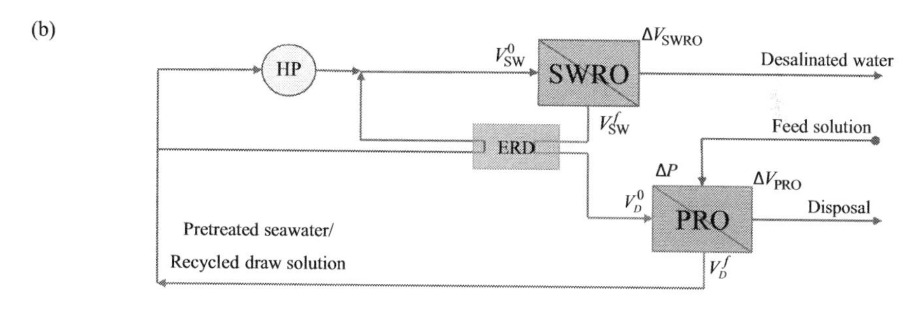

FIGURE 9.1
Schematic drawings of the integrated SWRO+ERD+PRO processes (a) $\Delta V_{PRO} < \Delta V_{SWRO}$, (b) $\Delta V_{PRO} = \Delta V_{SWRO}$.

effluent from ERD2 is split into two streams before ERD1: one is pressurized by ERD1 using the high-pressure brine from SWRO and the other is pressurized by a high-pressure pump HP2 to the operating pressure of the SWRO system. All three feed streams are then combined and enter the SWRO system, where pure water is produced. The high-pressure brine from the SWRO passes part of its energy to seawater via ERD1, and its pressure is reduced to a medium-high value. The medium-high pressure brine acts as the draw solution in the PRO system. In the PRO system, the concentrated brine gets diluted. As water permeates through the PRO membrane, the chemical potential is converted to the hydraulic pressure of the draw solution. Therefore, the draw solution is able to maintain its medium-high pressure at an increased volumetric flow rate. The diluted draw solution is then utilized in ERD2 to pressurize the pretreated seawater. When $\Delta V_{PRO} < \Delta V_{SWRO}$, the diluted draw solution from PRO has a salinity that is higher than the pretreated seawater and it is not economical to recycle it as the feed solution to SWRO, because this will require a higher energy input to achieve the same recovery in SWRO and also result in salt accumulation in the seawater feed. Hence, the diluted draw solution is discharged after pressure exchanging with the pretreated seawater feed. It is worth mentioning that the ERD does not have an efficiency of 100% and therefore a boost pump is always needed to make up the pressure loss, which is not shown in the process flow diagrams.

In the special case of $f = 1$, the general SWRO+ERD+PRO process can be simplified as presented in Figure 9.1(b). In this process, the permeate volume in PRO is the same as that in SWRO, i.e., $\Delta V_{PRO}=\Delta V_{SWRO}$ and $V^0{}_{SW} = V^f{}_D$. Therefore, the diluted draw solution from PRO can be directly used as the feed solution to SWRO. This process has several additional advantages: (1) recycling the draw solution as the feed to SWRO can tremendously cut down the pretreatment cost of SWRO, (2) it eliminates the need of an additional HP1 and ERD2 as shown in Figure 9.1(a), and (3) there is no pressure loss due to the less than 100% efficiency of ERD2. As presented, the pretreated seawater or the diluted draw solution with a medium-high pressure splits into two streams before ERD1: one is pressurized by ERD1 using the high-pressure brine from the SWRO system and the other is pressurized by an HP to the operating pressure of the SWRO system. The two streams are then combined, enter the SWRO system, and go through the same cycle as per the process shown in Figure 9.1(a). Though the recycled draw solution can be any type of draw solution (Efraty 2013, Han et al. 2014), but pretreatment seawater is preferred because it requires less modification of the existing and heavily optimized SWRO system.

9.2.2 Material Balance

In both processes, the seawater or its concentrated brine runs through the entire system. The recovery of SWRO and dilution factor of PRO can be conveniently defined as follows to facilitate further investigations of the processes. The recovery of the SWRO system, Rec, is defined as the flow rate ratio of the permeate water over the total seawater feed.

$$\text{Rec} = \frac{V^0_{\text{SW}} - V^f_{\text{SW}}}{V^0_{\text{SW}}} = \frac{\Delta V_{\text{SWRO}}}{V^0_{\text{SW}}} \tag{1}$$

where V^0_{SW} is the flow rate at which the feed seawater enters the SWRO system, V^f_{SW} is the flow rate at which the seawater brine exits the SWRO system and ΔV_{SWRO} is the flow rate of the permeate from the SWRO membrane.

The dilutive factor of the draw solution in the PRO system, DF, is defined as follows:

$$\text{DF} = \frac{V^0_D}{V^f_D} = 1 - \frac{\Delta V_{\text{PRO}}}{V^f_D} \tag{2}$$

where V_D^0 is the flow rate at which the draw solution (i.e., the concentrated seawater brine) enters the PRO system. Numerically, V_D^0 is equal to V_{SW}^f. V_D^f is the flow rate at which the diluted draw solution exits the PRO system and ΔV_{PRO} is the flow rate at which water permeates through the PRO membrane. The DF indicates the fraction of the concentrated brine in the diluted brine after PRO. A DF approaching 0% indicates that the draw solution is infinitely diluted, while a DF approaching 100% indicates that the draw solution is not diluted.

As previously defined, f is the fraction of the total seawater feed $V^0{}_{SW}$ that is pressurized by ERD2. It is essential to keep the flow rates of the diluted draw solution and the feed seawater entering ERD2 equal in order to maximize the efficiency of ERD1 (Hauge 1995). Therefore, the flow rate of the diluted draw solution is also a fraction of the total seawater feed.

$$f = \frac{V_D^f}{V_{SW}^0} \leq 1 \tag{3}$$

The following two equations can be obtained from eqns. 1 and 2 by carrying out mass balance over the entire system.

$$\Delta V_{PRO} = \frac{\text{Rec} + f - 1}{\text{Rec}} \Delta V_{SWRO} \tag{4}$$

$$\text{DF} = \frac{1 - \text{Rec}}{f} \tag{5}$$

9.2.3 Power Generation by PRO

The power density, PD, of the PRO membrane is proportional to the hydraulic pressure difference and the water flux as follows:

$$\text{PD} = J_w \Delta P \tag{6}$$

The mass transfer across the PRO membranes has been studied in detail (Achilli et al. 2009, Lee et al. 1981, Sivertsen et al. 2012, 2013, Zhang and Chung 2013). For an ideal semipermeable membrane, the water permeation flux, J_w, is proportional to the driving force across the membrane and the pure water permeability, A, of the membrane, while the reverse salt flux is zero as follows:

$$J_w = A(\Delta\pi - \Delta P) \tag{7}$$

$$J_s = 0 \tag{8}$$

Substituting eqn. 7 to eqn. 6, the power density can be expressed as a quadratic equation of ΔP, with a maximum value of $A\Delta\pi^2/4$ at $\Delta P = \Delta\pi/2$.

However, in reality, the effective osmotic pressure difference across the selective layer is less than the bulk osmotic pressure difference, due to the concentrative internal concentration polarization (cICP) and concentrative external concentration polarization on the feed solution side, and dilutive external concentration polarization (dECP) on the draw solution side and the reverse salt flux (J_s). Achilli et al. developed a model that described the effects of cICP, dECP, and J_s for flat-sheet membranes, which can be extended to hollow fiber membranes as follows (Achilli et al. 2009, Lee et al. 1981):

$$\Delta\pi_{\text{eff}} = \frac{\pi_D \exp(-\frac{J_w}{k}) - \pi_F \exp(\frac{J_w S}{D})}{1 + \frac{B}{J_w}[\exp(\frac{J_w S}{D}) - 1]} \tag{9}$$

$$J_w = A(\Delta\pi_{\text{eff}} - \Delta P) \tag{10}$$

$$J_s = \frac{B}{iRT}\left(\frac{J_w}{A} + \Delta P\right) \tag{11}$$

where $\Delta\pi_{\text{eff}}$ is the effective osmotic pressure difference across the selective layer, B is the salt permeability, and k is the mass transfer coefficient of the draw solution.

$$k = \frac{ShD}{d_i} \tag{12}$$

where D is the solute diffusivity, d_i is the inner diameter of the hollow fiber, Sh is the Sherwood number, and S is the structural parameter.

$$S = \frac{\tau\lambda}{\varepsilon} \tag{13}$$

where τ, ε, and λ are the tortuosity, porosity, and wall thickness of the hollow fiber membranes, respectively.

9.2.4 PRO Modeling

In a PRO system, the amount of energy that can be harvested is proportional to the mixing volume. Therefore, dilution of the draw solution is inevitable. On one hand, more energy can be harvested as dilution goes on; on the other hand, dilution causes reductions of the effective driving force and the average power density. These two effects result in an increased demand for membrane area in order to achieve a higher dilution of the draw solution and a higher power generation. Eventually, the total amount of harvestable energy and the average power density of a PRO system will be limited by the membrane area available (Banchik et al. 2014).

As shown in Figure 9.2, the feed and draw solutions flow countercurrently in the hollow fiber module. The draw solution is on the lumen side while the feed solution is on the shell side of the module. The flow direction of the draw solution is arbitrarily chosen as the positive direction.

During the design of a PRO system, it is important to find out the membrane area required to achieve a certain dilution of the draw solution or recovery of the feed solution and the corresponding optimal operating conditions. The dilutive factor of the draw solution, DF, and the recovery of the feed solution, $\mathrm{Rec}_{\mathrm{PRO}}$, are the two sides of the same coin and can be interconverted through the following equation:

$$\mathrm{Rec}_{\mathrm{PRO}} = \frac{V_D^0}{V_F^0}\left(\frac{1}{\mathrm{DF}} - 1\right) \tag{14}$$

Since seawater and its brine run through the entire SWRO+ERD+PRO process, the analyses will be more systematic by carrying out material balance of the seawater/brine streams. A one-dimensional mass transfer model with respect to a differential membrane area (M) can be developed as follows:

$$d(\rho_D V_D) = \rho_W J_W dM \tag{15}$$

$$d(\rho_F V_F) = \rho_W J_W dM \tag{16}$$

$$d(C_D V_D) = -J_S dM \tag{17}$$

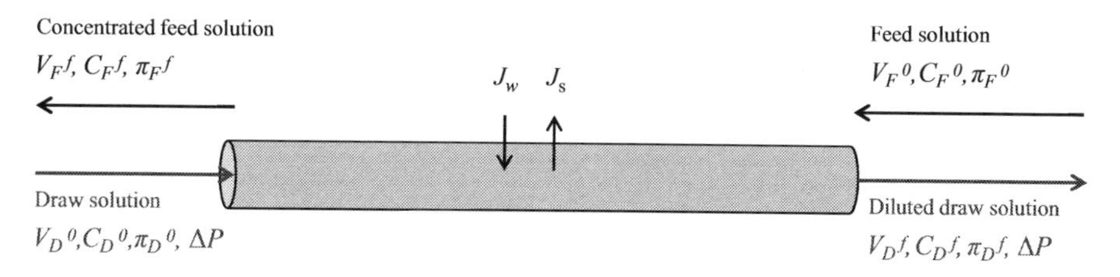

FIGURE 9.2
Schematic drawing of the PRO process in a single piece of hollow fiber.

$$d(C_F V_F) = -J_S dM \tag{18}$$

where ρ_D, ρ_F, and ρ_W are the densities of the draw solution, feed solution, and pure water, respectively. ρ_D and ρ_F vary as the concentrations of the draw solution and feed solution change. Since both the feed solution and the draw solution are diluted aqueous solutions, ρ_D and ρ_F can be estimated from the following equations.

$$\rho_D = \rho_W + C_D \tag{19}$$

$$\rho_F = \rho_W + C_F \tag{20}$$

Up to a salt concentration of 400 g/L, which is far beyond the salt concentration of the brine from conventional SWRO, the above equations can accurately predict the densities with less than 5% deviation from the correlations provided by Sharqawy et al. (2012). It is worth noting that the units of C_D and C_F are g/L in all the equations.

Equations 15–18 can be rearranged as functions of dV_D and integrated with the following boundary conditions to analyze the performance of PRO hollow fibers as dilution occurs. At the entrance of the draw solution where no dilution occurs (i.e., $V_D = V_D^0$), $M = 0$ and $C_D = C_D^0$. At the exit of the draw solution where the draw solution is diluted (i.e., $V_D = V_D^f = \frac{V_D^0}{\text{DF}}$), $V_F = V_F^0$ and $C_F = C_F^0$. The integration is carried out in this special way to evaluate the membrane area needed, the total amount of energy harvested, and the average power density obtained at a certain dilution of the draw solution.

The average water flux of the hollow fiber module, $\bar{J}_w$, can be defined as follows:

$$\bar{J}_w = \frac{V_D^f - V_D^0}{M} = \frac{V_F^0 - V_F^f}{M} \tag{21}$$

The pressure ratio, PR, is defined as the ratio of the hydraulic pressure difference over the maximum osmotic pressure difference between the draw and feed solutions in the PRO system.

$$\text{PR} = \frac{\Delta P}{\Delta \pi_{\text{max}}} \tag{22}$$

where $\Delta\pi_{\text{max}}$ is the maximum osmotic pressure difference between the draw and feed solutions at their respective inlets. If freshwater or solution of very low salinity is used as the feed solution, $\Delta\pi_{\text{max}}$ can be estimated as follows:

$$\Delta\pi_{\max} = \pi_D^0 - \pi_F^0 \approx \pi_D^0 = \frac{\pi_{SW}}{1 - \text{Rec}} \quad (23)$$

where π_{SW} is the osmotic pressure of the seawater feed to the SWRO system. $\frac{\pi_{SW}}{1-\text{Rec}}$ is the osmotic pressure of the draw solution (i.e., the concentration brine), which is also the minimum pressure that has to be overcome to achieve a recovery of Rec in SWRO.

The average power density of the module, $\overline{\text{PD}}$, at a constant ΔP is defined as:

$$\overline{\text{PD}} = \bar{J}_w \Delta P \quad (24)$$

The residence time of the draw solution in the hollow fiber module, t, is defined as the ratio of the volume enclosed by the hollow fibers over the flow rate of the draw solution.

$$t = \frac{M d_i}{4 V_D^0} \quad (25)$$

9.2.5 Energy Consumption of a SWRO+ERD+PRO System

The SEC of a stand-alone SWRO, that is, SEC_{SWRO} depends on the recovery ratio. A minimum hydraulic pressure equal to the osmotic pressure of the concentrated brine has to be applied on SWRO to achieve the desired recovery. In reality, the operating pressure is often 10–20% higher than this minimum value to compensate the pressure loss and enhance the SWRO water flux (Elimelech and Phillip 2011). The SEC to produce 1 m^3 of freshwater in the absence of an ERD is given by the following equation (Zhu et al. 2009a, 2009b):

$$\text{SEC}_{\text{SWRO}} = \frac{\pi_{SW}}{\eta_P \text{Rec}(1 - \text{Rec})} \quad (26)$$

where η_P is the efficiency of the HP. The SEC_{SWRO} has been significantly reduced due to the invention of ERD. The SEC of SWRO+ERD, that is, $\text{SEC}_{\text{SWRO}}^{\text{ERD}}$, can be calculated from the equation below (Zhu et al. 2009a, 2009b).

$$\text{SEC}_{\text{SWRO}}^{\text{ERD}} = \frac{[1 - \eta_E(1 - \text{Rec})]\pi_{SW}}{\eta_P \text{Rec}(1 - \text{Rec})} \quad (27)$$

where η_E is the efficiency of the ERD.

In the SWRO+ERD+PRO processes, more energy can be further recovered in the PRO system. Specific energy recovery is defined as the amount of energy that can be recovered from the brine for every cubic meter of freshwater produced in SWRO. If a process as shown in Figure 9.1 (a) is employed, then eqns. 3–5 and 21–24 can be combined as follows:

$$\mathrm{SER}_{\mathrm{PRO},a} = \frac{A_m\overline{PD}}{V_D^0}\frac{(1-\mathrm{Rec})}{\mathrm{Rec}} = \frac{\mathrm{PR}(\mathrm{Rec}+f-1)}{\mathrm{Rec}(1-\mathrm{Rec})}\pi_{\mathrm{SW}} \tag{28}$$

The factor of $\frac{(1-\mathrm{Rec})}{\mathrm{Rec}}$ in eqn. 28 is to convert the energy production of the PRO system from per cubic meter of the concentration brine basis to per cubic meter of the desalinated water basis.

It should be noted that not only the permeate flow from the PRO system, but also the draw solutions are used for energy recovery in ERD2. Thus, the amount of energy recovered in ERD2 is as follows:

$$\mathrm{SER}_{\mathrm{ERD2},a} = \frac{f\mathrm{PR}}{\mathrm{Rec}\,(1-\mathrm{Rec})}\pi_{\mathrm{SW}}\eta_E \tag{29}$$

The amount of energy recovered in ERD1 can be calculated from the following equation:

$$\mathrm{SER}_{\mathrm{ERD1},a} = \frac{(1-\mathrm{PR})}{\mathrm{Rec}}\pi_{\mathrm{SW}}\eta_E \tag{30}$$

The SEC for the integrated SWRO+ERD+PRO process in Figure 9.1 (a), SEC_a, is the difference between the energy consumed in the SWRO and those recovered in ERD1 and ERD2.

$$\begin{aligned}\mathrm{SEC}_a\eta_P &= \mathrm{SEC}_{\mathrm{SWRO}}\eta_P - \mathrm{SER}_{\mathrm{ERD1},a} - \mathrm{SER}_{\mathrm{ERD2},a} \\ &= \frac{[1-(1-\mathrm{Rec})(1-\mathrm{PR})\eta_E - f\mathrm{PR}\eta_E]\pi_{\mathrm{SW}}}{\eta_P\mathrm{Rec}(1-\mathrm{Rec})}\end{aligned} \tag{31}$$

Applying the same analysis for the process in Figure 9.1 (b), SEC_b, can be calculated from eqn. 32 by setting $f = 1$ and $\eta_{E2} = 100\%$ for ERD2 in eqn. 31, because the diluted draw solution is directly recycled to the SWRO.

$$\mathrm{SEC}_b = \frac{(1-\mathrm{PR})[1-(1-\mathrm{Rec})\eta_E]\pi_{\mathrm{SW}}}{\eta_P\mathrm{Rec}(1-\mathrm{Rec})} \tag{32}$$

As shown in eqns. 31 and 32, SEC of SWRO+ERD+PRO is a sole function of PR for given Rec and *f*. Therefore, finding the optimum PR is the key to determine the SEC of a SWRO+ERD+PRO system. The SECs of different SWRO-involved processes are summarized in Figure 9.3.

9.2.6 Operating Profit of a SWRO+ERD+PRO System

The operating profit of both the closed-loop and open-loop SWRO+ERD+PRO, OP ($/m^3 of seawater), can be defined as the profit generated for every cubic meter of seawater that enters the SWRO+ERD+PRO system.

$$\mathrm{OP} = \alpha \mathrm{Rec} - \beta \mathrm{SEC} \tag{33}$$

where α is the water price ($/m^3 of freshwater) and β is the electricity price ($/J).

The prices of water and electricity can be used as the weighting factors to determine the optimal operating condition of the SWRO+ERD+PRO system. Noticing that the unit of water price is in $/m^3 while the electricity price is in $/J, the following factors are defined to compare the two prices on the same basis.

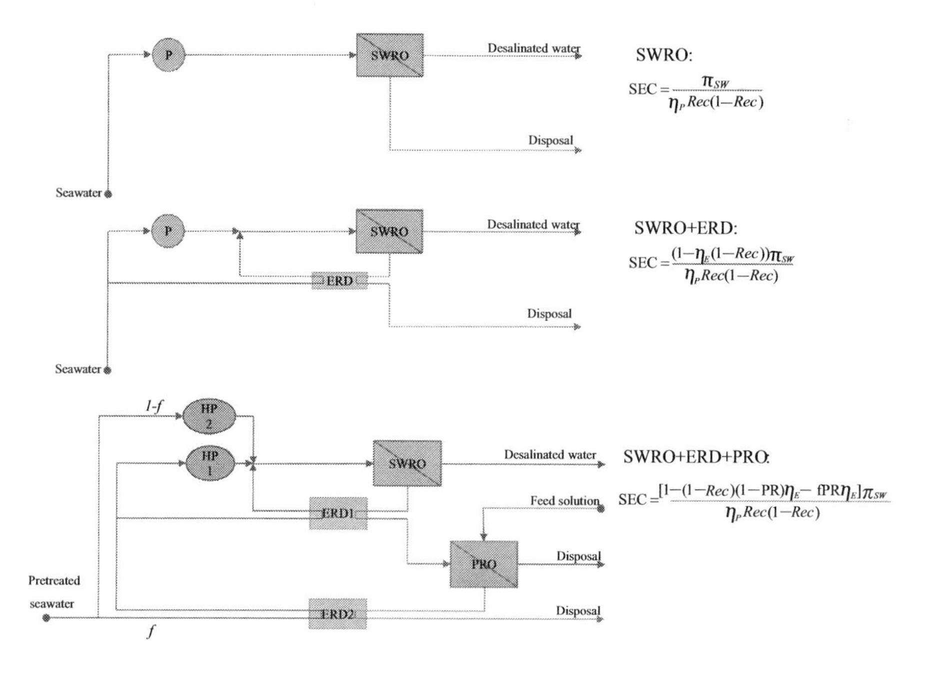

FIGURE 9.3
Summary of SECs on per cubic meter of the desalinated water basis in various SWRO-involved systems.

$$\gamma = \frac{\beta\pi_{\mathrm{SW}}}{\alpha} \tag{34}$$

where $\beta\pi_{sw}$ is the modified price of electricity ($/m^3 of seawater, 1 J/Pa = 1 m^3) and γ is the price ratio. Therefore, the dimensionless operating profit, ρ, can be expressed as follows:

$$\rho = \frac{\mathrm{OP}}{\alpha} = (\mathrm{Rec} - \gamma\mathrm{SEC}) \tag{35}$$

The operating profit of a conventional SWRO+ERD process can be calculated from the following equation.

$$\rho_{\mathrm{SWRO}} = \mathrm{Rec} - \frac{\gamma[1 - \eta_E(1 - \mathrm{Rec})]}{\eta_P(1 - \mathrm{Rec})} \tag{36}$$

In an ideal case where the HP and the ERD are 100% efficient, eqn. 36 can be reduced to the following equation.

$$\rho_{\mathrm{SWRO}} = \mathrm{Rec} - \frac{\gamma\mathrm{Rec}}{1 - \mathrm{Rec}} \tag{37}$$

If $\gamma < 1$, the above equation has the maximal value at the optimal recovery.

$$\rho_{\mathrm{SWRO,max}} = (1 - \sqrt{\gamma})^2 \tag{38}$$

$$R_{\mathrm{SWRO,opt}} = 1 - \sqrt{\gamma} \tag{39}$$

Similarly, in the SWRO+ERD+PRO process, the operating profit can be calculated from the following equation.

$$\rho_{\mathrm{SWRO-PRO}} = \mathrm{Rec} - \frac{\gamma[1 - \eta_E(1 - \mathrm{Rec} + \mathrm{PR} \times \mathrm{Rec})]}{\eta_P(1 - \mathrm{Rec})} \tag{40}$$

If the HP and the pressure exchanger are 100% efficient, eqn. 40 can be reduced to the following equation.

$$\rho_{\mathrm{SWRO-PRO}} = \mathrm{Rec} - \frac{\gamma(\mathrm{Rec} - \mathrm{PR} \times \mathrm{Rec})}{1 - \mathrm{Rec}} \tag{41}$$

For a fixed value of PR, the above equation has a maximal value at the optimal recovery.

$$\rho_{\text{SWRO-PRO,max}} = (1 - \sqrt{\gamma(1 - \text{PR})})^2 \quad (42)$$

$$R_{\text{SWRO-PRO,opt}} = 1 - \sqrt{\gamma(1 - \text{PR})} \quad (43)$$

By comparing eqns. 42 and 43 with eqns. 38 and 39, it was found that the advantages of the SWRO+ERD+PRO process over the conventional SWRO+ERD process are that SWRO+ERD+PRO can (1) be operated at a higher recovery and (2) achieve a higher operating profit for every cubic meter of seawater entering the system.

9.2.7 Optimization of the SWRO+ERD+PRO Process

The strategy to optimize the SWRO+ERD+PRO process is (1) to maximize the operating profit of the integrated process and (2) to simultaneously maintain the highest possible power density of PRO at the respective recovery of SWRO.

9.3 Results and Discussions

9.3.1 Simulation Inputs

The simulations in this study are based on the state-of-the-art PES-TFC hollow fiber membranes developed in our previous work, where a power density of 27 W/m^2 at 20 bar has been obtained in lab-scale tests using 1 M NaCl solution as the draw solution and deionized water as the feed solution (Wan and Chung 2015, Zhang et al. 2014). The characteristics of the membranes and the hypothetical module parameters are summarized in Table 9.1. The membranes in the modules are assumed to have the same characterizations, which in reality may vary.

Two scenarios are investigated in this study. In the first scenario, the recovery of the SWRO is 25%, which is a typical recovery of a one-stage SWRO system. In the second scenario, the recovery of the SWRO is 50%, which is a typical recovery of a multistage SWRO system. In each scenario, two types of feed solutions – freshwater (0 M) and wastewater (0.01 M) – are used in PRO. Wastewater, such as the typical secondary effluent from a wastewater plant or even the reject from a wastewater reverse osmosis plant, can be effectively used as the feed solution for PRO, due to its low total dissolved solids and low total organic carbon. In both the cases, the salt concentration of the seawater is 35 g/L and the temperature of the solutions is 25 °C.

TABLE 9.1

Summary of membrane characterizations and module design parameters

Membrane Characterizations		Module Parameters	
A (LMH/bar)	3.5	Diameter (cm)	25.4
B (LMH)	0.3	Packing density (%)	33
S (μm)	450	Number of hollow fibers	20,000
Outer diameter (μm)	1025	Total draw flow rate (L/min)	20–200
Inner diameter (μm)	575	Total feed flow rate (L/min)	eqn. (44)

Specifically, the results presented here are based on the total flow rate of the draw solution in the range of 20–200 L/min in each module. In a PRO process, it is advisable that the maximum recovery of the low salinity source should not exceed 80% in order not to aggravate fouling (Thorsen and Holt 2009). Therefore, the flow rate of the feed solution used in the simulations can be calculated as follows:

$$V_F^0 = \frac{V_D^0}{80\%}\left(\frac{1}{\mathrm{DF}} - 1\right) \tag{44}$$

9.3.2 Average Power Density of PRO Modules in the SWRO+ERD+PRO Process

The average power densities of the module at different conditions are presented in Figure 9.4 (a)–(d). In all of the four cases, the average power densities are functions of both the pressure ratio and the dilutive factor. Since the dilutive factor is only defined for the PRO system, the figures can be used extensively for any process that involves a PRO system with the same feed and draw solutions.

The diagonal line in each figure presents the operating condition that the hydraulic pressure difference equals the osmotic pressure difference at the inlet of the draw solution, above which the water flux and power density become negative. Therefore, it is not practical to operate the PRO system beyond this line, and the average power density is set to 0 by default. At a given dilutive factor, the average power density is a bell-shaped function of the pressure ratio. The average power density is maximized at the optimal pressure ratio. Either overpressurizing or underpressurizing from the optimal point will result in a reduced average power density. At a given pressure ratio, the average power density increases with the dilutive factor and reaches a maximum when there is minimal dilution (i.e., DF → 1).

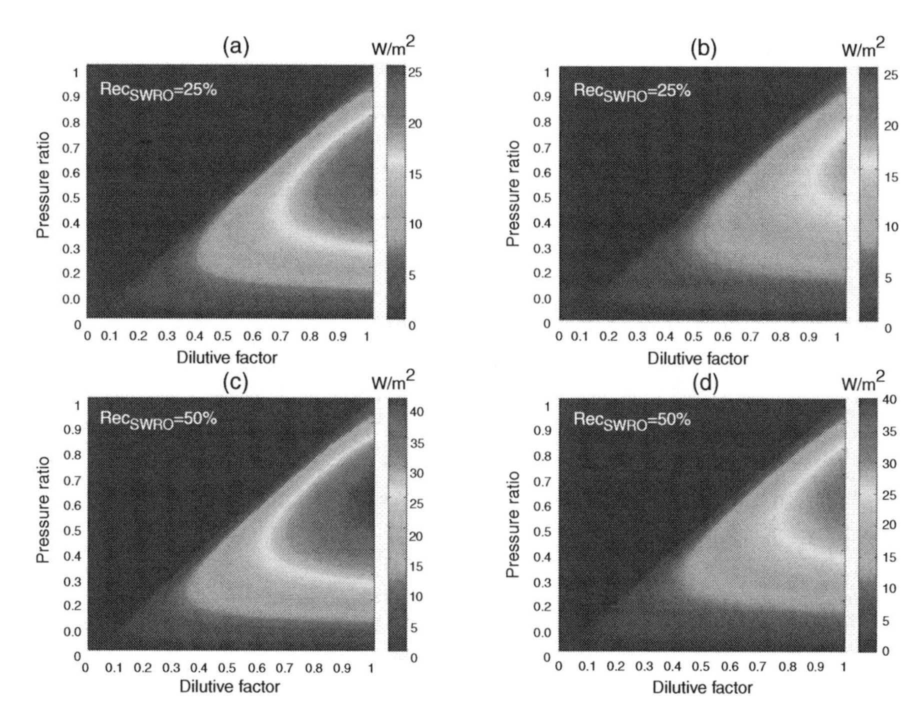

FIGURE 9.4
Average power density of a PRO hollow fiber module as functions of pressure ratio and dilutive factor. (a) Rec_{SWRO} = 25%, $C_D{}^0$ = 0.8 M, $C_F{}^0$ = 0 M, (b) Rec_{SWRO} = 25%, $C_D{}^0$ = 0.8 M, $C_F{}^0$ = 0.01 M, (c) Rec_{SWRO} = 50%, $C_D{}^0$ = 1.2 M, $C_F{}^0$ = 0 M, and (d) Rec_{SWRO} = 50%, $C_D{}^0$ = 1.2 M, $C_F{}^0$ = 0.01 M.

In Figure 9.4 (a) and (b), the recovery of the SWRO system is 25% and therefore the concentration of the draw solution entering the PRO system is 0.8 M. The average power density of the PRO system at a given dilutive factor and pressure ratio is always smaller when wastewater containing 0.01 M NaCl is used due to the enhanced ICP with increasing salt concentration in the feed solution (Achilli et al. 2009, She et al. 2012, Zhang and Chung 2013). The maximum average power densities appear as DF→1 (i.e., minimal dilution of the draw solution). The maximum average power densities are 23.8 and 18.0 W/m^2, respectively, when freshwater and wastewater are used as the feeds. In Figure 9.4 (c) and (d), the recovery of the SWRO system is 50% and therefore the concentration of the draw solution entering the PRO system is 1.2 M, and the maximum average power densities are boosted to 42.0 and 32.9 W/m^2, respectively, when freshwater and wastewater are used as the feeds.

Ideally, the maximum power density should increase quadratically with the draw solution concentration. However, the results show that when the bulk concentration of the draw solution is increased to 150%, the maximum average power density is increased to 176% instead of 225% when freshwater is used as the feed solution. This can be attributed to ICP – the main factor that compromises the PRO performance (Achilli et al. 2009, She et al. 2012), which becomes more severe due to the increased reverse salt flux at high draw solution concentrations (Achilli et al. 2009, She et al. 2012, Wan and Chung 2015, Zhang and Chung 2013). This results in a less than ideal increase in the maximum power density.

9.3.3 Optimal Operating Pressure of PRO

The operating lines of PRO systems, presented in Figure 9.5 (a) and (b), are obtained by connecting the points of the maximum average power densities at different dilutive factors in Figure 9.4 (a–d). In general, the optimal pressure ratio is higher at a higher draw solution concentration. This is because the ICP effect is more enhanced due to the increased reverse salt flux at a higher draw solution concentration. At 25% recovery, the optimal pressure ratios at DF→1 are 0.58 and 0.60, respectively, when freshwater and wastewater are used. When the draw solution concentration increases from 0.8 to 1.2 M, the optimal pressure ratios at DF→1 are increased to 0.60 and 0.63, respectively. As dilution occurs (i.e., decreasing DF), the driving force – the osmotic pressure difference – decreases and the corresponding optimal pressure ratio decreases as well.

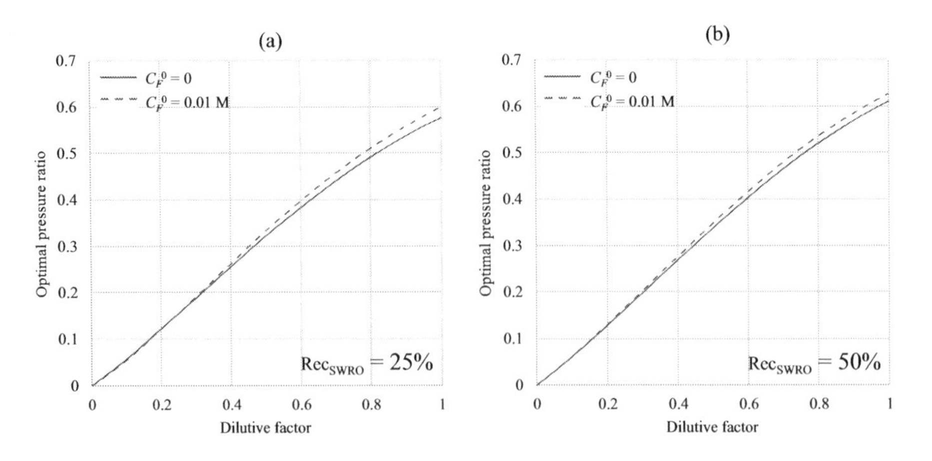

FIGURE 9.5
The optimal pressure ratio as a function of the dilutive factor. (a) $Rec_{SWRO} = 25\%$, $C_D^0 = 0.8$ M, $C_F^0 = 0$ or 0.01 M, (b) $Rec_{SWRO} = 50\%$, $C_D^0 = 1.2$ M, $C_F^0 = 0$ or 0.01 M.

9.3.4 Shortest Resident Time and the Minimum Membrane Area Required

The amount of energy that can be harvested from per cubic meter of draw solution in the PRO system at a given dilutive factor and pressure ratio is fixed by the following equation.

$$\frac{W_{\text{PRO}}}{V_{\text{D}}^{0}} = \frac{\text{PR}}{\text{DF}}\frac{\pi_{\text{SW}}}{1-\text{Rec}} \tag{45}$$

Therefore, the required membrane area can be minimized if PRO is operated at the optimal pressure ratio along the operating line where the average power density of the module is maximized. The normalized membrane area in terms of residence time (eqn. 25) is presented in Figure 9.6 (a) and (b). In both cases, the residence time increases more steeply when the dilutive factor approaches 0 or 100%. Therefore, it would be preferred to operate the PRO system with a dilutive factor in the range of 30–80% to enhance the stability of the system.

The minimum membrane area required can be affected by both concentrations of the feed and draw solutions. On one hand, the minimal membrane area is reduced by 15% on average due to the increased water flux when the draw solution concentration is increased from 0.8 to 1.2 M. On the other hand, the effect of increasing the salt concentration in the feed solution can be more significant. In other words, when the feed solution concentration increases from 0 to 0.01 M, the minimal

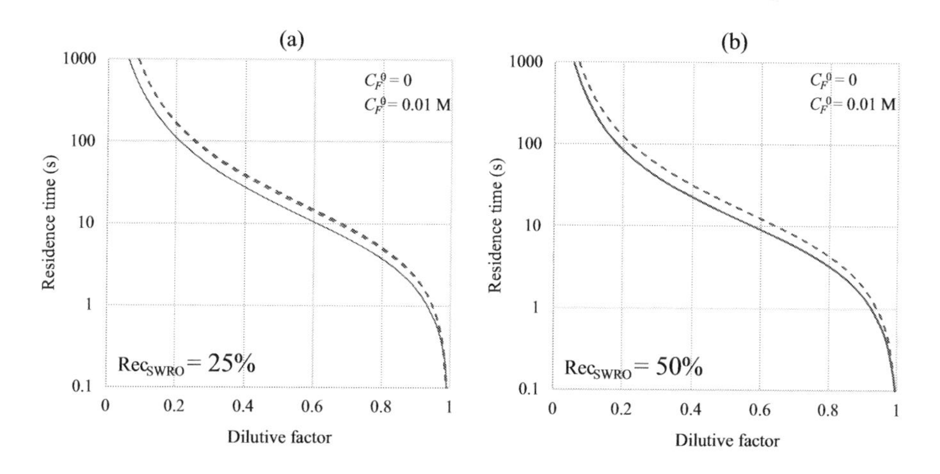

FIGURE 9.6
Shortest resident time of the draw solution inside the module as a function of the dilutive factor. (a) $\text{Rec}_{\text{SWRO}} = 25\%$, $C_D^{\,0} = 0.8$ M, $C_F^{\,0} = 0$ or 0.01 M, (b) $\text{Rec}_{\text{SWRO}} = 50\%$, $C_D^{\,0} = 1.2$ M, $C_F^{\,0} = 0$ or 0.01 M.

membrane area required is increased by more than 50% on average in both cases where 0.8 and 1.2 M NaCl solutions are used as the draw solution.

9.3.5 Comparisons of SECs in Various SWRO-Involved Systems

The SEC on the basis of per cubic meter of desalinated water for various processes at 25% SWRO recovery and 50% SWRO recovery are summarized in Figure 9.7 (a) and (b), respectively. According to eqn. 4, the following equation has to be satisfied, so that the SWRO+ERD+PRO process is physically possible.

$$\frac{f}{1-\text{Rec}}>1 \tag{46}$$

The value of $\frac{f}{1-\text{Rec}}$ indicates the ratio of flow rates of the draw solution exiting and entering the PRO system. A value of $\frac{f}{1-\text{Rec}}$ larger than unity indicates a positive ΔV_{PRO}. A larger f indicates that more water is drawn by the PRO system, and more energy can be recovered in ERD2. Therefore, the SEC is smaller at a larger f.

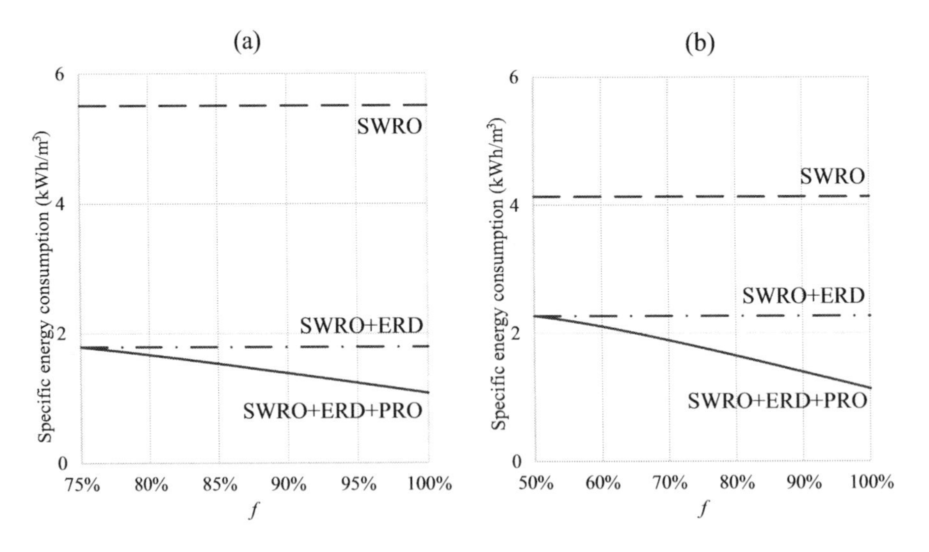

FIGURE 9.7
Comparison of SECs between SWRO, SWRO+ERD, and SWRO+ERD+PRO. While the SECs of SWRO and SWRO+ERD are independent of f, the SEC of SWRO+ERD+PRO can be further reduced because more energy can be recovered in PRO at a higher f. The values are calculated with η_P = 80% and η_E = 90%. (a) Rec_{SWRO} = 25% and (b) Rec_{SWRO} = 50%.

As shown in Figure 9.7 (a), at 25% recovery, it takes 5.51 kWh to produce 1 m^3 of desalinated water. An ERD can effectively reduce the SEC by 67.5% to 1.79 kWh/m^3. Currently, ERD is the dominant technology for energy recovery in SWRO plants (Hauge 1995, Migliorini and Luzzo 2004). By diluting the concentrated brine (i.e., the draw solution) to the seawater level, PRO is able to further decrease the SEC to 1.08 kWh/m^3 at f = 100%.

As shown in Figure 9.6 (b), at 50% recovery, the SEC of an SWRO plant without an ERD is 4.13 kWh/m^3, which is lower than that of a 25% recovery plant. This is because even though it takes lower pressure to achieve the 25% recovery, the energy used to pressurize the remaining 75% brine is wasted without an ERD. If energy in the brine is recovered by an ERD, the SEC of a 50% recovery SWRO plant becomes 2.27 kWh/m^3, 26.8% higher than that of a 25% recovery SWRO plant with an ERD. When PRO is used for energy recovery, the SEC of the SWRO+ERD+PRO process further drops to 1.14 kWh/m^3 at f = 100%.

Figure 9.8 summarizes the minimal SECs of various SWRO-involved processes with different efficiencies of HP and ERD. As shown, SECs are increased if the efficiencies of ERD and HP are reduced. Interestingly, the smaller the recovery of SWRO, the larger the increase in SECs due to the reduced efficiencies. This is because there is more energy carried by the concentrated brine at a smaller recovery. As more energy can be recovered by incorporating ERD and PRO, low efficiencies of HP and ERD will cause more energy losses. This is also why the energy loss

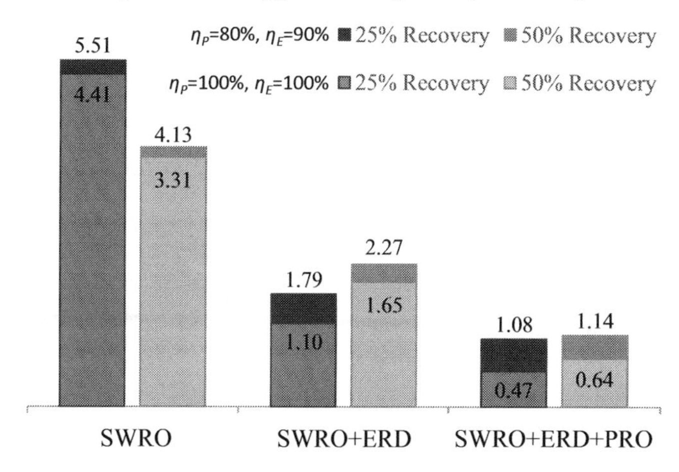

FIGURE 9.8
Comparison of the minimum SECs of SWRO, SWRO+ERD, and SWRO+ERD+PRO processes with different efficiencies of HP and ERD.

due to the reduced efficiencies in SWRO+ERD+PRO is larger than that in SWRO+ERD, which in turn is larger than that in SWRO.

9.3.6 SER and SEC of SWRO+ERD+PRO Integrated Processes

The relative amounts of specific energy recovery (SER) and SEC in an ideal process (i.e., η_P = 100%, η_E = 100%) are investigated and the results are presented in Figure 9.9 (a). The *y*-axis is the flow rate of the brine/seawater streams normalized by the flow rate of the pretreated seawater feed, and the *x*-axis is the pressure of the brine/seawater streams normalized by the operating pressure of SWRO. Therefore, the area enclosed by $x = y = 100\%$ represents the total energy required for the desalination step (eqn. 26). Area 1 represents the SER by ERD1 (eqn. 30), and area 2 represents the SER by ERD2 (eqn. 29). Figure 9.9 (a) also exposes the limitations of energy recovery by PRO in the SWRO+ERD+PRO integrated process. First, the concentration brine has enough energy to pressurize the feed seawater to the operating pressure of SWRO, but its flow rate cannot match that of the feed seawater because a fraction of Rec of the feed seawater is recovered as freshwater in SWRO. Second, the flow rate of the diluted brine is always less than or equal to the flow rate of the feed seawater, and it does not have the adequate energy to pressurize the feed seawater directly to the operating pressure of SWRO because the operating pressure of PRO is only a fraction of the operating pressure of SWRO. Therefore, areas 3 and 4

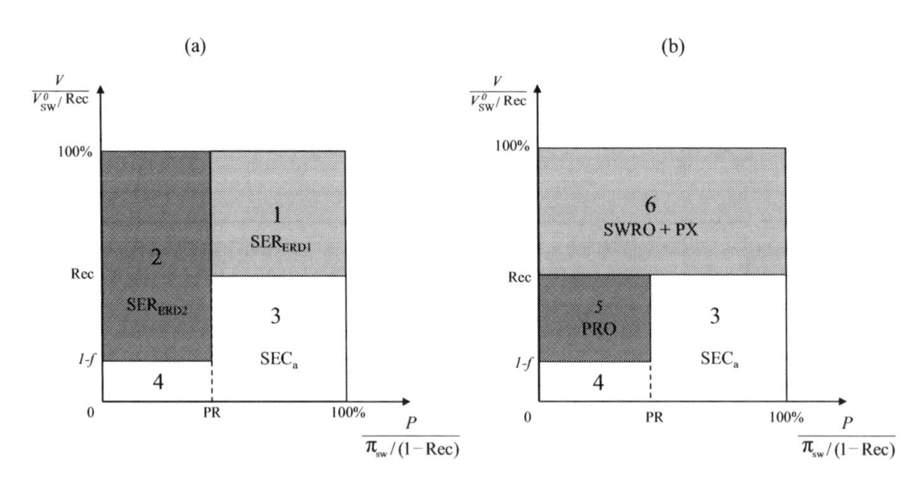

FIGURE 9.9
(a) The SER by ERD1 and ERD2, and SEC in a SWRO+ERD+PRO process as fractions of the total energy required for desalination – area 1: SER by ERD1, area 2: SER by ERD2, area 3+4: SEC of the integrated process. (b) The additional SER by PRO compared to the conventional SWRO+ERD process as fractions of the total energy required for desalination – area 5: additional SER by PRO, area 6: SER by the conventional SWRO+ERD process.

represent the dead energy losses that cannot be recovered. Therefore, the total area of areas 3 and 4 is the SEC of the process. Area 4 is indeed the energy loss due to inadequate dilution of the draw solution in PRO, which can be eliminated by increasing f to 1 as in Figure 9.1 (b). Area 3 is the energy loss due to the production of desalinated freshwater. It cannot be eliminated as long as freshwater is produced. The additional energy that can be recovered by PRO (eqn. 28) in the SWRO+ERD+PRO process is presented in area 5 in Figure 9.9 (b), which is only a fraction of Rec of the energy recovered by ERD1 (area 2). Area 6 in Figure 9.9 (b) represents the energy recovered in the SWRO+ERD process. As shown, area 6 is the sum of area 1 and the upper portion of area 2. This is because in the SWRO+ERD process, all the energy carried by the high-pressure brine is used for energy recovery, and therefore the pressure of the brine can be reduced to 0. However, in the SWRO+ERD+PRO process, the brine exiting ERD1 still has a medium high pressure as required by the PRO system.

Incorporating a PRO system into the existing SWRO–ERD plant will lead to an energy saving of 0.71–1.13 kWh/m^3. It is projected that 36 million m^3/day of desalinated water will be produced by SWRO by 2016 (GWI 2010). Therefore, it is estimated that 25.6–40.7 million kWh/day can be saved globally.

9.3.7 Optimization of SWRO+ERD+PRO

The price ratio between electricity and water reflects their relative demands. A smaller electricity-to-water price ratio implies that more energy can be consumed to produce water while maintaining the SWRO+ERD+PRO process profitable. On the contrary, a larger ratio implies that electricity is scarcer than water, and the SWRO+ERD+PRO process may no longer be profitable if it consumes too much energy. While the water price influences the total profit of the SWRO+ERD+PRO process, it is only the electricity-to-water price ratio that determines the optimal operating condition of the process according to eqn. 41. The electricity price, water price, and their ratios in different countries are listed in Table 9.2. As shown, the price ratio ranges from 0.01 to 0.5 in most countries. Interestingly, Jamaica and Denmark have similar electricity prices, but their water prices are 12 times different. This will significantly impact the optimal operating condition of SWRO+ERD+PRO.

The normalized operating profits obtained are presented in Figure 9.10. The maximal normalized operating profit is unity at the price ratio of 0. This means electricity is free and therefore the SWRO can operate at a recovery of 100% and can still make a good profit. As the price ratio increases, the optimal recovery decreases, and at some point the operating profit becomes 0 or even negative because electricity cost surpasses the profit of producing desalinated water. Eventually, as price ratio continues to increase, the operating profit will become negative at all recoveries. In the price ratio range between 0.01 and 0.5, the optimal recovery is between 0.38 and 0.9. The optimal operating curve of the

TABLE 9.2

Summary of water and electricity prices in different countries

Country	Water Price (α, US$/m^3)	Elec. Price (β, US$/kWh)	Mod. Elec. Price ($\beta\pi_{sw}$, US$/m^3)	Price Ratio ($\gamma = \beta\pi_{sw}/\alpha$)
Jamaica	0.76	0.45	0.375	0.489
Singapore	0.94	0.21	0.173	0.185
US	1.30	0.17	0.140	0.108
Dubai	2.64	0.06	0.049	0.019
Denmark	9.21	0.40	0.33	0.036

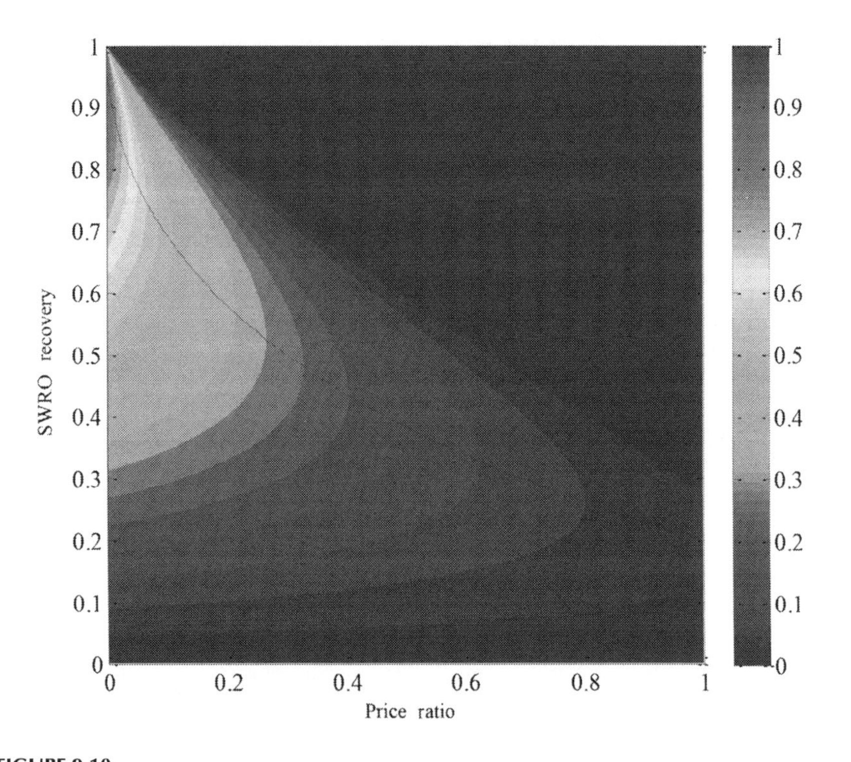

FIGURE 9.10
Normalized operating profit as a function of price ratio and SWRO recovery of the SWRO +ERD+PRO integrated process. The solid line is the optimal recovery of the SWRO+ERD +PRO integrated process as a function of the electricity–water price ratio.

SWRO unit is obtained by connecting the maximal operating profit at each price ratio in Figure 9.10. The maximal normalized operating profits of the conventional SWRO+ERD process and the SWRO+ERD+PRO process are shown in Figure 9.11. Since the price ratio in most countries are smaller than

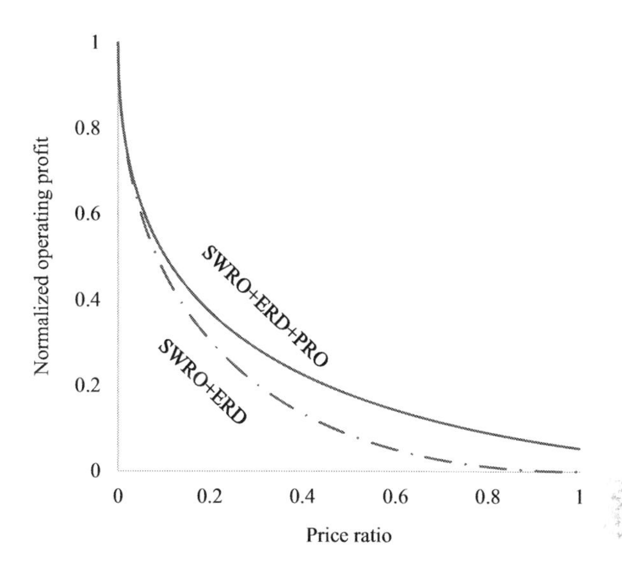

FIGURE 9.11
A comparison of the maximal normalized operating profits of SWRO+ERD and SWRO+ERD +PRO at different price ratios.

0.2, SWRO+ERD+PRO can harvest additional operating profits up to 22%. At a price ratio of 0.5, more than 100% of additional operating profit can be achieved.

9.4 Conclusion

In this study, detailed designs of novel SWRO–PRO processes are presented, with the option to form a closed-loop process that can substantially cut down the pretreatment costs of seawater. The SECs of various SWRO-involved processes are investigated. It shows that the SEC of the SWRO–PRO integrated process is the lowest among all the cases. When the SWRO is operated at 25% and 50% recovery and the brines are diluted to the seawater level, the SECs to produce 1 m^3 of desalinated water can be reduced to 1.08 and 1.14 kWh, respectively. It is evidenced from the SWRO–PRO model that finding the optimal operating pressure of the PRO system is of vital importance to maximize the average power density of the PRO system and minimize the SEC of the SWRO–PRO integrated process. While the integration of SWRO and PRO is possible, the development of new PRO membranes that can sustain the optimal operating pressure of the PRO system and maintain a high water flux is needed.

Moreover, the operating profit of a SWRO–PRO process is investigated. Maximization of the operating profit of SWRO can be a better approach than minimization of SEC in order to optimize SWRO–PRO, since it takes into account the relative demands of water and energy. The SWRO–PRO process can be optimized by finding the optimal recovery that maximizes the operating profit and the optimal pressure ratio that maximizes the power density at the respective recovery. The model shows that while the prices of water and electricity determine the operating profit of the system, only their ratio determines the optimal operating condition. As evidenced from the analyses, integration of SWRO with PRO will push the recovery above 50%, reduce the SEC by up to 35%, and increase the operating profit by up to 100%.

While integration of PRO with SWRO will reduce the specific energy of desalination and increase the operating profit, such integration requires additional capital investment in membranes, pumps, and ERDs. Detailed techno-economic and life cycle analyses are needed in the future to investigate in terms of capital expense (CAPEX) and operating expense (OPEX) in order to fully assess the feasibility of SWRO–PRO.

References

Achilli, A., Cath, T.Y., Childress, A.E., 2009. Power generation with pressure retarded osmosis: an experimental and theoretical investigation. *J. Membr. Sci.* 343, 42–52.

Banchik, L.D., Sharqawy, M.H., Lienhard, J.H., 2014. Limits of power production due to finite membrane area in pressure retarded osmosis. *J. Membr. Sci.* 468, 81–89.

Chung, T.S., Li, X., Ong, R.C., Ge, Q., Wang, H., Han, G., 2012. Emerging forward osmosis (FO) technologies and challenges ahead for clean water and clean energy applications. *Curr. Opin. Chem. Eng.* 1, 246–257.

Efraty, A., 2013. Pressure retarded osmosis in closed circuit: a new technology for clean power generation without need of energy recovery. *Desalin. Water Treat.* 51, 7420–7430.

Elimelech, M., Phillip, W.A., 2011. The future of seawater desalination: energy, technology and the environment. *Science* 333, 712–717.

Greenlee, L.F., Lawler, D.F., Freeman, B.D., Marrot, B., Moulin, P., 2009. Reverse osmosis desalination: water sources, technology, and today's challenges. *Water Res.* 43, 2317–2348.

GWI, 2010. *Desalination market 2010*, Global Water Intelligence, Oxford.

Han, G., Ge, Q., Chung, T.S., 2014. Conceptual demonstration of novel closed-loop pressure retarded osmosis process for sustainable osmotic energy generation. *Appl. Energy* 132, 383–393.

Hauge, L.J., 1995. The pressure exchanger – a key to substantial lower desalination cost. *Desalination* 102, 219–223.

Lee, K.L., Baker, R.W., Lonsdale, H.K., 1981. Membranes for power generation by pressure-retarded osmosis. *J. Membr. Sci.* 8, 141–171.

Liyanaarachchi, S., Shu, L., Muthukumaran, S., Jegatheesan, V., Baskaran, K., 2013. Problems in seawater industrial desalination processes and potential sustainable solutions: a review. *Rev. Environ. Sci. Bio.* 13, 203–214.

Migliorini, G., Luzzo, E., 2004. Seawater reverse osmosis plant using the pressure exchanger for energy recovery: a calculation model. *Desalination* 165, 289–298.

Peñate, B., García-Rodríguez, L., 2012. Current trends and future prospects in the design of seawater reverse osmosis desalination technology. *Desalination* 284, 1–8.

Sharqawy, M.H., Lienhard, J.H., Zubair, S.M., 2012. Thermophysical properties of seawater: a review of existing correlations and data. *Desalin. Water Treat.* 16, 354–380.

She, Q., Jin, X., Tang, C.Y., 2012. Osmotic power production from salinity gradient resource by pressure retarded osmosis: effects of operating conditions and reverse solute diffusion. *J. Membr. Sci.* 401–402, 262–273.

Sim, V.S., She, Q., Chong, T.H., Tang, C.Y., Fane, A.G., Krantz, W.B., 2013. Strategic co-location in a hybrid process involving desalination and pressure retarded osmosis (PRO). *Membranes* 3, 98–125.

Sivertsen, E., Holt, T., Thelin, W., Brekke, G., 2012. Modelling mass transport in hollow fibre membranes used for pressure retarded osmosis. *J. Membr. Sci.* 417, 69–79.

Sivertsen, E., Holt, T., Thelin, W., Brekke, G., 2013. Pressure retarded osmosis efficiency for different hollow fibre membrane module flow configurations. *Desalination* 312, 107–123.

Thorsen, T., Holt, T., 2009. The potential for power production from salinity gradients by pressure retarded osmosis. *J. Membr. Sci.* 335, 103–110.

Wan, C.F., Chung, T.S., 2015. Osmotic power generation by pressure retarded osmosis using seawater brine as the draw solution and wastewater brine as the feed. *J. Membr. Sci.* 479, 148–158.

Zhang, S., Chung, T.S., 2013. Minimizing the instant and accumulative effects of salt permeability to sustain ultrahigh osmotic power density. *Environ. Sci. Technol.* 47, 10085–10092.

Zhang, S., Sukitpaneenit, P., Chung, T.S., 2014. Design of robust hollow fiber membranes with high power density for osmotic energy production. *Chem. Eng. J.* 241, 457–465.

Zhu, A., Christofides, P.D., Cohen, Y., 2009a. Effect of thermodynamic restriction on energy cost optimization of RO membrane water desalinition. *Ind. Eng. Chem. Res.* 48, 6010–6021.

Zhu, A., Christofides, P.D., Cohen, Y., 2009b. Minimization of energy consumption for a two-pass membrane desalination: effect of energy recovery, membrane rejection and retentate recycling. *J. Membr. Sci.* 339, 126–137.

10

Operation and Maintenance of SWRO–PRO Pilot Systems

Tianshi Yang, Chun Feng Wan, and Esther Swin Hui Lee
Department of Chemical and Biomolecular Engineering
National University of Singapore
Singapore

CONTENTS

10.1 Introduction

Pressure retarded osmosis (PRO) is the promising technology to harvest sustainable osmotic energy using semipermeable membranes (Han et al. 2015, Loeb and Norman 1975, Skilhagen et al. 2008). Many researchers have studied this technology from concept validation to membrane and membrane module development in laboratories and made great progresses (Lee et al. 1981, Post et al. 2007, Seppälä and Lampinen 1999, Wan et al. 2017, Wang et al. 2019, Xiong et al. 2016, Zhang et al. 2014). However, in order to gain more understanding of the membrane performance and operating conditions outside laboratories, pilot studies employing large membrane modules and realistic water sources are urgently needed. Previous studies have shown that using seawater and river water as the feed pair was hardly energy efficient due to the low osmotic pressure difference (Loeb 2002, Skilhagen et al. 2008, Thorsen and Holt 2009). Thus, using the concentrated brine from seawater reverse osmosis (SWRO), denoted as SWBr, has been proposed for PRO because it has a higher salt concentration and better water quality than raw seawater (Chung et al. 2015, Yip et al. 2016). Therefore, the integration of PRO with SWRO plants becomes attractive because it not only reduces the overall energy consumption for seawater desalination but also mitigates the disposal and environment issues of SWBr (Chung et al. 2012, Li et al. 2017, Wan and Chung 2015).

10.2 Design and Operation of SWRO–PRO Pilot Systems

10.2.1 Design of the SWRO–PRO Pilot System at NUS

An SWRO–PRO integration pilot system was built at National University of Singapore (NUS) in 2016 to investigate the osmosis power generation, as shown in Figure 10.1 (Lee et al. 2017). This SWRO–PRO pilot system comprised (1) an SWRO system, (2) a PRO system, (3) a pressure exchanger to transfer the osmotic energy of the high-pressure draw solution from the PRO outlet stream to the low-pressure RO feed stream, and (4) two units of clean-in-place (CIP) systems. The SWRO system was purposely run at a recovery of 45%, so that the SWRO brine (i.e., SWBr) had a concentration of 0.8 M (46,750 mg/L). The SWBr was then sent to the PRO system and used as the draw solution in the pilot tests. The osmotic energy recovered from this SWBr by the PRO system was used to compensate the energy consumption for the SWRO system. In the PRO system, 1-inch inner-selective thin-film composite polyethersulfone (TFC-PES) hollow fiber membrane modules were employed as the PRO units and the total module capacity were 6 trains, with 10 modules in each train. The PRO

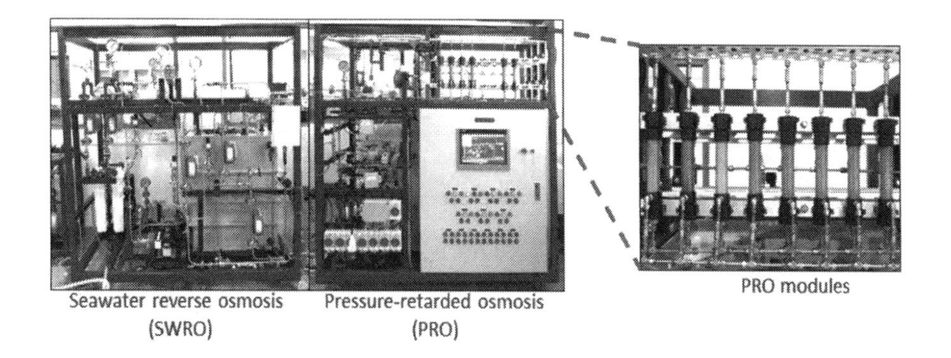

FIGURE 10.1
The NUS SWRO–PRO pilot with dimensions of 4.8 m × 1 m × 2 m ($L \times W \times H$) consisting of PRO membrane modules.

modules were developed in-house from the state-of-the-art inner selective (TFC-PES) hollow fiber membranes (Wan et al. 2017).

For performance comparison, both tap water and wastewater brine (WWBr) from a local water recycling plant were utilized as feed solutions in pilot tests, while two types of brines were employed as draw solutions. They were real SWBr from SWRO and synthetic brine (0.8 M NaCl), both of which have the same osmolality. The flow rates at the lumen side (i.e., draw solution side) and the shell side (i.e., feed solution side) were set and maintained at 1 liter/min.

10.2.2 The Operation and Results of the SWRO–PRO Pilot System at NUS

Testing and commissioning of the pilot were carried out prior to the actual operation in order to assure that all components were designed, installed, and operated in accordance to the operational requirements. The commissioning data were obtained by using a 0.8 M synthetic brine as the draw solution and tap water as the feed solution at a constant applied pressure of 15 ± 0.5 bar. The operating parameters remained unchanged for a certain period of time to assess the stability of the pilot plant. Successful commissioning of the pilot was achieved because of its operability in terms of performance, reliability, safety, and information traceability.

The PRO pilot study was divided into four stages depending on the operational purposes and the feed pair. They were (1) stage I: system and process stabilization; (2) stage II: pilot study using 0.8 M synthetic brine and tap water as the feed pair; (3) stage III: pilot study using real SWBr and tap water as the feed pair; and (4) stage IV: pilot study using real SWBr and WWBr as the feed pair. Figure 10.2 illustrates the variations of

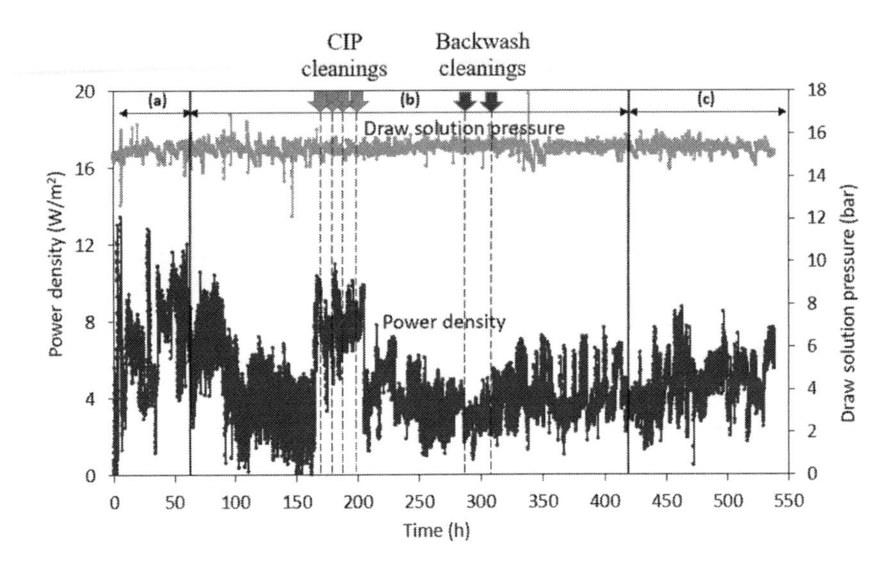

FIGURE 10.2
Variations of power density and draw solution pressure as a function of time in (a) stage I: system and process stabilization, (b) stage II: pilot study using 0.8 M synthetic brine and tap water as the feed pair, (c) stage III: pilot study using real SWBr and tap water as the feed pair with indications of cleaning timelines.

power density as a function of time for the first three stages. The pressure at the draw solution side was kept within the range of 15 ± 0.5 bar with occasional pressure drops and surges.

10.2.2.1 Stage I: System and Process Stabilization

In the first 63-h stage, various performance indicators, namely (1) the stability of the SWRO–PRO pilot, which included all systems and components; (2) robustness and performance of the PRO membrane modules; and (3) process parameters such as operating pressure and power density, were assessed. In order to minimize the effects of fouling issues, 0.8 M synthetic brine and tap water were used as the feed pair. The applied pressure at the draw solution side was controlled within 15 ± 0.5 bar. Results indicated that there were some fluctuations in power density in the first 37 h during the system stabilization stage. After that, the power density became stable and an average value of 8.4 W/m^2 was obtained in this stage. The operating pressure was stable with minimal fluctuations throughout the entire time period. A high and stable pressure exchanger efficiency, which was another critical performance indicator in the SWRO–PRO pilot system, was obtained with an average value of 92.9%.

10.2.2.2 Stage II: 0.8 M Synthetic Brine and Tap Water as the Feed Pair

In the second stage, the SWRO–PRO pilot study was run for a longer time (i.e., 355 h) in order to investigate the process trend using the same testing methods and feed pair. Several membrane cleaning strategies, such as CIP and backwash cleanings with tap water, were carried out to recover the permeate flux as well as the power density. The details about these membrane cleaning strategies would be discussed in the following sections. The timelines of membrane cleaning were shown in stage (b) in Figure 10.2. At 90 h operating time, a decline in power density was observed until CIP cleaning was carried out at 165 h. Subsequent CIP cleanings with tap water were carried out at 175, 185, and 195 h, at every 10 h interval. The duration of the first CIP cleaning was 4 h while the durations of the subsequent three CIP cleanings were reduced to 1 h each. Results showed that four rounds of CIP cleanings with tap water, regardless of the cleaning duration, were able to fully recover the power density to an average value of around 8.4 W/m^2.

Backwash cleanings with tap water were also conducted at 280 and 305 h, as illustrated in Figure 10.2. Each of backwash cleanings took place for 20 min and 2.5 h, respectively. However, different from the CIP cleaning, only 55% of the power density was recovered using backwash cleanings regardless of the cleaning duration. Considering that there was no flushing on the shell side of the membrane, the only permeate stream from the lumen side was not strong enough to loosen the foulants that were accumulated inside the porous membrane substrates and clogged the flow channels. The average power density of this stage was 5.7 W/m^2.

10.2.2.3 Stage III: Real SWBr and Tap Water as the Feed Pair

In this stage, real SWBr was used as the draw solution instead of the synthetic brine in the PRO process. The variations of power density and pressure as a function of operating time are shown in stage (c) in Figure 10.2. Over the 120 h of the pilot operation started from 416 h onwards, a stable trend of power density was observed and its average value was 5 W/m^2, which was close to the value obtained in stage II. Thus, it could be concluded that fouling induced by both synthetic brine and real SWBr on the selective layer is relatively insignificant in the PRO process.

10.2.2.4 Stage IV: Real SWBr and WWBr as the Feed Pair

As illustrated in Figure 10.3, a feed pair consisting of SWBr and WWBr was used in the PRO process in this stage, and the power density generated by this pair was compared with that using SWBr and tap water as the feed pair. In the first operational cycle of 2.3 h, an average power density of 5.7 W/m^2 was obtained using WWBr as the feed solution, while 6.7 W/m^2 was obtained using tap water. However, the power density using the feed pair of SWBr and

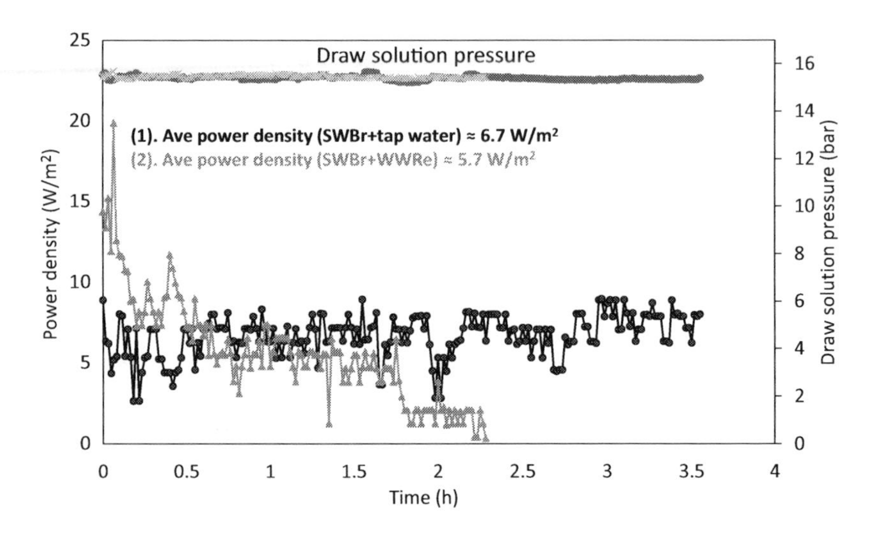

FIGURE 10.3
Variations of power density and draw solution pressure as a function of time using (1) 0.8 M SWBr and tap water as the feed pair and (2) real SWBr and wastewater retentate (WWBr) as the feed pair.

WWBr declined to near 0 after 2.3 h of operation due to severe fouling and scaling induced by the inorganic and organic species of WWBr. Therefore, the operational duration of 2.3 h was used to calculate the aforementioned average power densities. Membrane cleaning became necessary in this scenario, and the details would be discussed in the following sections.

10.2.3 Design of the SWRO–PRO Pilot System in Korea

In order to explore the effectiveness of the SWRO–PRO hybrid system, a Korean R&D project called Global MVP (GMVP) was launched in 2013 (Lee et al. 2019, Park et al. 2017). As shown in Figure 10.4, its SWRO–PRO pilot plant consisted of SWRO and PRO systems, as well as a control office. The PRO had a treatment capacity of 20 m^3/day, and it consisted essentially of SWRO and PRO systems. Different from the SWRO–PRO pilot system at NUS, where the PRO system was used to compensate the energy consumption rate for the RO system, a Pelton turbine was used in the Korean SWRO–PRO pilot system to directly generate electricity. The pressure, flow rate, concentration, conductivity, and temperature were monitored and recorded every 10 s using the Human Machine Interface of the SWRO–PRO system. Four 8-inch spiral wound PRO membrane modules manufactured by Toray Chemical Korea (Seoul, Korea) were employed in this pilot system. The modules were 40 inches in length and 18 m^2 in total membrane area. A synthetic SWRO feed solution of 35,000 mg/L was injected into RO modules. The SWBr of

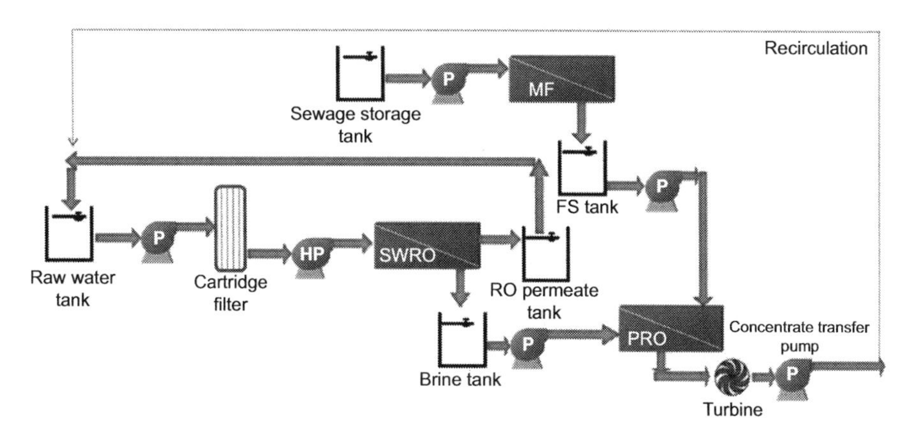

FIGURE 10.4
A Korean SWRO–PRO pilot plant with a PRO capacity of 20 m^3/d.

70,000 mg/L was used as the high-salinity draw solution and the SWRO permeate was used as the low-salinity feed solution. The flow rates of the draw and feed solution were maintained at 10 L/min, respectively. The applied hydraulic pressure difference (ΔP) was fixed at 20 bar after testing the system from 5 to 35 bar.

10.2.4 The Operation and Results of the SWRO–PRO Pilot System in Korea

Several generations of 8-inch PRO modules (CSM-PRO-1, CSM-PRO-2, CSM-PRO-3, and CSM-PRO-4) with different structure parameters were tested under the applied pressure of 20 bar. The first-generation PRO module (CSM-PRO-1) showed a permeate flux of 2.3 L/m^2 h and a feed solution recovery rate of 9.2%. These performance values were improved greatly with the second and third versions: 300% and 574% in permeate flux and 224% and 440% in feed solution recovery, respectively. The overall power density achieved by CSM-PRO-4 was within the range of 6–8 W/m^2, and the feed solution recovery was 49.3% in the Korean SWRO–PRO pilot plant (Lee et al. 2019, Park et al. 2017).

10.3 Membrane Cleaning Strategies

10.3.1 Commonly Used Membrane Cleaning Strategies

Membrane fouling is one of the major challenges in PRO processes because it not only reduces membrane performance and shortens membrane life, but also imposes extra operating costs in terms of energy and chemical consumption due to frequent membrane cleaning (Han et al. 2016, Li et al. 2017, She et al. 2013, Thelin et al. 2013). Therefore, membrane cleaning becomes significantly important.

There were several commonly used membrane cleaning methods, such as physical and chemical rinsing, osmotic backwash, and hydraulic backwash. During physical and chemical rinsing, one of the following solutions was circulated through the shell side of the membrane modules: (1) deionized (DI) water, (2) aqueous citric acid solution (pH = 3.6), (3) NaOH solution (pH = 11.2), (4) aqueous NaOCl solution (200 ppm, pH = 8.3), (5) commercially available acid membrane cleaner Genesol 38 (3 wt%, pH = 3.6), and (6) alkaline membrane cleaner Genesol 704 (1 wt%, pH = 11.2). Osmotic backwash was conducted by circulating DI water along the polyamide selective layer and 2 M NaCl solution along the porous substrate; it aimed to induce the osmotic-driven permeate from the polyamide side to the substrate side so that the foulant was washed out from the porous substrate. The hydraulic backwash by means of a hydraulic pressure was carried out similarly to typical RO operations, where only DI water was circulated on the polyamide side and pressurized to 1.0–1.5 bar. The permeate across the membrane could flush away the foulants underneath the membrane surface as well as within the membrane substrates.

Han et al. (2016) adopted these cleaning strategies when the lab-scale TFC PES hollow fiber membranes were fouled by a real WWRe during PRO tests. Figure 10.5 summarizes the normalized initial water fluxes of the fouled and regenerated TFC-PES membranes by different cleaning strategies. The initial water flux $J_{w,DI}$ was measured using DI water as the feed and then normalized by the initial flux ($J_{w0,DI}$) of a fresh membrane. As illustrated in Figure 10.5(a), water flux dropped to 46% of the initial value after fouling tests. Neither physical rinsing with freshwater nor osmotic backwash could effectively recover the water flux due to the irreversible fouling and scaling inside the membrane substrates. On the contrary, hydraulic backwash could considerately clean the membranes and recover the water flux. Because the cross-membrane flow of freshwater induced by the hydraulic pressure carried the accumulated salts and foulants away from the substrates, 90% of the initial water flux was recovered.

Figure 10.5(b) shows a comparison of cleaning effectiveness among various chemical cleaning solutions. Aqueous citric acid solution (pH = 3.6) and the commercially available acid cleaner Genesol 38 (pH = 3.6) could recover the water flux to 53–60% of the initial value, which was only slightly higher than that of DI water rinse (i.e., 52%). However, the alkaline flushing using a NaOH solution (pH = 11.2), NaOCl solution (pH = 8.3), or commercially available cleaner Genesol 704 (pH = 11.2) showed outstanding effectiveness in removing foulants. The NaOH solution was able to recover the water flux to 83% of the initial value

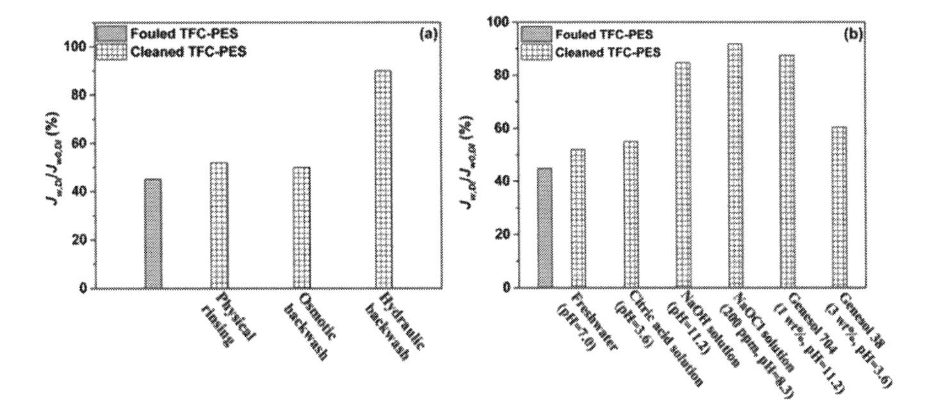

FIGURE 10.5
Normalized initial water fluxes, $J_{w,DI}/J_{w0,DI}$, of the fouled and cleaned TFC-PES hollow fiber membranes. (a) Cleaned by physical rinsing of DI water, osmotic backwash, and hydraulic backwash, separately; and (b) flushed by various chemical solutions for 12 h, separately. $J_{w0,DI}$ and $J_{w,DI}$ were obtained under the PRO mode at $\Delta P = 0$ bar using DI water as the feed and 2 M NaCl as the draw solution. $J_{w0,DI}$ was the water flux of the fresh membrane.

and Genesol 704 recovered the water flux to 87%. Among the three cleaning solutions, the NaOCl solution with the highest pH value provided the most effective cleaning performance with a water flux recovery of 90%.

Similarly, Li et al. (2017) used backwash strategy with different solutions to clean PRO membranes fouled by the same WWRe. In their work, (1) DI water, (2) EDTA solution at pH 10.9 (dosage per membrane area of 203 g/m^2), (3) HCl solution at pH 2.8 (dosage per membrane area of 11 g/m^2), and (4) NaOH solution at pH 11.8 (dosage per membrane area of 77 g/m^2) were chosen as the cleaning agents. During the cleaning process, DI water was pumped into the lumen side of PRO hollow fiber membranes at 1.2 m/s under 15 ± 1.0 bar while one of the four cleaning agents was circulated in the shell side at 0.1 m/s under 0 bar. Figure 10.6 summarizes the fouling and cleaning behavior of charged hyperbranched polyglyceral-grafted TFC (CHPG-TFC) membranes. The CHPG modification was conducted on the membrane surface with aims to mitigate fouling from the feed solution. The first PRO runs in four different cases, which were performed under identical conditions where the trans-membrane pressure was 15.0 ± 1.0 bar. All four runs showed a sharp decline in the normalized water flux, which indicated that the CHPG-TFC membranes were fouled quickly by the real WWRe. The normalized water flux dropped to around 75% of its initial value at the end of the first PRO run. Compared with the work by Han et al., in which the normalized water flux of unmodified PRO membranes dropped to 45%, this CHPG surface modification showed certain effectiveness to

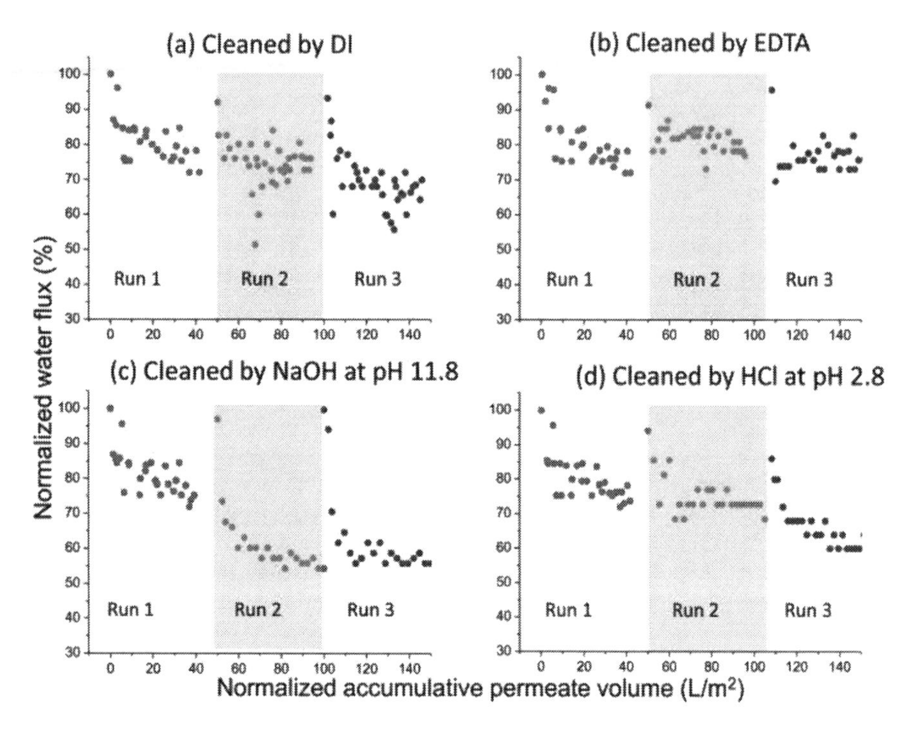

FIGURE 10.6
Fouling behaviors of CHPG-TFC membranes with different cleaning agents. The draw solution initially contained 0.81 mol/L NaCl, and a wastewater effluent was used as the feed solution. The hydraulic pressure difference between feed and draw solutions was 15.0 ± 1.0 bar. The first data point in each run was tested in DI water other than in wastewater. Runs 1, 2, and 3 are testing cycles after cleaning.

mitigate membrane fouling to a certain level. After the first round of cleaning, all four cleaning solutions were able to recover the water flux to more than 90% of the initial value. However, it was noted that only the EDTA solution could preserve the anti-fouling effect of CHPG-TFC membranes as indicated by the flux patterns, while the anti-fouling effect was gradually weakened by DI water cleaning and HCl cleaning. Although NaOH cleaning was effective in recovering the water flux, the water flux in the second PRO run dropped even further to less than 60% after NaOH cleaning as shown in Figure 10.6(c). These results implied that the cleaning agent might decrease the stability of surface coating and impair the structure of porous hollow fiber substrates. The second round cleaning and the third PRO runs further proved that the EDTA solution was more suitable to clean the CHPG-TFC membranes. This work also reminded us that the choice of the most suitable cleaning agent highly depended on the type of PRO membranes.

10.3.2 Membrane Cleaning Strategies for SWRO–PRO Pilot System Using Synthetic Brine and Tap Water as the Feed Pair

Several online cleaning approaches were performed to evaluate the cleaning efficiency of the fouled membrane modules in the NUS pilot (Section 2.2; Lee et al. 2017). Two units of CIP systems, which are (1) shell side CIP and (2) lumen side CIP systems, were utilized to clean the PRO membrane modules. The pressure of the lumen side solution was fixed at 8 bar while the pressure of the shell side solution was fixed at 2.5 bar. The PRO operation was resumed after completing each cleaning. All online cleanings were manually triggered through the Human Machine Interface touchscreen at a designated interval of operating time. The efficiencies of different cleaning methods were evaluated by comparing their percentages of average water flux recovery before and after each cleaning method.

CIP cleaning was used to enhance the membrane performance by regaining its short-term permeability, increasing its maintenance cleaning (MC) interval or operational duration. No chemical agent was added during the cleaning process. The outer surface (i.e., shell side) and the inner-selective layer (i.e., lumen side) of the fouled membranes were cleaned by continuously circulating the tap water with the aid of the two CIP units. Because the applied pressures at the lumen and shell sides were 8 and 2.5 bar, respectively, it created a crossflow across the membrane. During the intermittent CIP cleaning, the cleaning duration was kept within 30 min with minimum stoppage of the pilot. Chemical-free backwash with tap water was also performed to remove foulants from the porous support layer by passing the water flow from the polyamide-selective layer of the lumen side to the shell side at an applied pressure of 8 bar to dislodge the foulants. The dislodged foulants were drained through the shell side ports. Results from Figure 10.2 indicated that CIP cleaning was effective in water flux recovery and it was more favorable than backwash.

10.3.3 Membrane Cleaning Strategies for SWRO–PRO Pilot System Using SWBr and WWRe as the Feed Pair

As discussed in Section 2.2, when WWRe was directly used as the feed solution to the PRO process, severe fouling quickly occurred, and the power density of the fouled membrane dropped to 0 within 2.5 h. Thus, effective membrane cleaning strategies became necessary, especially at this pilot level. Two units of CIP systems were connected to the lumen and shell side inlet ports to perform membrane cleaning and the cleaning solutions were cycled back to the CIP tanks. Whenever the PRO power density declined to near 0 W/m^2, cleaning methods were conducted and evaluated at the end of each operational cycle.

Backwash, CIP, and MC were the membrane cleaning strategies applied in this pilot study using SWBr and WWBr as the feed pair. MC was performed by adding either a 200 ppm HCl solution at pH 2.3 or a 200 ppm NaOCl solution at pH 10.5 to the shell side CIP system, while tap water was used to clean the lumen side or the inner-selective layer of the fouled membranes concurrently. Besides physical and chemical flushing during MC, a few rounds of soaking and recirculation steps were involved as well, which allowed more chemical contact time between the cleaning agents and membranes. Therefore, the MC took a longer time than backwash or CIP cleaning. The entire cleaning process ended with one final step of CIP cleaning with tap water to wash off the residual chemicals on PRO membrane modules before resuming to the normal PRO operation.

Figure 10.7 showed the typical power density and pressure variations in 19-h pilot trials using SWBr and WWBr as the feed pair. Six operational runs were conducted to evaluate the cleaning efficiencies of various cleaning methods. The vertical dotted lines on the plot indicated the time when the cleanings were carried out. A 3-h backwash was conducted after the first cycle, where a partial water flux recovery of 42%

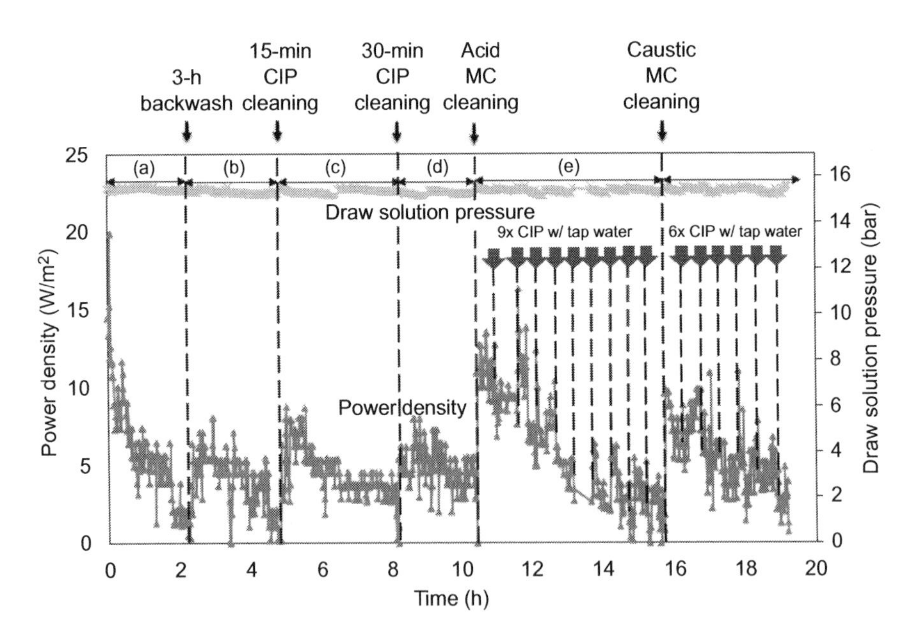

FIGURE 10.7
Variations of power density and draw solution pressure as a function of time using SWBr and WWBr as the feed pair in stage IV. Vertical dotted lines indicate the time when various types of cleanings were conducted. The process can be divided into several sections: (a) baseline, (b) after backwash at 8 bar, (c) after 15-min CIP with tap water, (d) after 30-min CIP with tap water, (e) after acid MC and with intermittent CIP cleanings, and (f) after caustic MC and with intermittent CIP cleanings.

was achieved. Two CIP cleanings, one with a cleaning duration of 15 min and the other with 30 min, were carried out; they achieved flux recovery of 47% and 51%, respectively. Stages (a), (b), (c), and (d) in Figure 10.7 illustrate the effectiveness of power density restoration with different cleaning methods. Table 10.1 tabulates the details of cleaning methods and their corresponding power densities, operational durations, and % water flux recoveries.

As summarized in Table 10.1, backwash and two CIP cleanings could only recover around half of the initial water flux for the feed pair of real SWBr and WWBr. Since WWBr contained many complicated foulants, it was easier for those foulants to accumulate within the membrane porous support layer, and they were more difficult to be washed out (Chen et al. 2015, She et al. 2013, Thelin et al. 2013). Therefore, it was deduced that

TABLE 10.1

Cleaning methods, average power density per cycle, and water flux recovery using real SWBr and wastewater retentate (WWBr) feed pair in stage IV

Cleaning Method	Average Power Density/Cycle (W/m^2)	Operational Duration (h)	Water Flux Recovery (%)
Before cleaning	5.7	2.3	–
After 3-h backwash with tap water @ 8 bar	4.0	2.5	42
After 30-min CIP with tap water	4.5	3.3	47
After 15-min CIP with tap water	4.8	2.2	51
After acid MC	10.9	0.5	~100
After 15-min CIP with tap water – 1	9.1	0.5	95
After 15-min CIP with tap water – 2	9.4	0.5	98
After 15-min CIP with tap water – 3	7.6	0.5	80
After 15-min CIP with tap water – 4	5.0	0.5	53
After 15-min CIP with tap water – 5	3.0	0.5	31
After 15-min CIP with tap water – 6	3.4	0.5	35
After 15-min CIP with tap water – 7	3.7	0.5	39
After 15-min CIP with tap water – 8	3.0	0.5	31
After 15-min CIP with tap water – 9	2.2	0.5	23
After caustic MC	7.7	0.5	80
After 15-min CIP with tap water – 1	8.0	0.5	83
After 15-min CIP with tap water – 2	5.7	0.5	60
After 15-min CIP with tap water – 3	4.2	0.5	44
After 15-min CIP with tap water – 4	4.9	0.5	51
After 15-min CIP with tap water – 5	4.6	0.5	48
After 15-min CIP with tap water – 6	2.2	0.5	23

cleaning methods by chemical-free backwash and CIP were not effective enough when using WWBr as the feed.

Starting from Stage (e) in Figure 10.7, MC with the addition of an acid or caustic cleaning agent was employed at the shell side and the effects of intermittent CIP with tap water were investigated. The intermittent CIP was introduced at every 30-min interval. The power density diminished to 0 W/m^2 at 10.5 h and a 200 ppm HCl solution at pH 2.3 was used for membrane cleaning. Results indicated that this acid cleaning could fully recover the water flux. At 11 h, the first intermittent CIP cleaning was introduced and subsequent eight rounds of CIP cleanings were conducted thereafter within an operational cycle. Due Because of the frequent cleaning, the operational duration was sustained and prolonged from 2.3 to 5 h compared with that without CIP cleaning. Considering that the intermittent CIP cleaning only used tap water under a nominal pumping pressure with lesser operational disruption and without using any chemicals, it was a good option to sustain the energy production in SWRO–PRO operations.

On the other hand, the caustic MC using a 200 ppm NaOCl, solution at pH 10.5 showed a flux recovery of 80%, which was slightly lower than that of acid MC as shown in Figure 10.7 and Table 10.1. Similarly, intermittent CIP cleanings were introduced six times until the power density completely declined. The total operational duration after caustic MC was 3.5 h, indicating that acid MC was more effective than caustic MC to sustain the pilot operation. The reason why acid MC was superior to caustic MC was possibly due to the unique chemistry of the local WWBr and its corresponding scaling mechanism that dominated the flux reduction of the PRO membranes (Chen et al. 2016, Xiong et al. 2016).

10.4 Summary

The integration of SWRO and PRO pilot systems is an effective method to harvest osmotic energy from the mixing of SWBr and various feed water. Because of the severe fouling and scaling on membrane surface, strategies to cleaning membranes and regenerating their performance are necessary to prolong the system life. In real applications, based on the nature of membranes, the chemistry of feed streams and their corresponding fouling mechanisms, it is possible to combine CIP, acid, and caustic MC methods in one operational cycle in order to improve the sustainability of SWRO–PRO operations. However, the frequency of MC strongly depends on feed water quality and its variations.

References

Chen, S.C., Amy, G.L., Chung, T.S., 2016. Membrane fouling and anti-fouling strategies using RO retentate from a municipal water recycling plant as the feed for osmotic power generation. *Water Res.* 88, 144–155.

Chen, S.C., Wan, C.F., Chung, T.S., 2015. Enhanced fouling by inorganic and organic foulants on pressure retarded osmosis (PRO) hollow fiber membranes under high pressures. *J. Membr. Sci.* 479, 190–203.

Chung, T.S., Luo, L., Wan, C.F., Cui, Y., Amy, G., 2015. What is next for forward osmosis (FO) and pressure retarded osmosis (PRO). *Sep. Purif. Technol.* 156, 856–860.

Chung, T.S., Zhang, S., Wang, K.Y., Su, J., Ling, M.M., 2012. Forward osmosis processes: Yesterday, today and tomorrow. *Desalination* 287, 78–81.

Han, G., Zhang, S., Li, X., Chung, T.S., 2015. Progress in pressure retarded osmosis (PRO) membranes for osmotic power generation. *Prog. Polym. Sci.* 51, 1–27.

Han, G., Zhou, J., Wan, C., Yang, T., Chung, T.S., 2016. Investigations of inorganic and organic fouling behaviors, antifouling and cleaning strategies for pressure retarded osmosis (PRO) membrane using seawater desalination brine and wastewater. *Water Res.* 103, 264–275.

Lee, E.S.H., Xiong, J.Y., Han, G., Wan, C.F., Chong, Q.Y., Chung, T.S., 2017. A pilot study on pressure retarded osmosis operation and effective cleaning strategies. *Desalination* 420, 273–282.

Lee, K.L., Baker, R.W., Lonsdale, H.K., 1981. Membranes for power generation by pressure-retarded osmosis. *J. Membr. Sci.* 8(2), 141–171.

Lee, S., Choi, J., Park, Y.-G., Shon, H., Ahn, C.H., Kim, S.-H., 2019. Hybrid desalination processes for beneficial use of reverse osmosis brine: Current status and future prospects. *Desalination* 454, 104–111.

Li, X., Cai, T., Amy, G.L., Chung, T.S., 2017. Cleaning strategies and membrane flux recovery on anti-fouling membranes for pressure retarded osmosis. *J. Membr. Sci.* 522, 116–123.

Loeb, S., 2002. Large-scale power production by pressure-retarded osmosis, using river water and sea water passing through spiral modules. *Desalination* 143(2), 115–122.

Loeb, S., Norman, R.S., 1975. Osmotic power plants. *Science* 189(4203), 654–655.

Park, Y.G., Chung, K., Yeo, I.H., Lee, W.I., Park, T.S., 2017. Development of a SWRO-PRO hybrid desalination system: Pilot plant investigations. *Water Supply* 18, 473–481.

Post, J.W., Veerman, J., Hamelers, H.V.M., Euverink, G.J.W., Metz, S.J., Nymeijer, K., Buisman, C.J.N., 2007. Salinity-gradient power: Evaluation of pressure-retarded osmosis and reverse electrodialysis. *J. Membr. Sci.* 288(1), 218–230.

Seppälä, A., Lampinen, M.J., 1999. Thermodynamic optimizing of pressure-retarded osmosis power generation systems. *J. Membr. Sci.* 161(1), 115–138.

She, Q., Wong, Y.K.W., Zhao, S., Tang, C.Y., 2013. Organic fouling in pressure retarded osmosis: Experiments, mechanisms and implications. *J. Membr. Sci.* 428, 181–189.

Skilhagen, S.E., Dugstad, J.E., Aaberg, R.J., 2008. Osmotic power – Power production based on the osmotic pressure difference between waters with varying salt gradients. *Desalination* 220, 476–482.

Thelin, W.R., Sivertsen, E., Holt, T., Brekke, G., 2013. Natural organic matter fouling in pressure retarded osmosis. *J. Membr. Sci.* 438, 46–56.

Thorsen, T., Holt, T., 2009. The potential for power production from salinity gradients by pressure retarded osmosis. *J. Membr. Sci.* 335(1), 103–110.

Wan, C.F., Chung, T.S., 2015. Osmotic power generation by pressure retarded osmosis using seawater brine as the draw solution and wastewater retentate as the feed. *J. Membr. Sci.* 479, 148–158.

Wan, C.F., Li, B., Yang, T., Chung, T.S., 2017. Design and fabrication of inner-selective thin-film composite (TFC) hollow fiber modules for pressure retarded osmosis (PRO). *Sep. Purif. Technol.* 172, 32–42.

Wang, Q., Zhou, Z., Li, J., Tang, Q., Hu, Y., 2019. Investigation of the reduced specific energy consumption of the RO-PRO hybrid system based on temperature-enhanced pressure retarded osmosis. *J. Membr. Sci.* 581, 439–452.

Xiong, J.Y., Cheng, Z.L., Wan, C.F., Chen, S.C., Chung, T.S., 2016. Analysis of flux reduction behaviors of PRO hollow fiber membranes: Experiments, mechanisms, and implications. *J. Membr. Sci.* 505, 1–14.

Yip, N.Y., Brogioli, D., Hamelers, H.V.M., Nijmeijer, K., 2016. Salinity gradients for sustainable energy: Primer, progress, and prospects. *Environ. Sci. Technol.* 50, 12072–12094.

Zhang, S., Sukitpaneenit, P., Chung, T.S., 2014. Design of robust hollow fiber membranes with high power density for osmotic energy production. *Chem. Eng. J.* 241, 457–465.

11

Techno-Economic Evaluation of Various RO+PRO and RO+FO Integrated Processes

Chun Feng Wan

Department of Chemical and Biomolecular Engineering
National University of Singapore
Singapore

CONTENTS

11.1 Introduction

Utilization of wastewater or wastewater retentate as the feed solution and seawater reverse osmosis (RO) brine as the draw solution has received increasing interests (Chung et al. 2015, Prante et al. 2014, Saito et al. 2012, Sakai et al. 2016, Skilhagen et al. 2008, Wan and Chung 2015, 2016). The new feed pair can potentially produce more osmotic energy, generate a higher power density and mitigate the environmental issue of discharging concentrated seawater brine. Sharqawy et al. and Feinberg et al. conducted thermodynamic analyses on the hybrid RO–PRO process and concluded that a 38% reduction in specific energy consumption (SEC) and a 20% increase in second law efficiency can be achieved (Feinberg et al. 2013, Sharqawy et al. 2011). In Chapter 9, the calculated SECs dropped to 1.08 and 1.14 kWh/m^3 for RO–PRO systems with 25% and 50% RO recovery, respectively (Wan and Chung 2016), meanwhile (Prante et al. 2014) obtained a similar SEC of 1.2 kWh/m^3 for a RO–PRO with 50% RO recovery.

Lab and pilot systems have been developed to investigate the technical feasibility of RO–PRO. Achilli et al. demonstrated the RO–PRO system but only achieved a power density of 1.1–2.3 W/m^2 and an energy saving of 1 kWh/m^3 due to the limited RO recovery and lack of effective PRO membranes (Achilli et al. 2014). The Japan Mega-ton water system had a prototype PRO plant to generate electricity from a 50% recovery RO brine. The reported power density from a 10-inch PRO module at 27 bar was 10.1–13.5 W/m^2 (Saito et al. 2012, Sakai et al. 2016). In the Korea GMVP project, RO was the first step to generate freshwater, which was followed by a membrane distillation (MD) process to increase the freshwater production and generate a more saline brine for PRO. The RO–MD–PRO process achieved a higher freshwater recovery with a reduced SEC (Kim et al. 2016). Lee et al. (2017) obtained a high power density of 5.7 W/m^2 in a PRO plant using brine from a 25% recovery RO unit as the draw solution and wastewater retentate from municipal water plants as the feed solution. However, most studies on RO–PRO focused solely on the osmotic energy recovery by PRO, but did not include the potential energy saving from (1) reduced seawater pretreatments, if the diluted brine can be recycled as the seawater feed to RO, (2) reduced pretreatments of the PRO draw solution (RO brine) since it has already been pretreated, and (3) reduced brine discharge.

Though the reduction in SEC by PRO is significant and attractive, the integration incurs additional costs that may offset the benefit of SEC reduction. There is limited information on the economic feasibility of RO–PRO. The breakeven power density of 5 W/m^2 is usually used as the benchmark for RO–PRO, but it was originally derived from the feed pair of river water and seawater (Gerstandt et al. 2008, Skilhagen et al. 2008, Thorsen and Holt 2009). On the other hand, forward osmosis (FO) can utilize seawater or seawater brine as the draw solution to extract water from impaired sources to (1) dilute the seawater feed before entering RO and therefore reduce the RO operating pressure or (2) dilute and recycle RO brine as the seawater feed to RO. Therefore, osmotic dilution by FO also offers great potential for energy and cost savings (Blandin et al. 2015, Chekli et al. 2016, Coday et al. 2015, Kim et al. 2015, Valladares Linares et al. 2016).

In this chapter, detailed process flow diagrams of various RO–PRO and RO–FO integrations are presented. Their technical performances are evaluated via well-established mass transfer models. Superior technical performances do not always translate into commercial successes. Beyond the technical evaluations, this chapter also presents the first economics-driven designs of RO–PRO and RO–FO integrated processes. The operating expenditure (OpEx) and capital expenditure (CapEx) of the integrated systems are investigated in the cost model to validate the economic feasibilities of the integrations and to guide the design of the hybrid systems.

11.2 Methodology

11.2.1 Process

Figure 11.1 shows a widely used conventional seawater RO design with an ERD (Hauge 1995, Zhu et al. 2009). A high-pressure pump (HP) pressurizes $R \times 100\%$ of seawater to the RO operating pressure. The remaining $(1 - R) \times 100\%$ of seawater is pressurized by the ERD and a booster pump (BP) to the RO operating pressure. The pretreated high-pressure seawater

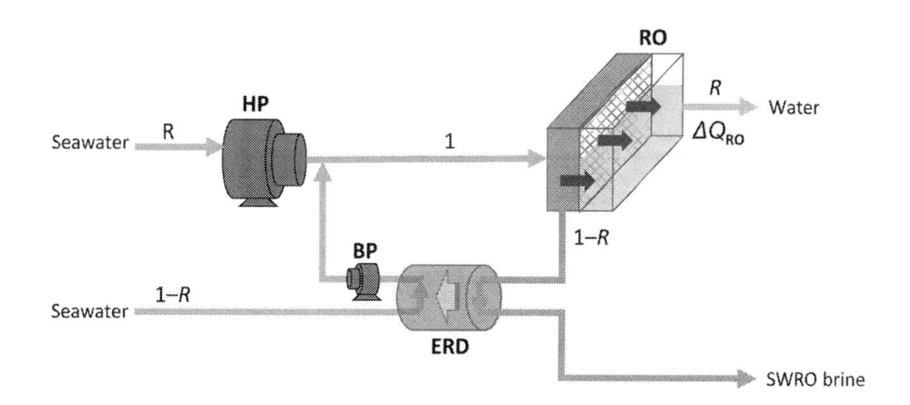

FIGURE 11.1
Seawater RO with an ERD. The denoted flow rates are normalized by the total seawater flow rate to RO.

enters the RO system, $R \times 100\%$ of which is recovered as freshwater. The remaining concentrated RO brine with a high pressure is then utilized in the ERD to pressurize $(1 - R) \times 100\%$ of the seawater feed. Because of the limited efficiency of the ERD, a BP is always required to make up the pressure loss in the ERD. The BP is left out in the process flow diagrams in Figure 11.2 (a)–(d) to simplify the drawings.

The concentrated brine from RO is an ideal source of draw solution for PRO. Because of (1) the requirement of a huge membrane area to transport the same amount of water in PRO as in RO processes and (2) the limited supply of wastewater or wastewater retentate from municipal water plants, the permeation volume in PRO is usually an x fraction of that in RO (i.e., $\Delta Q_{PRO} = x\Delta Q_{RO}$, $x \leq 1$) in actual applications. Figure 11.2(a) shows the integration of RO with an open-loop PRO, referred to as RO+oPRO thereafter. The high-pressure brine from RO passes through ERD1 and part of the energy carried by the high-pressure brine is used to pressurize the medium-pressure seawater in ERD1. The RO brine exiting ERD1 carries a medium pressure, the same as the operating pressure of PRO. In PRO, the brine draws water from the wastewater feed and becomes diluted. The diluted brine then releases its energy in ERD2 to pressurize part of the seawater feed to a medium pressure. Because $\Delta Q_{PRO} \leq \Delta Q_{RO}$, the flow rate of the diluted brine from PRO is less than the fresh seawater feed. The remaining seawater feed that cannot be pressurized by ERD2 needs to be pressurized by the low-pressure pump (LP) to the same medium pressure. The seawater outflow from ERD2 splits into two streams. One is pressurized by ERD1 and the other is pressurized by the HP to the RO operating pressure. The seawater from HP and ERD1 combines as the total seawater feed to RO. This design utilizes all the RO brine in PRO but requires additional ERD2 to harvest the osmotic energy.

Figure 11.2(b) shows an alternative integration of RO with a closed-loop PRO, referred to as RO+cPRO thereafter. The RO brine splits into two streams: $x \times 100\%$ goes through a lower loop similar to that of the RO+oPRO design, becomes diluted to the same concentration of seawater, and then is recycled as a feed to RO. Therefore, no additional ERD is required in the RO+cPRO design to harvest the osmotic energy. The remaining $(1 - x) \times 100\%$ of the RO brine goes through an upper loop similar to a conventional RO in Figure 11.1 to directly pressurize part of the seawater feed to the RO operating pressure and then is discharged without recovering its osmotic energy. It is worth noting that the total capacity of ERD1 and ERD2 in Figure 11.2(b) is the same as the capacity of ERD in Figure 11.1 and ERD1 in Figure 11.2(a).

In Figure 11.2(c), the RO brine completely releases its pressure in ERD and hence the FO unit operates at an ambient pressure, referred to as RO+FO thereafter. $x \times 100\%$ of the brine from ERD acts as the draw solution in FO, becomes diluted to the seawater concentration, and gets recycled to RO. The remaining brine that cannot be utilized in FO is

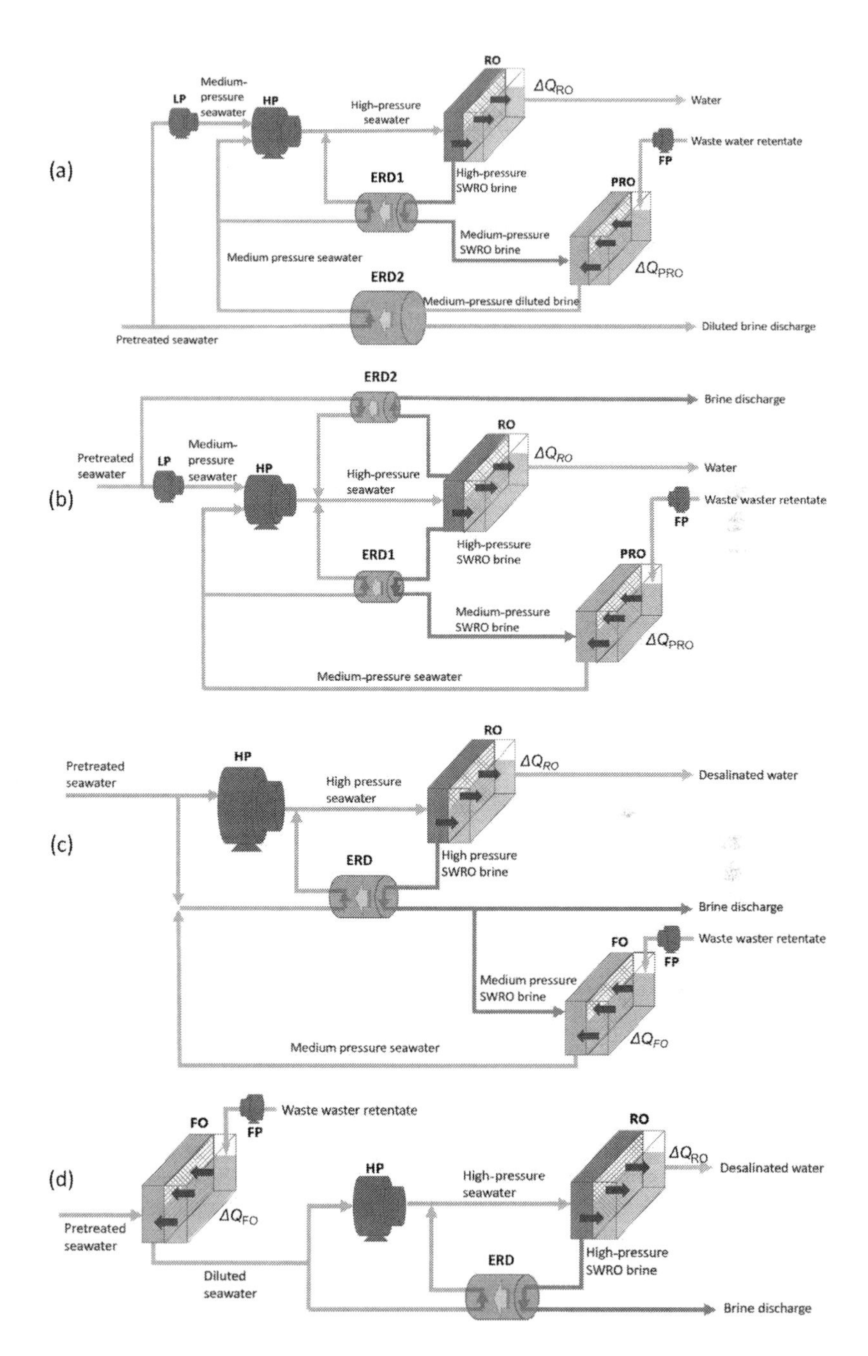

FIGURE 11.2
(a) RO and open-loop PRO integrated process (RO+oPRO). (b) RO and closed-loop PRO integrated process (RO+cPRO). (c) RO and FO integrated process (RO+FO). (d) FO and RO integrated process (FO+RO).

directly discharged. If 100% of the brine is used as the draw solution in FO, there may not be enough wastewater feed to fully dilute the brine to the seawater concentration. If such insufficiently diluted brine with a high salt concentration is recycled to RO, it will cause salt accumulation within the system and therefore cannot maintain a steady-state operating condition. Because of the brine reuse, RO+FO and RO+cPRO are able to reduce CapEx on seawater intake and pretreatment (SWIP), and brine discharge.

Another possible integration of FO with RO, as presented in Figure 11.2(d), is to use FO to dilute the seawater feed before entering RO, referred to as FO+RO thereafter. Therefore, RO can operate at a reduced pressure and achieve the same recovery. This leads to a reduced SEC and a lower operating cost. The brine from RO is directly used to pressurize the seawater by an ERD as in a conventional RO and then is discharged.

The capacity of each unit normalized by the total seawater flow rate into RO is tabulated in Table 11.1 and denoted in Figure 11.1. The analysis is conducted for an RO with a total seawater flow rate of 200,000 m^3/day. At a given RO recovery, the capacities of HP and RO remain the same in all cases, as shown in Table 11.1. This eliminates the complexity and uncertainty arisen from different RO systems and simplifies the cost model. The following analyses will focus on the savings on CapEx and OpEx of the additional equipment required for the integrations.

TABLE 11.1

Capacity of each unit normalized by the total seawater feed to RO (Q_{RO})

Unit	Abbreviation	RO	RO+oPRO	RO+cPRO	RO+FO	FO+RO
Seawater intake and pretreatment	SWIP	1	1	$1 - x$	$1 - x$	$1 - xR$
Low-pressure pump	LP		$R - xR$	$R - xR$		
High-pressure pump	HP	R	R	R	R	R
Energy recovery device 1	ERD1	$1 - R$	$1 - R$	$x(1 - R)$	$1 - R$	$1 - R$
Energy recovery device 2	ERD2		$1 - R + xR$	$(1 - x)(1 - R)$		
Booster pump 1	BP1	$1 - R$	$1 - R$	$x(1 - R)$	$1 - R$	$1 - R$
Booster pump 2	BP2			$(1 - x)(1 - R)$		
Pressure retarded osmosis/forward osmosis	PRO/FO		xR	xR	xR	xR
Feed pump	FP		xR/80%	xR/80%	xR/80%	xR/80%
Brine discharge	BD	$1 - R$	$1 - R + xR$	$(1 - x)(1 - R)$	$(1 - x)(1 - R)$	$1 - R$

11.2.2 Technical and Economic Evaluations

11.2.2.1 Seawater Intake and Pretreatments

A correlation of combined SWIP capital expenditure (CE, in 1995 USD) has been widely used as follows (Kim et al. 2009, Malek et al. 1996, Marcovecchio et al. 2005, Vince et al. 2008):

$$CE_{SWIP} = 12,659.84Q_{SWIP}^{0.8} \tag{1}$$

where Q_{SWIP} is the flow rate of the seawater feed in m^3/h.

The pressure (ΔP_{SWIP}) required for RO pretreatment is about 0.2–1 bar (Avlonitis et al. 2003, Straub et al. 2016). An average value of 0.6 bar is used to calculate the energy consumption (E) of pretreatments.

$$E_{SWIP} = \frac{\Delta P_{SWIP}Q_{SWIP}}{36\eta_P} \tag{2}$$

where η_P is the pump efficiency of 85%, the factor 36 is to convert the unit of E to kW.

11.2.2.2 Energy Recovery Devices

The following cost correlation (in 1995 USD) between flow rate (Q_{ERD}) and operating pressure (P_{ERD}) of an ERD was initially developed for a reversely running centrifugal pump. The correlation can be extended to estimate the cost of other ERDs, such as a pressure exchanger (Choi et al. 2016, Kim et al. 2009, Malek et al. 1996, Marcovecchio et al. 2005, Vince et al. 2008).

$$CE_{ERD} = \frac{Q_{ERD}}{450}(393,000 + 10,710P_{ERD}) \tag{3}$$

The amount of energy recovered by an ERD is

$$E_{ERD} = \frac{\Delta P_{ERD}Q_{ERD}\eta_{ERD}}{36} \tag{4}$$

ERD usually requires equal flow rates of the low-pressure and high-pressure streams in order to minimize mixing and maximize its efficiency (Migliorini and Luzzo 2004, Stover 2007). With equal flow rates, the energy balance inside the ERD can be simplified as follows:

$$\Delta P_{ERD\_LP} = -\Delta P_{ERD\_HP}\eta_{ERD} \tag{5}$$

The subscript ERD_LP stands for the low-pressure stream, such as the seawater feed into the ERD, while ERD_HP stands for the high-pressure stream, such as the RO brine into the ERD. ΔP is the pressure change of each stream and η_{ERD} is the efficiency of the ERD. The reported efficiency of ERD is in the range of 90–96%, and an average value of 93% is used in the calculation.

11.2.2.3 Pumps

Centrifugal pumps are widely used in desalination plants. Their costs are calculated as:

$$\mathrm{CE}_P = (F_m)(F_{\mathrm{pr}})(\text{base cost}) \tag{6}$$

where F_{m} and F_{pr} are the material adjustment factor and pressure adjustment factor, respectively. Stainless steel with a F_{m} of 2.4 is the most popular material for seawater pumps. The correlation between the base cost and flow rate and the correlation between F_{pr} and P are interpolated from the data provided in Peters et al. (2003).

$$(\text{base cost}) = 844.31 {Q_P}^{0.3726} \tag{7}$$

$$F_{\mathrm{pr}} = 1, \ P_P \leq 10.35 \text{ bar} \tag{8a}$$

$$F_{\mathrm{pr}} = 0.3968 P^{0.4132}, \ 10.35 \text{ bar} < P_P \leq 300 \text{ bar} \tag{8b}$$

where the base cost is in 2002 USD, Q_P is the capacity of the centrifugal pump in m^3/h, P_p is the operating pressure of the centrifugal pump in bar. These correlations can be used to estimate the costs of LPs for RO, FPs for PRO, and the BPs for ERDs with the appropriate pressure and material factors.

The energy consumption of the centrifugal pump can be calculated from the following equation.

$$E_P = \frac{\Delta P_P Q_P}{36 \eta_P} \tag{9}$$

11.2.2.4 Brine Discharge

Most large seawater desalination plants worldwide dispose their concentrates through surface water discharge to an open water body. The correlation in 2013 USD is extracted from the data report in Voutchkov (2013).

$$CE_{DIS} = -0.1152Q_{Dis}^{2} + 4396.8Q_{Dis} + 117,480 \quad (10)$$

The RO brine discharged from the ERD usually carries a residual pressure of 2 bar and therefore no additional pumping is required for surface water discharge (Voutchkov 2013). However, this residual pressure is a loss of energy that cannot be recovered by the ERD.

11.2.2.5 PRO and FO

PRO and FO are driven by the osmotic pressure difference. However, due to the detrimental effects of concentrative internal concentration polarization, concentrative external concentration polarization, dilutive external concentration polarization, and salt leakage, the effective osmotic pressure difference ($\Delta\pi_{eff}$) is much smaller. The following model describes such effects on $\Delta\pi_{eff}$ for flat-sheet membranes, which can be further extended for hollow fiber membranes (Achilli et al. 2009, Cheng and Chung 2017, Han et al. 2015, Lee et al. 1981, Sivertsen et al. 2012, 2013, Thorsen and Holt 2009, Xiong et al. 2017).

$$\Delta\pi_{eff} = \frac{\pi_D \exp(-\frac{J_w}{k}) - \pi_F \exp(\frac{J_w S}{D})}{1 + \frac{B}{J_w}[\exp(\frac{J_w S}{D}) - 1]} \quad (11)$$

The water flux (J_w in L/m^{2}h) and reverse salt flux (J_s in g/m^{2}h) can be calculated as follows:

$$J_W = A(\Delta\pi_{eff} - \Delta P_{PRO}) \quad (12)$$

$$J_S = \frac{B}{iRT}\left(\frac{J_W}{A} - \Delta P_{PRO}\right) \quad (13)$$

where A is the pure water permeability, B is the salt permeability, k is the mass transfer coefficient of the draw solution, and S is the structural parameter. Table 11.2 summarizes the values of these membrane properties used in calculations.

The above equations can be applied to FO applications with the active layer facing the draw solution (AL-DS) by setting ΔP_{PRO} to 0. The AL-DS orientation (i.e., the PRO mode) usually generates a high flux due to the less severe concentration polarizations. However, this orientation is more prone to fouling because the porous support layer is facing wastewater or wastewater retentate (She et al. 2012). Studies show that fouling from wastewater and wastewater retentate can be effectively mitigated by ultrafiltration and chemical treatments (Chen et al. 2015, 2016, Saito et al. 2012,

TABLE 11.2

Summary of membrane characteristics

Membrane Characteristics	
A (LMH/bar)	3.5
B (LMH)	0.3
S (μm)	450
Outer diameter (μm)	1025
Inner diameter (μm)	575

Sakai et al. 2016). With appropriate pretreatments, it is more favorable to operate FO in the AL-DS orientation to yield a higher water flux.

To describe the mass transfer inside the PRO module, the one-dimensional counter-current flow PRO model in Chapter 9 is employed. Modules consisting of 8-inch elements with characteristics listed in Table 11.3 and a total membrane area (M) of 30 m^2 are used in PRO (Wan et al. 2017). The membrane cost is assumed to be similar to that of RO membranes of \$25/m^2 with a lifetime of 5 years. Six elements are used in one pressure vessel with a unit vessel price of \$1000. The cost of PRO modules (CE_{PRO}) and membrane replacement cost (C_{MEM}) are then calculated as follows:

$$CE_{PRO} = 25M + \frac{1000M}{30 \times 6} \tag{14}$$

$$C_{MEM} = \frac{1}{5} \times 25M \tag{15}$$

11.2.2.6 Chemical Pretreatment and Cleaning

Chemical pretreatment and cleaning have demonstrated great effectiveness in reducing fouling and recovering membrane performance (Chen et al. 2015,

TABLE 11.3

Summary of membrane module characteristics

Module Characteristics	
Diameter (inch)	8
Length (inch)	40
Packing density (%)	42
Number of hollow fibers	16,500
Membrane area (m^2)	30

2016, Han et al. 2016, Lee et al. 2017, Li et al. 2017, Saito et al. 2012, Sakai et al. 2016, Thelin et al. 2013). Based on chemical pretreatment and cleaning protocols established in our previous studies, fouling can be reduced if the feed solution is acidified with a 0.115 ppm HCl solution (to mitigate the calcium phosphate scaling) (Chen et al. 2016) and then maintenance cleaning is performed by flushing the supportive layer with a 200 ppm HCl solution for 60 min for every 12 h of operation (Lee et al. 2017). Therefore, the chemical cost for 8400 working hours every year is calculated at a price of $250/ton for the 31% HCl solution.

$$C_{CHE} = \left(0.115Q_F + \frac{200Q_F}{13}\right) \times 8400 \times \frac{250}{31\% \times 10^6} \tag{16}$$

11.2.3 Economic Analysis

The CapEx saving arising from the system integration is calculated as the summation of the change in CapEx (ΔCE) of each unit as compared to the conventional RO.

$$\Delta CE = \Delta CE_{SWIP} + \Delta CE_{LP} + \Delta CE_{ERD} + \Delta CE_{BP} + \Delta CE_{DIS} + \Delta CE_{PRO} + \Delta CE_{FP} \tag{17}$$

The annualized CapEx Saving (ACS) in $/year is calculated at an interest rate of i =3% for a plant life (LT) of 20 years.

$$ACS = (\Delta CE)\frac{i(1+i)^{LT}}{(1+i)^{LT} - 1} \tag{18}$$

The annual OpEx saving (AOS) in $/year is calculated at an electricity cost of 0.12 $/kWh for 8400 working hours every year. The AOS includes (1) the energy consumption, (2) membrane replacement, and (3) chemical costs (Avlonitis et al. 2003).

$$AOS = 0.12 \times 8400 \times (\Delta E_{SWIP} + \Delta E_{LP} + \Delta E_{ERD} + \Delta E_{BP} + \Delta E_{DIS} + \Delta E_{PRO} + \Delta E_{FP}) + C_{MEM} + C_{CHE} \tag{19}$$

The annual total saving (ATS) is the sum of ACS and AOS.

$$ATS = ACS + AOS \tag{20}$$

The change in the RO desalination cost (cent/m^3) is calculated by dividing the ATS by the total flow rate of RO permeate every year. The negative sign indicates an increase in RO costs.

$$\Delta(\text{RO cost}) = \frac{-\text{ATS} \times 100}{Q_{\text{RO}} \times 8400} \tag{21}$$

11.3 Results and Discussions

11.3.1 Optimal Operating Pressure and Maximum Power Density of PRO

Figure 11.3(a) and (b), respectively, show the optimal operating pressures and maximum average power densities of an oPRO and a cPRO with draw solutions from 25% and 50% recovery ROs, as a function of ΔQ_{PRO} to ΔQ_{RO} ratio (referred to as the permeation ratio, x, thereafter). As the permeation ratio increases, the RO brine draws more water from the feed and becomes more diluted. Because of the reduced average osmotic pressure difference, the optimal operating pressure and maximum average power density of oPRO decrease as well. For oPRO with the brine from a 25% recovery RO as the draw solution, the optimal operating pressure and maximum average power density of oPRO decrease from 20.6 bar and 18.0 W/m^2 to

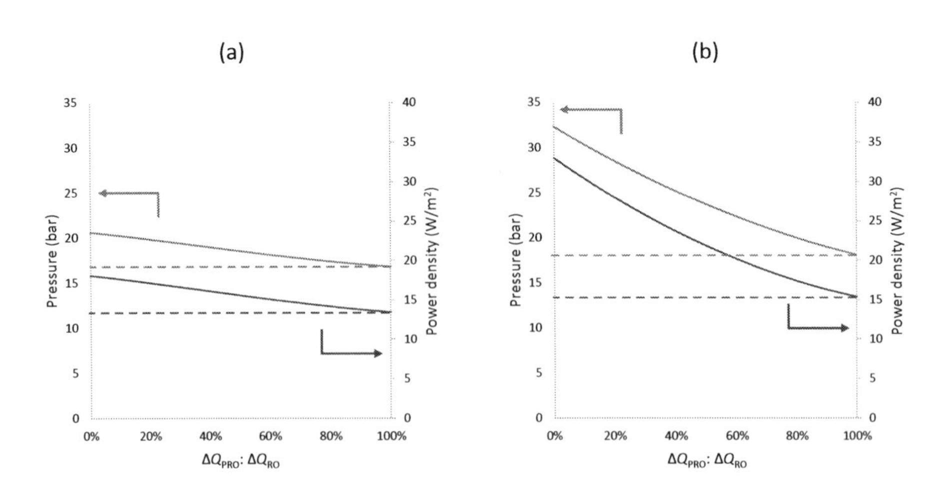

FIGURE 11.3
The optimal operating pressures and maximal average power densities of PRO with brines from (a) a 25% recovery RO and (b) a 50% recovery RO as the draw solution and 0.01 M NaCl solution as the feed solution in an open-loop PRO (—) and closed-loop PRO (—).

16.8 bar and 13.4 W/m^2, respectively. The brine from a 50% recovery RO has a higher osmotic pressure and can potentially generate a higher power density. With the brine from a 50% recovery RO as the draw solution, the optimal operating pressure and maximum average power density of oPRO decrease from 32.0 bar and 32.9 W/m^2 to 18.2 bar and 15.3 W/m^2, respectively.

In cPRO, the brine is always diluted to the same salinity of seawater. This is equivalent to the case of a permeation ratio of 100% in oPRO when the same amount of water is recovered from RO and then added to PRO. Therefore, cPRO generates a smaller power density and a smaller total energy output as compared to oPRO.

11.3.2 Specific Energy Recovery by PRO

The total amount of energy released from PRO can be calculated as the product of the optimal operating pressure of PRO and ΔQ_{PRO}. At the same permeation ratio, oPRO has a higher optimal operating pressure as shown in Figure 11.3 and therefore a higher total energy output. ERDs are utilized to harvest the osmotic energy released from oPRO. However, each ERD has a limited efficiency due to the friction loss and mixing of streams. Moreover, the high-pressure outflow from ERD2 of Figure 11.2(a) carries a residual pressure of 2 bar that cannot be recovered. Therefore, a significant amount of energy is lost in this ERD. On the contrary, cPRO recycles the diluted brine directly as the seawater feed to RO, as shown in Figure 11.2(b). It therefore does not suffer from the energy loss of brine discharge at 2 bar.

As shown in Figure 11.4(a) with a 25% recovery RO, the specific energy saved from oPRO per m^3 of product water increases with the permeation ratio and reaches a maximum of 0.47 kWh/m^3. However, only 82% of the energy is effectively harvested on average because of the efficiency losses from ERD2 and brine discharge at 2 bar. Though less osmotic energy is saved from cPRO, all osmotic energy is effectively harvested. The specific energy saving of cPRO increases linearly to 0.47 kWh/m^3 and exceeds that of oPRO from a permeation ratio of 20% onwards. With a 50% recovery RO in Figure 11.4(b), the maximum specific energy saved from oPRO is 0.51 kWh/m^3 and 84% can be harvested on average. Initially, more osmotic energy can be harvested from oPRO up to a permeation ratio of 70%. After that, cPRO achieves a higher specific energy saving.

11.3.3 PRO/FO Water Flux and Membrane Area

The required PRO and FO membrane areas in hybrid systems with 25% and 50% recovery RO as a function of permeation ratio are presented in Figure 11.5 (a) and (b), respectively. RO+FO generates the highest water flux because of the high salinity of the draw solution and its ambient operating pressure, and hence requires the least membrane area. FO+RO takes seawater instead of

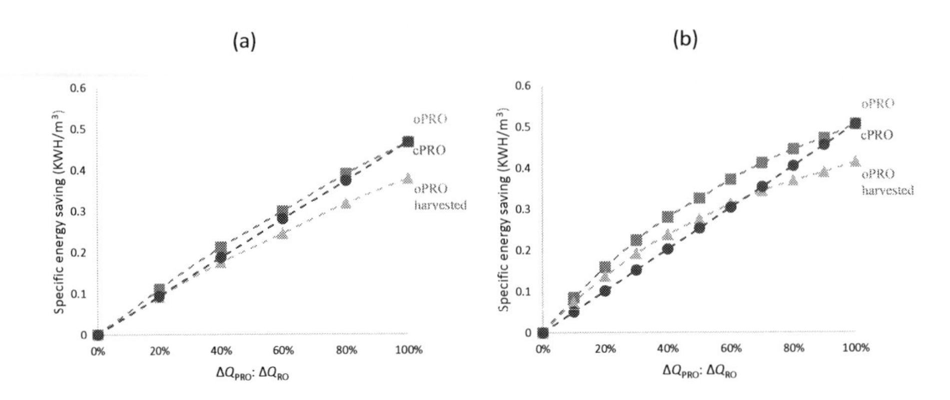

FIGURE 11.4
Specific energy savings of oPRO and cPRO integrated with (a) a 25% recovery RO and (b) a 50% recovery RO.

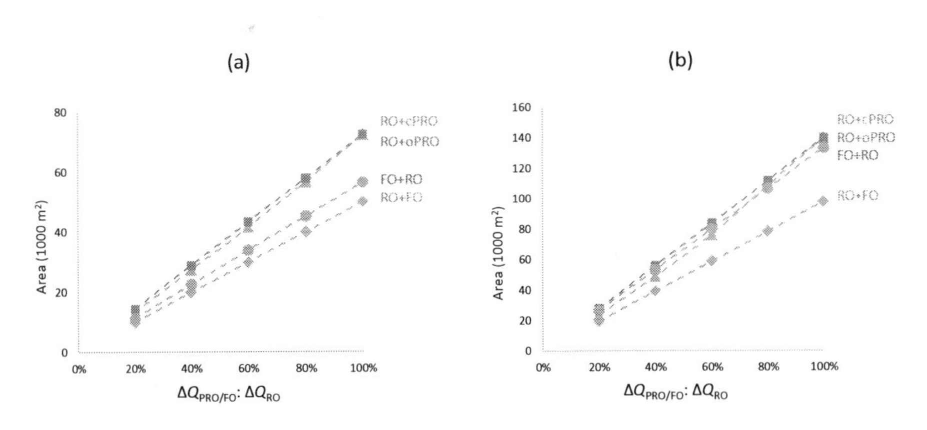

FIGURE 11.5
The required PRO and FO membrane areas when integration with (a) a 25% recovery RO and (b) a 50% recovery RO.

seawater brine as the draw solution, which results in a smaller flux and a higher membrane area than RO+FO. PRO requires a relatively higher membrane area than FO because of the reduced water flux at elevated operating pressures. RO+oPRO utilizes all the brine as the draw solution. Therefore, it has a higher water flux and requires less membrane areas than RO+cPRO.

ΔQ_{PRO} and ΔQ_{FO} are doubled in the 50% recovery RO integrations. A 50% recovery RO generates a more concentrated brine and therefore induces higher average PRO and FO water fluxes. Therefore, compared to the 25% recovery RO integrations, the increases of membrane areas in RO+oPRO,

RO+cPRO, and RO+FO are less than 100% (84.5%, 92.8%, and 95.5%, respectively). The seawater pre-dilution by FO has a higher dilution in a 50% recovery FO+RO integration, which results in a lower average water flux and 135% more membrane area than a 25% recovery FO+RO.

11.3.4 Energy Cost Saving

Figure 11.6(a) and (b) summarize the monetary value of energy saving including energy savings from SWIP and HP. RO+cPRO and RO+FO recycle the diluted brine as the seawater feed to RO and eliminate the need of SWIP for the recycled seawater. In FO+RO, FO also functions as the pretreatment unit that draws clean water to the seawater feed. Therefore, as tabulated in Table 11.1, only $(1 - x) \times 100\%$ of the seawater feed needs to be pretreated in RO+cPRO and RO+FO, and only $(1 - xR) \times 100\%$ of the seawater feed needs to be pretreated in FO+RO.

The major energy cost saving comes from the reduced energy consumptions of HPs. There are two ways to reduce the HP energy consumptions. oPRO and cPRO recover the osmotic energy from the concentrated brine to compensate the HP energy consumptions. As discussed in Section 3.2, though cPRO saves less energy, it is more efficient to deliver the energy to the seawater feed. As a result, compared to RO+oPRO, RO+cPRO almost doubles the energy saving in the 25% recovery RO integration. The latter also achieves up to 50% more energy saving than the former from a permeation ratio of 30% onwards in the 50% recovery RO integration, as shown in Figure 11.6(b).

Since RO+cPRO has energy savings from SWIP and PRO, while RO+FO only has energy saving from SWIP, the difference between RO+cPRO and RO+FO is the energy saving from cPRO. FO+RO reduces the HP energy

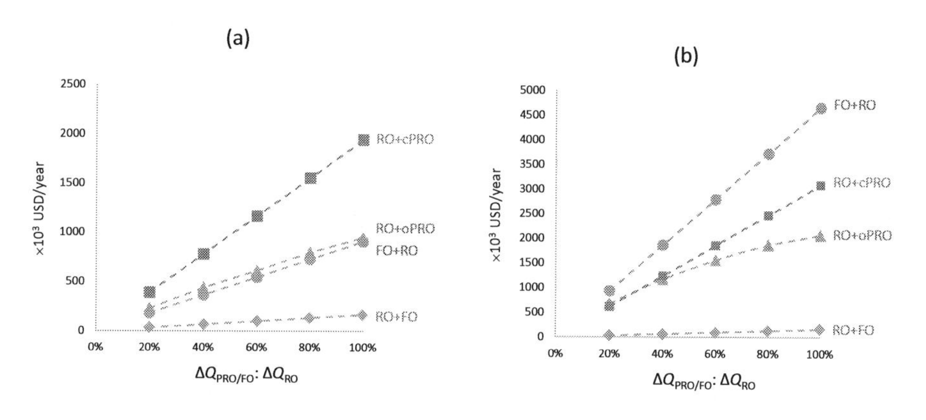

FIGURE 11.6
Energy saving of various PRO and FO integrations with (a) a 25% recovery RO and (b) a 50% recovery RO. The RO system has a fixed seawater input of 200,000 m^3/day.

consumption by diluting the seawater feed and enabling RO to operate at a lower pressure to achieve the same freshwater recovery. In a 25% recovery RO, the operating pressure is reduced from 42 to 31 bar, resulting in a saving up to $905,000/yr. This saving becomes more significant when the RO recovery is increased to 50%. FO pre-dilution halves the operating pressure of the 50% recovery RO and achieves a saving in energy cost up to $4,664,000/yr. Among the 50% recovery RO integrations, the saving from FO+RO is the greatest and even 50.8% more than that of RO+cPRO.

The saving in energy cost can be also reflected in terms of SEC value for seawater desalination. As illustrated in Figure 11.7(a) and (b), the SEC of a 25% recovery RO can be reduced from 1.95 to 1.83, 1.47, 1.46, and 0.98 kWh/m^3 by RO+FO, FO+RO, RO+oPRO, and RO+cPRO, respectively. The 50% recovery RO has a higher SEC of 2.28 kWh/m^3 due to its higher operating pressure. RO+FO, RO+oPRO, RO+cPRO, and FO+RO can effectively reduce the SEC to 2.24, 1.79, 1.55, and 1.17 kWh/m^3, respectively. In both cases, the most effective integrations are able to reduce SECs of seawater desalination by up to 50% at a permeation ratio of 100%.

11.3.5 Annual OpEx Saving

The AOS includes the energy, membrane replacement, and chemical costs. Since energy cost saving is the major part of AOS, the trend of AOS in Figure 11.8(a) and (b) agrees with the trend in Figure 11.6(a) and (b) but is shifted downwards by the respective costs of membrane replacement and chemical consumption. Since cPRO and oPRO require relative more membrane areas, their integrations incur the highest membrane replacement costs. In both cases, RO+cPRO, FO+RO, and RO+oPRO produce positive

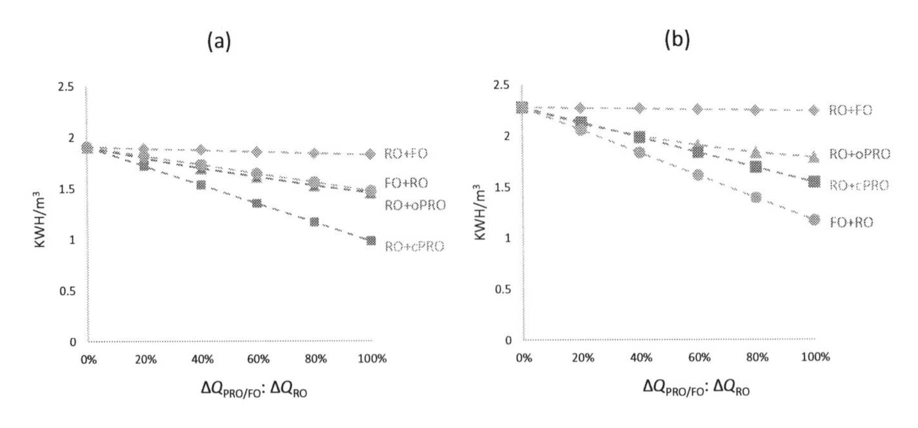

FIGURE 11.7
SEC of (a) 25% recovery RO integrations and (b) 50% recovery RO integrations.

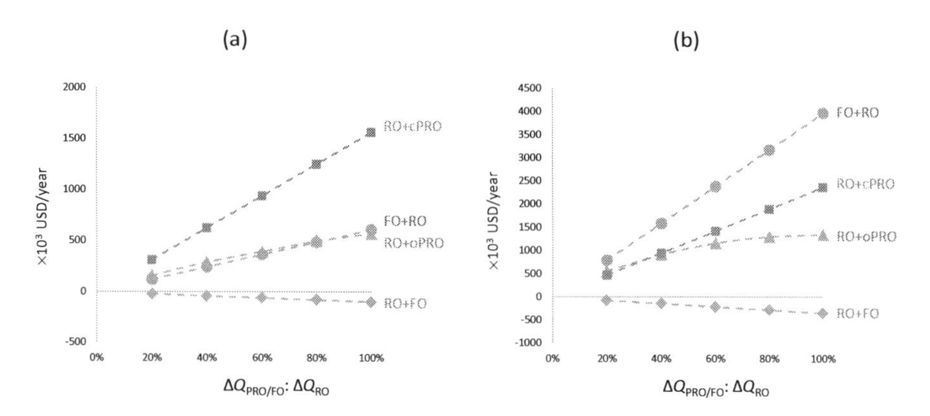

FIGURE 11.8
AOS of (a) 25% recovery RO integrations and (b) 50% recovery RO integrations. The RO system has a fixed seawater input of 200,000 m^3/day.

AOS. Among the 25% recovery RO integrations, RO+cPRO and FO+RO can save up to $1,562,000/yr and $605,000/yr, respectively. Among the 50% recovery RO integrations, FO+RO achieves the highest saving up to $3,971,000/yr, followed by RO+cPRO up to $2,367,000/yr. However, due to the low energy saving of RO+FO, the total AOS becomes negative when the costs of membrane replacement and chemical consumption are taken into consideration. Based on the cost model, RO+FO can achieve positive AOS only if the FO membrane cost drops from $25/m^2 to $12/m^2 for a 25% recovery RO+FO, and to $6/m^2 for a 50% recovery RO+FO.

11.3.6 Annualized CapEx Saving

The integrated processes require additional pumps, ERDs, and membrane modules. This incurs additional CapEx, denoted by the negative sign, as shown in Figure 11.9(a) and (b). RO+FO requires only FO modules and FPs for the process integration. The additional CapEx of FO+RO is slightly higher because its FO unit has a lower flux by using seawater as the draw solution and therefore requires more FO modules. RO+cPRO requires a LP to pressurize the seawater feed to a medium high pressure and therefore it further increases the cost. The additional CapEx of RO+oPRO is the highest because of the requirement of additional ERD2. The additional capacity of ERD2 is $(1 + x) \times 100\%$ of ERD1. Therefore, a bigger ERD2 is needed as the capacity of oPRO increases. The ACS of ERD2 in RO+oPRO increases from −$807,000/yr to −$954,000/yr in a 25% recovery RO+oPRO, and −$699,000/yr to −$979,000/yr in a 50% recovery RO+oPRO.

Other components that contribute to ACS are the SWIP and brine discharge units. As shown in Figure 11.10(a) and (b), RO+oPRO has an

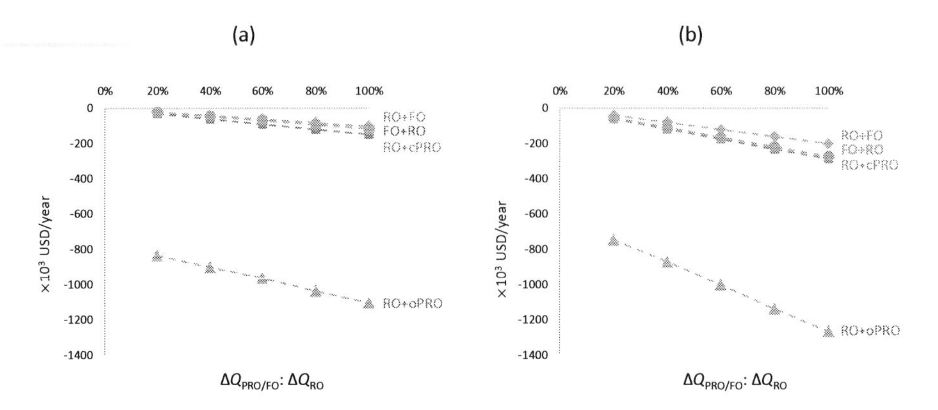

FIGURE 11.9
ACS of pumps (LP, FP, and BP), ERDs, and membrane modules of (a) 25% recovery RO integrations and (b) 50% recovery RO integrations. The RO system has a fixed seawater input of 200,000 m^3/day.

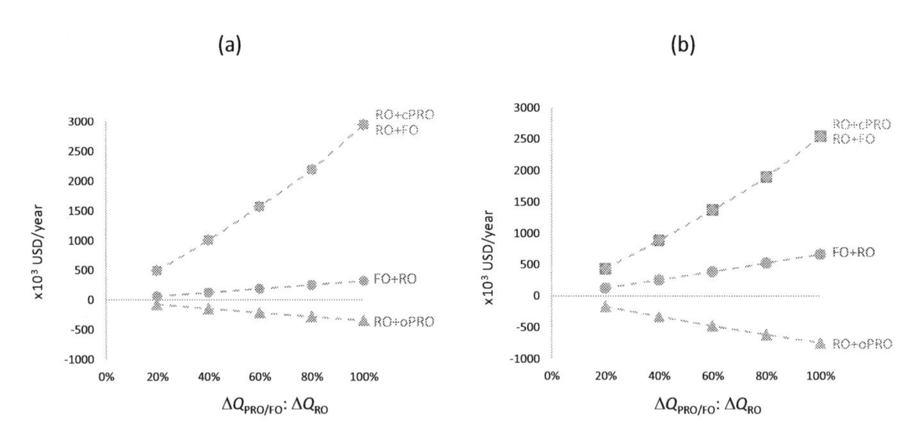

FIGURE 11.10
ACS of SWIP and brine discharge units of (a) 25% recovery RO integrations and (b) 50% recovery RO integrations. The RO system has a fixed seawater input of 200,000 m^3/day.

increased volume of brine to be discharged, which increases the CapEx by up to $342,000/yr and $745,000/yr in the 25% and 50% recovery processes, respectively. FO+RO draws wastewater or wastewater retentate as the feed to RO and therefore downsizes SWIP. RO+cPRO and RO+FO recycle some or even all of the brine to replace the seawater feed to RO, which significantly downsizes both SWIP and the brine discharge unit. Therefore, RO+cPRO and RO+FO can achieve a positive ACS up to $2,500,000/yr.

11.3.7 ATS and Change in RO Desalination Cost

Two scenarios are considered when calculating ATS. In the first scenario, PRO and FO are integrated with an existing RO to form a hybrid process. For RO+cPRO, RO+FO, and FO+RO, only the ACS of additional pumps, ERDs, and membrane modules, as shown in Figure 11.9, are included. However, the previous costs to build the oversized SWIP and discharge units cannot be recovered. On the other hand, the negative ACS to expand the SWIP and discharge units in RO+oPRO needs to be included in its cost analysis, as shown in Figure 11.10. In general, ATS in Figure 11.11(a) and (b) follows the trend of AOS in Figure 11.8, except for the case of RO+oPRO. There is a significant drop in ATS due to the additional requirements of ERDs, SWIP, and discharge units. RO+FO is the only integration system with a negative AOS and an even higher negative ATS when an existing RO is modified to RO+FO. Therefore, it is not economic to add oPRO or FO post dilution to an existing RO. Among the 25% recovery RO integrated processes, the highest annual saving up to $1,413,000 is achieved by RO+cPRO, followed $489,000 by FO+RO.

Figure 11.12(a) and (b) show the variations of cost saving for 25% and 50% recovery RO, respectively. For the former, the increases of RO desalination cost are approximately 3.1–3.9 cent/m^3 for RO+oPRO and 0.3–1.2 cent/m^3 for RO+FO, while the decreases of RO desalination cost are approximately 1.6–8.1 cent/m^3 for RO+cPRO and 0.5–2.8 cent/m^3 for FO+RO. For the latter, the highest annual saving up to $3,698,000 is achieved by FO+RO, followed by $2,081,000 achieved by RO+cPRO. Therefore, the increases of RO desalination cost are approximately 1.0–1.9 cent/m^3 for RO+oPRO and 0.3–1.6 cent/m^3 for RO+FO. In contrast, the RO desalination costs decrease approximately 1.2–5.9 cent/m^3 for RO+cPRO and 2.1–10.6 cent/m^3 for FO+RO.

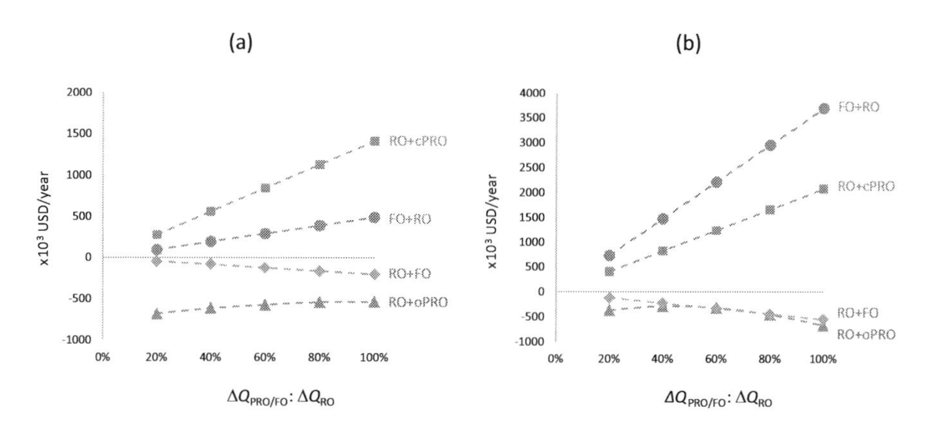

FIGURE 11.11
ATS by integrating PRO and FO with an existing (a) 25% recovery RO and (b) 50% recovery RO. The RO system has a fixed seawater input of 200,000 m^3/day.

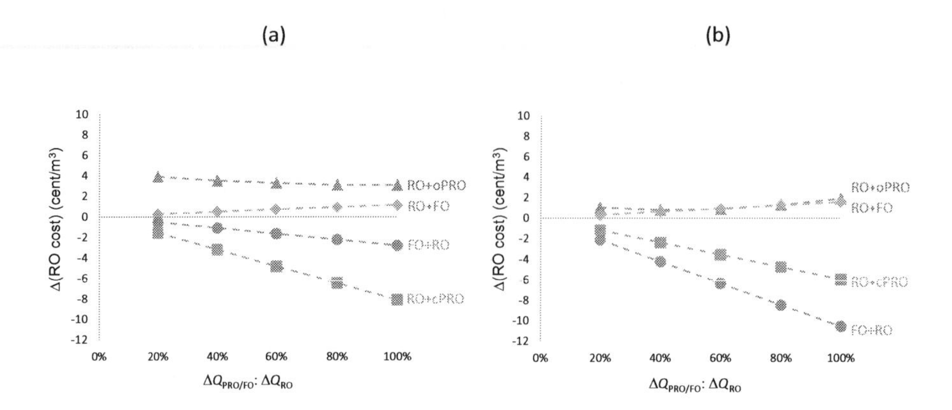

FIGURE 11.12
Changes in RO desalination costs by integrating PRO and FO with an existing (a) 25% recovery RO and (b) 50% recovery RO. The RO system has a fixed seawater input of 200,000 m^3/day.

In the second scenario, new integrated processes are to be built from scratch. Both ACS of pumps, ERDs, and membrane modules in Figure 11.9 and ACS of SWIP and brine discharge units in Figure 11.10 are included in the cost model for all processes. Among the 25% recovery RO integrations as shown in Figures 11.13(a) and 11.14(a), RO+cPRO achieves the highest cost saving up to $4,362,000/yr and 24.9 cent/m^3, due to the huge saving on ACS. RO+FO achieves the second highest cost savings up to $2,744,000/yr and 15.7 cent/m^3, due to the huge saving on ACS. The difference in savings between RO+cPRO and RO+FO can be attributed to

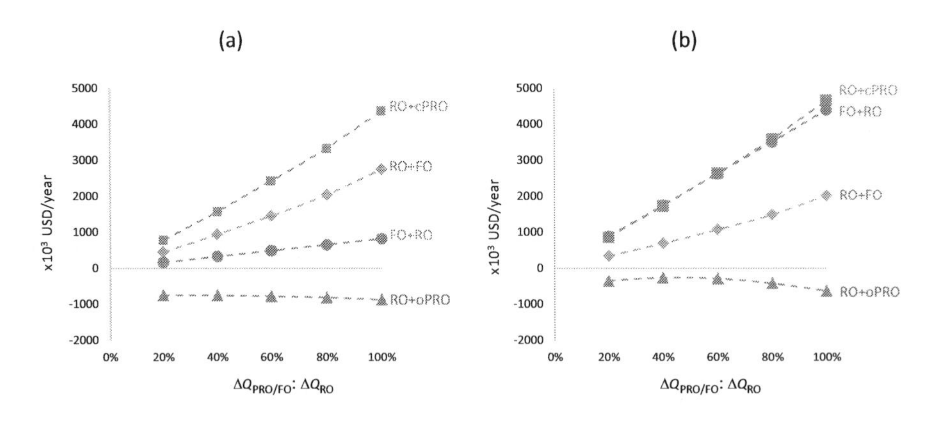

FIGURE 11.13
ATS by building new (a) 25% recovery RO integrated systems and (b) 50% recovery RO integrated systems. The RO system has a fixed seawater input of 200,000 m^3/day.

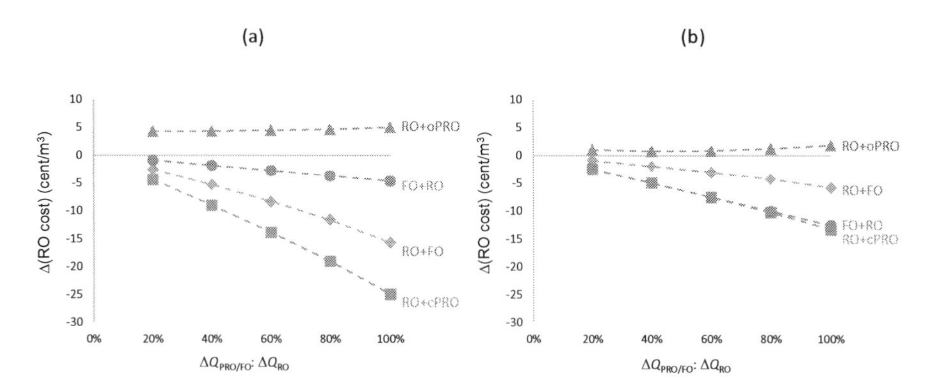

FIGURE 11.14
Changes in RO desalination costs by building new (a) 25% recovery RO integrated systems and (b) 50% recovery RO integrated systems. The RO system has a fixed seawater input of 200,000 m^3/day.

the energy saving by PRO in RO+cPRO. A higher cost saving is possible with a higher recovery RO integration. In Figures 11.13(b) and 11.14(b) of 50% recovery RO integrations, the highest cost saving up to \$4,655,000/yr and 13.3 cent/m^3 is achieved by RO+cPRO, closely followed by FO+RO with a cost saving up to \$4,390,000/yr and 12.5 cent/m^3. The ATS of FO+RO increases from \$810,000/yr and 4.63 cent/m^3 with a 25% recovery RO to \$4,390,000/yr and 12.5 cent/m^3 with a 50% recovery RO because of the significant reductions in RO operating pressure and energy consumption. However, the ATS of RO+oPRO remains negative despite of its potential for high osmotic energy recovery, causing 1–1.8 cent/m^3 increases in RO desalination costs.

11.3.8 Influence of PRO and FO Membrane Prices

So far, the prices of PRO and FO membranes are assumed to be the same as commercial RO membranes. However, as emerging technologies, PRO and FO may initially command higher membrane prices. The membrane prices will come down as the technologies mature and may eventually become lower than RO membranes because of the relatively lower operating pressures of PRO and FO. Figure 11.15(a) and (b) show the influence of PRO and FO membrane prices on ATS when building new 25% and 50% recovery RO integrated processes, respectively. The PRO/FO to RO permeation ratio is fixed at 50%. In both figures, ATS decreases due to the increased module cost and membrane replacement cost. Therefore, it is crucial to make PRO and FO membranes economical and durable with prolonged lifetimes. With a 25% recovery RO, RO+cPRO and RO+FO, respectively, maintain positive ATS of \$1,267,000/yr and \$685,000/yr up

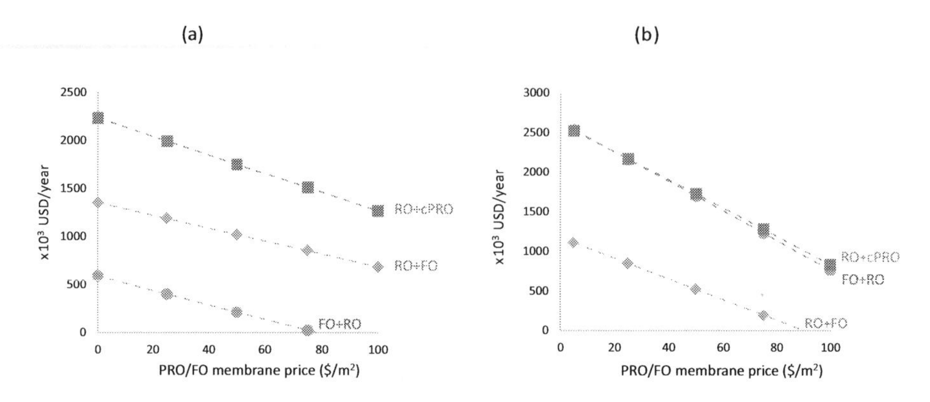

FIGURE 11.15
Influence of PRO and FO membrane prices on ATS when building new (a) 25% recovery RO integrations and (b) 50% recovery RO integrations with a permeation ratio of 50%. The RO system has a fixed seawater input of 200,000 m^3/day.

to a membrane price of \$100/m^2, while the ATS of FO+RO drops to negative at a membrane price of \$57/m^2. With a 50% recovery RO, RO+cPRO and FO+RO continue producing positive ATS of \$834,000/yr and \$763,000/yr, respectively, even at the increased membrane price of \$100/m^2. However, RO+FO becomes no longer feasible when the membrane price hits \$90/m^2.

11.4 Conclusions

Techno-economic analyses have been conducted for four potential RO integrated processes; namely, RO+oPRO, RO+cPRO, RO+FO, and FO+RO, with a fixed RO seawater input of 200,000 m^3/day, to investigate the savings on OpEx and CapEx.

For the 25% recovery RO integrations, RO+cPRO is the most cost-effective integration that achieves up to \$1,413,000/yr and 8.1 cent/m^3 saving when an existing RO is to be modified and up to \$4,362,000/yr and 24.9 cent/m^3 when a new integrated process is to be developed. FO+RO reduces SEC and hence achieves positive savings in both cases. RO+FO increases both OpEx and CapEx when modifying an existing RO, but it significantly reduces the CapEx of SWIP and discharge units. Therefore, it would achieve a positive saving when a new RO+FO process is to be built. Positive savings can still be achieved by RO+cPRO and RO+FO even at an increased membrane cost of \$100/m^2. However, RO+oPRO dramatically increases CapEx in both cases and therefore its integration is not economic with 25% and 50% recovery ROs.

50% recovery RO integrations can achieve higher cost savings in most cases. FO+RO and RO+cPRO are the two most effective processes to reduce OpEx of RO and therefore achieve the highest savings up to 10.6 and 5.9 cent/m^3, respectively, when an existing RO is integrated. When it comes to develop new seawater desalination plants, RO+cPRO and FO+RO can effectively reduce both OpEx and CapEx of desalination and achieve tremendous savings up to 13.3 and 12.5 cent/m^3, respectively. RO+cPRO and FO+RO maintain positive savings even if the membrane price is increased to \$100/m^2.

Most of the studies on RO integrations focus on the utilization of RO brines in PRO or FO, so that the operating conditions of the existing RO systems will remain unchanged. RO+cPRO offers great advantages over RO+oPRO and RO+FO because it can (1) most efficiently harvest the osmotic energy without utilizing additional energy recovery or generation devices and (2) dilute and recycle the RO brines. FO+RO can dilute the seawater feed and lower the operating pressure of the existing RO systems, which may be less favorable because the existing RO equipment and control systems need to be modified to accommodate the lowered RO operating pressure. Nonetheless, such limitation of FO+RO will be lifted if a new integrated process will be designed and built from scratch. This techno-economic analysis offers the guidelines for engineers to design the integrated processes based on the recovery of RO and the strategies to lower the cost of desalination.

References

Achilli, A., Cath, T.Y., Childress, A.E., 2009. Power generation with pressure retarded osmosis: an experimental and theoretical investigation. *J. Membr. Sci.* 343, 42–52.

Achilli, A., Prante, J.L., Hancock, N.T., Maxwell, E.B., Childress, A.E., 2014. Experimental results from RO-PRO: a next generation system for low-energy desalination. *Environ. Sci. Technol.* 48, 6437–6443.

Avlonitis, S.A., Kouroumbas, K., Vlachakis, N., 2003. Energy consumption and membrane replacement cost for seawater RO desalination plants. *Desalination* 157, 1–3.

Blandin, G., Verliefde, A.R.D., Tang, C.Y., Le-Clech, P., 2015. Opportunities to reach economic sustainability in forward osmosis–reverse osmosis hybrids for seawater desalination. *Desalination* 363, 26–36.

Chekli, L., Phuntsho, S., Kim, J.E., Kim, J., Choi, J.Y., Choi, J.-S., Kim, S., Kim, J.H., Hong, S., Sohn, J., Shon, H.K., 2016. A comprehensive review of hybrid forward osmosis systems: performance, applications and future prospects. *J. Membr. Sci.* 497, 430–449.

Chen, S.C., Amy, G.L., Chung, T.S., 2016. Membrane fouling and anti-fouling strategies using RO retentate from a municipal water recycling plant as the feed for osmotic power generation. *Water Res.* 88, 144–155.

Chen, S.C., Wan, C.F., Chung, T.S., 2015. Enhanced fouling by inorganic and organic foulants on pressure retarded osmosis (PRO) hollow fiber membranes under high pressures. *J. Membr. Sci.* 479, 190–203.

Cheng, Z.L., Chung, T.S., 2017. Mass transport of various membrane configurations in pressure retarded osmosis (PRO). *J. Membr. Sci.* 537, 160–176.

Choi, Y., Shin, Y., Cho, H., Jang, Y., Hwang, T.-M., Lee, S., 2016. Economic evaluation of the reverse osmosis and pressure retarded osmosis hybrid desalination process. *Desalin. Water Treat.* 57, 26680–26691.

Chung, T.S., Luo, L., Wan, C.F., Cui, Y., Amy, G., 2015. What is next for forward osmosis (FO) and pressure retarded osmosis (PRO). *Sep. Sci. Technol.* 156, 856–860.

Coday, B.D., Miller-Robbie, L., Beaudry, E.G., Munakata-Marr, J., Cath, T.Y., 2015. Life cycle and economic assessments of engineered osmosis and osmotic dilution for desalination of Haynesville shale pit water. *Desalination* 369, 188–200.

Feinberg, B.J., Ramon, G.Z., Hoek, E.M., 2013. Thermodynamic analysis of osmotic energy recovery at a reverse osmosis desalination plant. *Environ. Sci. Technol.* 47, 2982–2989.

Gerstandt, K., Peinemann, K.V., Skilhagen, S.E., Thorsen, T., Holt, T., 2008. Membrane processes in energy supply for an osmotic power plant. *Desalination* 224, 64–70.

Han, G., Zhang, S., Li, X., Chung, T.S., 2015. Progress in pressure retarded osmosis (PRO) membranes for osmotic power generation. *Prog. Polym. Sci.* 51, 1–27.

Han, G., Zhou, J., Wan, C., Yang, T.S., Chung, T.S., 2016. Investigations of inorganic and organic fouling behaviors, antifouling and cleaning strategies for pressure retarded osmosis (PRO) membrane using seawater desalination brine and wastewater. *Water Res* 103, 264–275.

Hauge, L.J., 1995. The pressure exchanger – a key to substantial lower desalination cost. *Desalination* 102, 219–223.

Kim, D.I., Kim, J., Shon, H.K., Hong, S., 2015. Pressure retarded osmosis (PRO) for integrating seawater desalination and wastewater reclamation: energy consumption and fouling. *J. Membr. Sci.* 483, 34–41.

Kim, J., Park, M., Shon, H.K., Kim, J.H., 2016. Performance analysis of reverse osmosis, membrane distillation, and pressure-retarded osmosis hybrid processes. *Desalination* 380, 85–92.

Kim, Y.M., Lee, Y.S., Lee, Y.G., Kim, S.J., Yang, D.R., Kim, I.S., Kim, J.H., 2009. Development of a package model for process simulation and cost estimation of seawater reverse osmosis desalination plant. *Desalination* 247, 326–335.

Lee, E.S.H., Xiong, J.Y., Han, G., Wan, C.F., Chong, Q.Y., Chung, T.S., 2017. A pilot study on pressure retarded osmosis operation and effective cleaning strategies. *Desalination* 420, 273–282.

Lee, K.L., Baker, R.W., Lonsdale, H.K., 1981. Membranes for power generation by pressure-retarded osmosis. *J. Membr. Sci.* 8, 141–171.

Li, X., Cai, T., Amy, G.L., Chung, T.S., 2017. Cleaning strategies and membrane flux recovery on anti-fouling membranes for pressure retarded osmosis. *J. Membr. Sci.* 522, 116–123.

Malek, A., Hawlader, M.N.A., Ho, J.C., 1996. Design and economics of RO seawater desalination. *Desalination* 105, 245–261.

Marcovecchio, M.G., Aguirre, P.A., Scenna, N.J., 2005. Global optimal design of reverse osmosis networks for seawater desalination: modelling and algorithm. *Desalination* 184, 259–271.

Migliorini, G., Luzzo, E., 2004. Seawater reverse osmosis plant using the pressure exchanger for energy recovery: a calculation model. *Desalination* 165, 289–298.

Peters, M.S., Timmerhaus, K.D., West, R.E., 2003. *Plant design and economics for chemical engineers*, fifth edition, McGrawHill, US.

Prante, J.L., Ruskowitz, J.A., Childress, A.E., Achilli, A., 2014. RO-PRO desalination: an integrated low-energy approach to seawater desalination. *Appl. Energy* 120, 104–114.

Saito, K., Irie, M., Zaitsu, S., Sakai, H., Hayashi, H., Tanioka, A., 2012. Power generation with salinity gradient by pressure retarded osmosis using concentrated brine from SWRO system and treated sewage as pure water. *Desalin. Water Treat.* 41, 114–121.

Sakai, H., Ueyama, T., Irie, M., Matsuyama, K., Tanioka, A., Saito, K., Kumano, A., 2016. Energy recovery by PRO in seawater desalination plant. *Desalination* 389, 52–57.

Sharqawy, M.H., Zubair, S.M., Lienhard, J.H., 2011. Second law analysis of reverse osmosis desalination plants: an alternative design using pressure retarded osmosis. *Energy* 36, 6617–6626.

She, Q., Jin, X., Tang, C.Y., 2012. Osmotic power production from salinity gradient resource by pressure retarded osmosis: effects of operating conditions and reverse solute diffusion. *J. Membr. Sci.* 401–402, 262–273.

Sivertsen, E., Holt, T., Thelin, W., Brekke, G., 2012. Modelling mass transport in hollow fibre membranes used for pressure retarded osmosis. *J. Membr. Sci.* 417–418, 69–79.

Sivertsen, E., Holt, T., Thelin, W., Brekke, G., 2013. Pressure retarded osmosis efficiency for different hollow fibre membrane module flow configurations. *Desalination* 312, 107–123.

Skilhagen, S.E., Dugstad, J.E., Aaberg, R.J., 2008. Osmotic power – power production based on the osmotic pressure difference between waters with varying salt gradients. *Desalination* 220, 476–482.

Stover, R.L., 2007. Seawater reverse osmosis with isobaric energy recovery devices. *Desalination* 203, 168–175.

Straub, A.P., Deshmukh, A., Elimelech, M., 2016. Pressure-retarded osmosis for power generation from salinity gradients: is it viable? *Energy Environ. Sci.* 9, 31–48.

Thelin, W.R., Sivertsen, E., Holt, T., Brekke, G., 2013. Natural organic matter fouling in pressure retarded osmosis. *J. Membr. Sci.* 438, 46–56.

Thorsen, T., Holt, T., 2009. The potential for power production from salinity gradients by pressure retarded osmosis. *J. Membr. Sci.* 335, 103–110.

Valladares Linares, R., Li, Z., Yangali-Quintanilla, V., Ghaffour, N., Amy, G., Leiknes, T., Vrouwenvelder, J.S., 2016. Life cycle cost of a hybrid forward osmosis – low pressure reverse osmosis system for seawater desalination and wastewater recovery. *Water Res* 88, 225–234.

Vince, F., Marechal, F., Aoustin, E., Bréant, P., 2008. Multi-objective optimization of RO desalination plants. *Desalination* 222, 96–118.

Voutchkov, N., 2013. *Desalination engineering: planning and design*, McGrawHill, US.

Wan, C.F., Chung, T.S., 2015. Osmotic power generation by pressure retarded osmosis using seawater brine as the draw solution and wastewater brine as the feed. *J. Membr. Sci.* 479, 148–158.

Wan, C.F., Chung, T.S., 2016. Energy recovery by pressure retarded osmosis (PRO) in SWRO–PRO integrated processes. *Appl. Energy* 162, 687–698.

Wan, C.F., Li, B., Yang, T., Chung, T.S., 2017. Design and fabrication of inner-selective thin-film composite (TFC) hollow fiber modules for pressure retarded osmosis (PRO). *Sep. Sci. Technol.* 172, 32–42.

Xiong, J.Y., Cai, D.J., Chong, Q.Y., Lee, S.H., Chung, T.S., 2017. Osmotic power generation by inner selective hollow fiber membranes: an investigation of thermodynamics, mass transfer, and module scale modelling. *J. Membr. Sci.* 526, 417–428.

Zhu, A., Christofides, P.D., Cohen, Y., 2009. Effect of thermodynamic restriction on energy cost optimization of RO membrane water desalination. *Ind. Eng. Chem. Res.* 48, 6010–6021.

12

Design of Other Pressure Retarded Osmosis Hybrid Processes (Pressure Retarded Osmosis–Membrane Distillation and Pressure Retarded Osmosis–Forward Osmosis)

Zhen Lei Cheng and Gang Han

Department of Chemical and Biomolecular Engineering
National University of Singapore
Singapore

CONTENTS

12.1 Introduction

Pressure retarded osmosis (PRO) has received growing attention during the last decade as an emerging technology to harvest the renewable salinity-gradient energy. In this process, water transports from a low-salinity feed solution across a semipermeable membrane to a high-salinity draw solution against a hydraulic pressure. The increased volume of the diluted draw solution is a form of mechanical energy, which can be converted to electricity through hydro turbines or recovered via other energy recovery devices (Gerstandt et al., 2008; Lee et al., 1981; Loeb and Norman, 1975; Ramon et al., 2011). The estimated global osmotic energy that can be generated from the mixing of ocean and fresh river water is approximately 1750–2000 TWh per year (Thorsen and Holt, 2009). Additional osmotic energy is projected when using high-salinity water sources that provide greater salinity gradients (Achilli et al., 2014; Chung et al., 2012; Kim et al., 2012). For instance, recent research focuses on the use of the feed pair consisting of wastewater retentate (WWRe) from municipal water recycling plants and seawater brine (SWBr) from seawater reverse osmosis (RO) desalination plants (Chung et al., 2015; Kim et al., 2015; Straub et al., 2016; Touati and Tadeo, 2017). By utilizing these two waste effluents, one can not only generate larger specific osmotic power than the conventional feed pair of river water and seawater, but also can mitigate the disposal and environmental issues of SWBrs (Chung et al., 2015; Kim et al., 2015; Straub et al., 2016; Touati and Tadeo, 2017). Despite of these merits, (1) the relatively high costs of pretreatment and pumping, (2) the challenges of removing membrane fouling and scaling, (3) a low extractable osmotic energy due to the small salinity gradient, and (4) potential environmental impacts of discharging the brackish water may render PRO difficult to generate a sizable net power via conventional mixing of seawater and river water (Efraty, 2013; Lin et al., 2014a; Sikdar, 2014; Wan and Chung, 2015).

On the other hand, membrane distillation (MD) is a thermally driven membrane process where the hot feed stream evaporates and transports across the membrane, and then condenses on the cold permeate side (Curcio and Drioli, 2005; Edwie and Chung, 2013; Francis et al., 2014; Gryta et al., 2001). Compared to RO, MD is less sensitive to the salt concentration or the osmotic pressure of the feed. Thus, it can treat feed streams with higher total dissolved solids (TDS). In addition, the modest operating temperature allows MD to utilize sustainable or alternative energy resources such as solar energy, geothermal energy, or waste heat. Nonetheless, the use of MD for direct seawater desalination is still challenging because of its relatively high energy intensity and difficulties to mitigate membrane fouling and wetting (Curcio and Drioli, 2005; Edwie and Chung, 2013; Francis et al., 2014; Gryta et al., 2001).

By integrating PRO and MD processes, the proposed PRO–MD system enables the sustainable generation of osmotic power and clean water (Han et al., 2015b; Lee et al., 2015; Lin et al., 2014a). Such a hybrid system may not only avoid the shortcomings of each individual process but also strengthen their advantages. In contrast to a typical MD process, the net energy consumption for desalination is lowered by the power generated from PRO, while the circulated draw solution becomes almost foulant-free due to the design of an integrated-loop process. Thus, the membrane regeneration and replenishment costs can be significantly reduced. From the perspective of PRO, the draw solution is an abundantly available and low-cost residual from the MD unit but with a high purity and salinity. Therefore, an ultrahigh osmotic energy is achievable for the hybrid PRO–MD system and the membrane fouling on the draw solution side could be well controlled. In addition, the system integration could mitigate the disposal problems of the MD-concentrated brine and the resultant brackish water from the PRO unit could be discharged easily. Furthermore, the elimination of additional chemical usage in the pretreatment and possible reuse of the impaired wastewater make this integrated process more environmentally friendly. Overall, the PRO–MD hybrid process possesses the merits of (1) lower operation costs for desalination, (2) higher water recovery, (3) less membrane fouling, and (4) minimal negative environmental impacts.

Another primary challenge to prevent the widespread use of PRO technology is membrane fouling. In order to have a higher power output and mechanical stability, the active layer (AL) of PRO membranes usually faces the draw solution (DS) (i.e., AL-DS mode) in PRO processes. As a consequence, the foulants of the feed solution (e.g., WWRe) can be easily brought to the porous support and accumulate underneath the active layer, which may substantially lower both the water flux and power generation (She et al., 2013, 2016; Thelin et al., 2013; Yip and Elimelech, 2013; Zhang et al., 2014a). Since the foulants are trapped inside the porous support, it is hard to remove them and clean up the fouled membrane by varying the hydrodynamic conditions (Mi and Elimelech, 2008; She et al., 2016).

To effectively control the membrane fouling in PRO, two strategies are normally adopted, namely, pretreatment of feed solutions and development of antifouling PRO membranes. However, the first strategy involves extra energy and chemical consumption as well as operating costs (Chun et al., 2017; She et al., 2016; Wan and Chung, 2015). While, for the second strategy, tailoring antifouling PRO membranes with balanced characteristics is not trivial because one must carefully examine every counterbalanced factor such as high hydrophilicity versus good mechanical strength, strong repulsion to foulants versus low resistance to water, and multi-functionality versus easy processing (Han et al., 2017; Hu et al., 2016; Le et al., 2017; Li et al., 2016; Nguyen et al., 2013). Thus, to explore a more general fouling control strategy, an additional forward osmosis (FO) unit is proposed to extract water from the feed solution of WWRe to the inter-loop solution as the pretreatment step. Being analogous to PRO, FO shares the same transport mechanism except for the absence of pressure on the draw solution side (i.e., the inter-loop solution) and the option of operating as the AL facing the feed solution (FS) mode (i.e., AL-FS mode). By using the AL-FS mode, FO membranes are able to provide a stable and relatively higher water flux, as the AL could effectively reject the foulants of WWRe with minimal fouling deterioration (Chun et al., 2017; Klaysom et al., 2013; Mi and Elimelech, 2008; She et al., 2013, 2016). Subsequently, the diluted inter-loop solution would serve as a clean feed to the PRO unit. The resultant PRO–FO hybrid system aims for sustainable osmotic power generation by having the advantages of FO such as (1) low fouling propensity, (2) easy membrane cleaning, and (3) minimal external energy requirements.

In summary, this chapter presents the aforementioned two emerging PRO hybrid processes. For the PRO–MD hybrid, two types of process configurations generally found in the literature and pilot studies; namely, close-loop and open-loop, will be classified in Chapter 12.2.1. While the details of process modeling, membrane design, and operation as well as economic analysis are elaborated in Chapter 12.2.2 to 12.2.4. For the PRO–FO hybrid, its process configuration and modeling are discussed in Chapter 12.3.1 and 12.3.2, respectively. In order to validate this proposed hybrid process, the concentration of the inter-loop solution is first determined at the local mass transport scenario followed by carrying out the corresponding FO/PRO bench-scale experiments in Chapter 12.3.3. Finally, full-scale analyses are conducted in Chapter 12.3.4 to predict the performance of the PRO–FO system using large-size modules. The last part, Chapter 12.4 highlights the challenges encountered and future prospects of both PRO–MD and PRO–FO hybrids.

12.2 PRO–MD

12.2.1 Process Configuration

12.2.1.1 Close-Loop Process

The close-loop PRO–MD hybrid process, also known as osmotic heat engine (OHE), is a promising technology that can produce electricity from waste heat sources (Logan and Elimelech, 2012). This process comprises a power generation stage (i.e., PRO) that converts the energy released from the mixing of two solutions with different concentrations into useful work, and a thermal separation stage (i.e., MD) where waste heat is used to regenerate the low and high concentration solutions, which are then recirculated back to the energy production stage (i.e., PRO) (McGinnis et al., 2007). A representative study was carried out by Lin et al. (2014b) and a simplified schematic of their proposed PRO–MD hybrid system is shown in Figure 12.1, where HX and PX represent the heat exchanger and pressure exchanger that are used to prompt the thermal/mechanical efficiency of the respective operating stage. A hydroturbine (TB) in parallel with PX is to convert the energy of mixing to useful work. It is noted that the additional heat source (i.e., the block "H") and the heat sink (i.e., the block "C") are associated with MD to reflect the working principle of a generic heat engine, which absorbs heat from the heat source, converts part of the absorbed heat to useful work through the turbine and releases the remaining heat to the heat sinks.

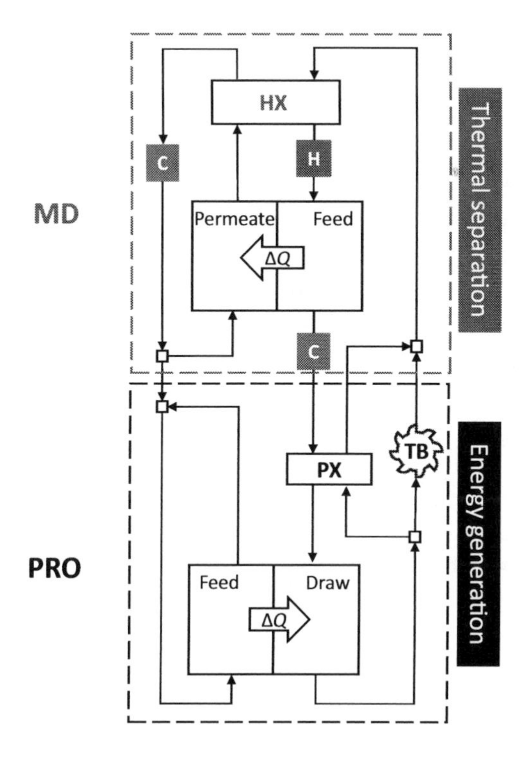

FIGURE 12.1
Schematic diagram of a PRO–MD close-loop hybrid system (Lin et al., 2014b).

To achieve a more holistic assessment of the viability of this hybrid process, Lin et al. conducted a systematic analysis by performing a module-scale modeling to couple the heat and mass transfer between the MD and PRO stages from a waste heat source ranging from 40 to 80°C and working concentrations of 1, 2, and 4 M NaCl (Lin et al., 2014b). Subsequently, the potential of this hybrid system is assessed to identify the major factors controlling the system performance and to examine the thermodynamic limit of the energy conversion efficiency, under both limited and unlimited mass and heat transfer kinetics. In both cases, the relative flow rate between the MD permeate and feed streams is identified as an important operation parameter and there is an optimal relative flow rate that maximizes the overall energy efficiency of the PRO–MD process for given operating temperatures and concentrations. In the case of unlimited mass and heat transfer kinetics, the energy efficiency of the system can be analytically determined based on thermodynamics. Their study indicates that the hybrid PRO–MD system can theoretically achieve an energy efficiency of 9.8% (81.6% of the Carnot efficiency) with hot and cold operating temperatures of 60 and 20°C, respectively, and a working solution of 1.0 M NaCl. When mass and heat transfer kinetics are limited, conditions that more closely represent actual operations, the practical energy efficiency will be lower than the theoretically achievable efficiency. In such practical operations, utilizing a higher working concentration will yield greater energy efficiency.

12.2.1.2 Open-Loop Process

Different from the close-loop process of solely converting waste heat into useful energy, the open-loop PRO–MD process provides a dual function of producing both clean water and osmotic energy. A conceptual experimental demonstration was first conducted by Han et al. (2015b) and the corresponding schematic of their proposed process is shown in Figure 12.2. The hybrid system consists of (1) one open-end loop using freshwater or wastewater as the feed solution for the PRO unit and (2) one closed loop including a PRO unit to generate osmotic power through a TB and a MD unit to re-concentrate the diluted draw solution and then circulate it back to the PRO unit. Meanwhile, the MD unit also produces potable water. HX is used for heat exchange between the diluted draw solution and the hot concentrated MD brine, further increasing the energy saving of the hybrid system, while a high-pressure pump (HP) is used to pressurize the concentrated MD brine before entering the PRO unit.

By employing the state-of-the-art polyethersulfone (PES) thin-film composite (TFC) PRO hollow fiber membrane (Han et al., 2016; Zhang and Chung, 2013) and polyvinylidene fluoride (PVDF) MD flat-sheet membrane (Han et al., 2015b) and using synthetic freshwater and real

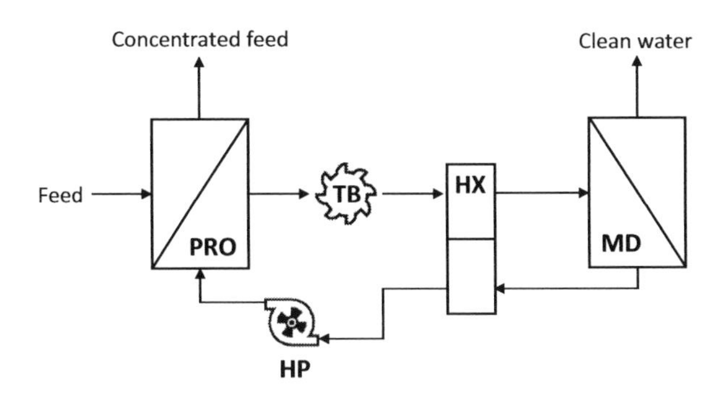

FIGURE 12.2
Schematic diagram of a PRO–MD open-loop hybrid system (Han et al., 2015b).

wastewater as feeds, the technical feasibility of the PRO–MD hybrid system and the key parameters governing the system have been systematically investigated. More specifically, the main operation variables are (1) the salinity and quality of the feed water and draw solution, (2) the power density of the PRO membrane and its salt rejection, (3) water recovery rate of the feed and the dilution effects of the draw solution (or the degree of mixing) in the PRO unit, (4) water flux and salt rejection in the MD unit, and (5) the fouling and scaling behaviors of each unit. These parameters are correlated with one another and could be optimized over a wide operational range to enhance the effectiveness of the hybrid system. As a result, it is found that when employing a 2 M NaCl MD concentrate as the draw solution, ultrahigh power densities of 31 and 9.3 W/m^2 can be achieved by the PRO unit using deionized water and real wastewater as feeds, respectively. Simultaneously, high-purity (i.e., ~100% rejection) potable water with a flux of 32.5–63.1 LMH can be produced by the MD unit at 40–60 °C almost without any detrimental effects of fouling. In case of using wastewater as the feed, the water flux decreases fast with increasing the percentage of water recovery because of membrane fouling. As a result, power density drops from 7.8 to 4.5 W/m^2 and further to 2.7 W/m^2 when the water recovery is increased to 10% and 70%, respectively. These results suggest that the water recovery cannot be too high in order to achieve the power density of 5 W/m^2, set by Statkraft as the commercially viable benchmark (Gerstandt et al., 2008; Skilhagen, 2010; Thorsen and Holt, 2009). Thus, regular cleaning is required to maintain the productivity of the PRO membrane.

Another process configuration of the open-loop hybrid system is to link the feed inlet of MD with the discharged brine from RO. In this case, MD produces more clean water by achieving a greater water recovery than the stand-alone RO. While its highly concentrated brine could be used as a draw solution in the subsequent PRO unit. The resultant RO–MD–PRO hybrid system has been adopted in a Korean R&D project entitled "Global MVP" (M for MD, V for valuable resource recovery, and P for PRO), which was launched in 2013 (Kim et al., 2016; Lee et al., 2019). Figure 12.3 illustrates its process diagram where the corresponding HX and HP components are not included for simplicity. The division stream from the RO brine exiting from its PX, shown as the dashed line, is optional based on the operating conditions.

Lee et al. (2015) pioneered a simulation study of the hybrid system by only integrating a multi-stage vacuum membrane distillation (MVMD) and PRO without considering the upstream RO unit. The MVMD unit employs a recycling flow scheme for the continuous production of both distillate water and highly concentrated brine. By theoretically assessing the distillate and power production of the MVMD–PRO system with respect to the inlet feed flow rate and recycling flow ratio in MVMD, it is found that the production of distillate water slightly increases with a decrease in the recycling flow when the inlet feed flow rate is constant. The maximum possible brine concentration from the MVMD unit is 1.9 M NaCl at an inlet feed flow rate of 3 kg/min and a 90% recycling flow. Under this condition, a peak power density of 9.7 W/m^2 is achieved in the PRO unit at 13 bar using river water as the feed solution with flow rates of 0.5 kg/min for both feed and draw streams.

A more comprehensive study including both the upstream RO unit and the downstream MD-PRO integration was later introduced by Kim's group (Kim et al., 2016). They analyzed the commercial feasibility of such a RO–MD–PRO hybrid configuration by evaluating the performance indicators, such as power generation and specific energy consumption (SEC), and the design and operating factors, such as the RO discharged brine division ratio (BDR), the plant dimension ratio of MD

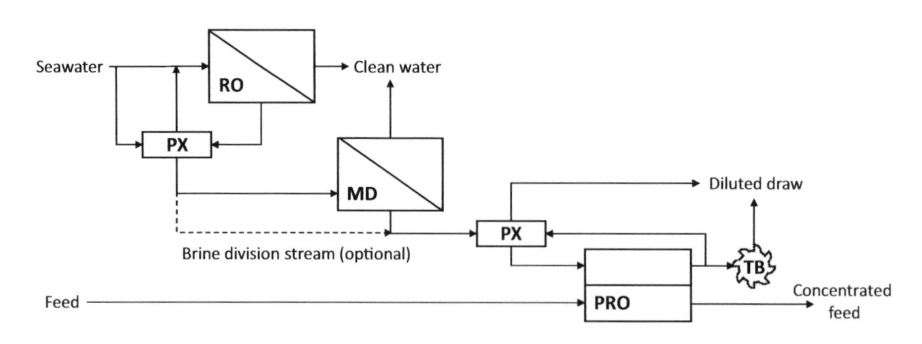

FIGURE 12.3
Schematic diagram of a RO–MD–PRO open-loop hybrid system (Kim et al., 2016).

+ PRO to RO as well as the cost of the MD heating source. Overall, the RO–MD–PRO hybrid system outperforms a stand-alone RO in terms of reducing SEC and alleviating the disposal and environmental impacts of the concentrated RO brine. An increase in BDR has positive effects on the efficiency of the hybrid system by raising the water production of MD and the energy generation of PRO. At lower BDR values, the SEC is greatly influenced by the PRO plant dimension. Nonetheless, both the MD and PRO plant dimensions almost equally contribute to the efficiency of the hybrid system with the increase of BDR. Last but not the least, the cost of the MD heating source has a dominant role in determining the overall process efficiency based on their analysis. If the required heat supply cannot be reimbursed, the hybrid system will be unfavorable at any BDR value. Thus, it is crucial to set the RO–MD–PRO hybrid system in locations where free or low-cost thermal energy is easily accessible. Similar findings have also been acquired in another study (Choi et al., 2017).

Few years later, the same group revised the previously proposed RO–MD–PRO hybrid configuration (i.e., Figure 12.3) by setting BDR equal to 1 and replacing the TB with a PX in order to transfer the pressure to the seawater intake stream in favor of greater mechanical efficiency (Chae et al., 2018). To provide a holistic understanding of the hybrid process, they proposed a new dimensionless performance index ($N_{r/c}$) to assess the energy efficiency of RO–PRO (i.e., RO–MD–PRO omitting the MD unit) and RO–MD–PRO under different water recovery rates of RO/MD and dilutive factors of PRO. This performance index ($N_{r/c}$) physically quantifies the ratio of the total energy generated by PRO to the total energy consumed by RO/MD. It is found that the typical water recovery rate of 0.4–0.6 for RO is not effective in terms of the energy efficiency of hybrid processes ($N_{r/c} < 0.3$). Both higher water recovery rate of RO/MD and larger dilutive factor of PRO, raised by a corresponding more concentrated inlet brine, favor $N_{r/c}$. However, this may impose a challenge in membrane design and process operation, as a high water recovery easily causes fouling/scaling of RO/MD membranes, and subsequently more robust PRO membrane modules with a less pressure drop are needed to handle the brine. Regular cleaning/maintenance and special precautions have to be paid when operating the hybrid system at such harsh conditions. It also should be noted that a higher water flux across the PRO membrane does not always guarantee a higher energy efficiency when it comes to PRO-hybridized processes, as opposed to the stand-alone PRO. Lastly, the simulation results once more confirm that RO–MD–PRO generally has a higher energy efficiency than RO–PRO if an inexpensive heating source, such as waste heat, is used. The heating cost of MD is found to have a great negative impact on $N_{r/c}$ under the low water recovery rate range (<0.6), while it is somewhat insensitive to $N_{r/c}$ under the high water recovery rate range (>0.7).

12.2.2 Process Modeling

Process modeling developed through the fundamentals of mass and heat transport is an essential pillar to design and optimize the PRO–MD hybrid process. For the MD unit, detailed mathematical modelings of both local and module scenarios in typical MD processes have been developed in our previous study (Lu et al., 2019; Wang and Chung, 2012). For the PRO unit, Chapter 4 presents a universal platform to analytically evaluate the mass transport of various membrane configurations of flat sheet, single-, and double-skinned hollow fibers. Nonetheless, the temperature effect has not been taken into account when modeling the PRO process. By measuring the PRO performance of commercial FO membranes supported by woven mesh/spacers, She et al. (2012) observed a growth of ~34% in water flux when increasing the testing temperature from 25°C to 35°C using 1 M NaCl and 10 mM NaCl as the feed pair while Anastasio et al. (2015) reported to double the power density by raising the temperature from 20°C to 40°C using 1.5 M NaCl and deionized water as the feed pair. As mentioned earlier, the elevated temperature could enhance both water flux and reverse solute flux at the same time. This trade-off relationship should be optimized through a model coupling mass and heat transport simultaneously across the membrane. Recently, Chowdhury and McCutcheon (2018) developed such a model for FO processes and validated it by carrying out experiments at controlled temperature conditions. Although FO and PRO processes share many similarities besides the application of hydraulic pressures on the draw solution side, future studies could extend their work to PRO processes.

On the other hand, the elevated temperature also affects the membrane transport properties (i.e., water permeability coefficient (A) and salt permeability coefficient (B)). Nonetheless, this aspect has not been fully analyzed for PRO membranes yet. In analogy, RO membrane systems have already established a temperature-correction factor (TCF) to calculate A and B at a desired temperature T (°C) (Ruiz-Saavedra et al., 2015). This could serve as a guideline for the PRO application.

$$\text{For } T \leq 25^\circ\text{C}: \text{TCF} = \exp\left[3480 \times \left(\frac{1}{298} - \frac{1}{273 + T}\right)\right] \tag{1}$$

$$\text{For } T > 25^\circ\text{C}: \text{TCF} = \exp\left[2640 \times \left(\frac{1}{298} - \frac{1}{273 + T}\right)\right] \tag{2}$$

$$A = A_0 \times \text{TCF} \tag{3}$$

$$B = B_0 \times \text{TCF} \tag{4}$$

where A_0 and B_0 are water and salt permeability coefficients, respectively, measured at 25°C.

Apart from traditional PRO processes, another important factor that should be considered is the nonideality of the hypersaline draw solution generated from the high recovery of the upstream MD unit. By assuming a relatively dilute draw solution, most PRO studies rely on the van't Hoff equation to calculate the osmotic pressure (π) (Cheng and Chung, 2017).

$$\pi = iCRT \tag{5}$$

where i, R, C, and T are the van't Hoff's factor, universal gas constant, concentration, and absolute temperature of the draw solution, respectively. Generally speaking, this simple equation gives a reasonable estimation of osmotic pressure for a salinity of up to 2 M (Bajraktari et al., 2017; Lee and Kim, 2016). For higher salinities, the osmotic pressure can be calculated from the water activity (γ_w) and molar volume (V) to get more accurate results (Bajraktari et al., 2017; Madsen et al., 2017).

$$\pi = \frac{RT}{V}\ln(\gamma_w) \tag{6}$$

Data of γ_w and V could be either obtained using Pitzer correlations or found in the literature (Pitzer et al., 1984). From an engineering point of view, the OLI software is often used to calculate π based on the activity model for quick and precise predictions, which has been practiced by many researchers (Bajraktari et al., 2017; Lee and Kim, 2016; McCutcheon and Elimelech, 2007; Phuntsho et al., 2011; Yip et al., 2011).

12.2.3 Membrane Design and Operation

For the hybrid process, the membrane design and operation of PRO and MD units generally follow the guidelines in their respective applications. In terms of MD operations, the aforementioned close-loop process (i.e., OHE) adopts the configuration of direct contact MD. While other operation configurations such as vacuum MD (VMD), air gap MD, and sweeping gas MD are also applicable to the open-loop process.

The representative membrane morphologies (Figure 12.4) and properties (Table 12.1) can be found in the study by Han et al. (2015b), where they employed a PES-TFC hollow fiber membrane for PRO and a PVDF flat-sheet membrane for MD. Figure 12.4(a)–(c) shows the morphology of the PRO membrane. It consists of a thin polyamide-selective layer on the inner surface of a porous PES hollow fiber substrate. Macroscopically, the PES-TFC membrane possesses a good concentricity and a medium dimension (i.e., outer diameter (OD) = 973 μm) with a relatively thick cross-section of

~200 μm. Microscopically, this membrane has a straight finger-like macrovoid structure near the inner cross-section and a highly porous open-cell microstructure in the outer surface that is capable of reducing water transport resistance. As a result, this PES-TFC membrane not only possesses outstanding mechanical strength but also exhibits promising transport characteristics. Figure 12.4(d)–(f) shows the images of the PVDF flat-sheet membrane for the MD unit. It possesses a relatively thin and dense top surface, an open-pore bottom surface, and a porous cross-section full of macrovoids. The thin top layer provides the separation function, while the porous cross-section could offer both a low transport resistance and a small thermal conductivity across the membrane due to a large porosity of up to ~78%. The porous bottom surface provides uninterrupted transport channels for water vapor transport. In order to maintain the vacuum of the membrane module at the permeate side and provide better wetting resistance, the PVDF membrane has a relatively small mean pore size of ~168 nm. Besides, the MD membrane is relatively hydrophobic with a water contact angle of 86.3° due to the hydrophobic nature of the PVDF material and the membrane surface morphology.

As the PRO–MD hybrid process often handles concentrated brines at elevated temperatures, special measures should be taken during the membrane design and operation. Owing to the nature of the close-loop system (i.e., OHE) and the pre-treatment in the upstream of the open-loop process (e.g., RO–MD–PRO), the brine stream entering MD/PRO is theoretically foulant-free. This saves a significant amount of efforts in tackling various

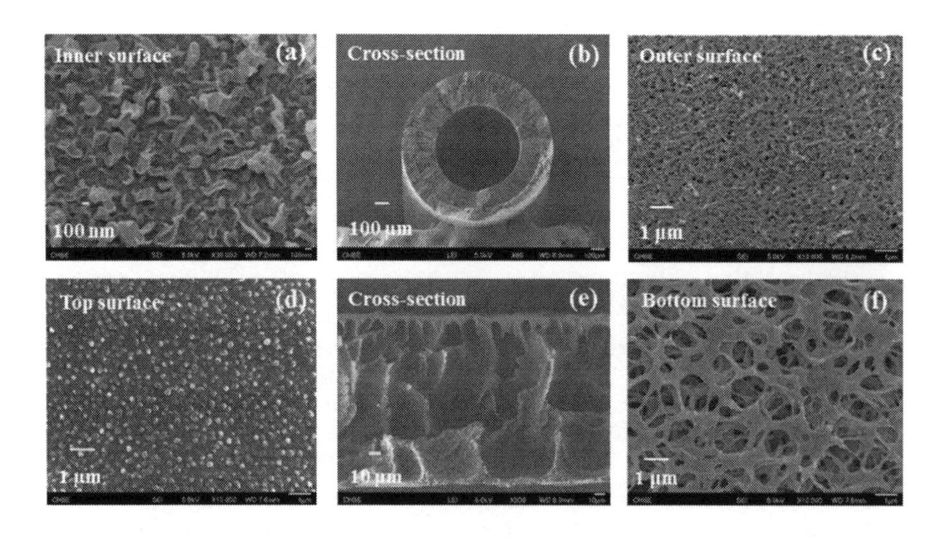

FIGURE 12.4
The surface and cross-section morphology of the PES-TFC PRO hollow fiber membrane (a–c), and the PVDF MD flat-sheet membrane (d–f) (Han et al., 2015b).

TABLE 12.1

The specifications of the PES-TFC PRO hollow fiber membrane and PVDF MD flat-sheet membrane (Han et al., 2015b)

	Water Permeability Coefficient, *A* (L/m^2 h bar)	**Salt Rejection, *R* (%) (2000 ppm @ 5 bar)**	**Salt Permeability Coefficient, *B* (L/m^2 h)**	**Structural Parameter, *S* (μm)**	**OD/ ID (μm)**	**Burst pressure (bar)**
PES-TFC PRO hollow fiber membrane	3.3 ± 0.3	97.5 ± 0.2	0.31 ± 0.3	450	973/ 562	> 20
	Mean Pore Size (μm)	**Water Contact Angle (°)**	**Porosity (%)**	**Selective layer**		**Effective area per module (cm^2)**
PVDF MD flat-sheet membrane	168.4 ± 1.6	86.3 ± 1.3	78.6 ± 2.3	Top surface		15.2

types of fouling, such as biological, particulate, and colloidal fouling, and is considered to be an advantage of the hybrid process as mentioned earlier in Chapter 12.1. However, inorganic scaling is particularly concerned when operating MD at a high water recovery and feed salinity. It is noted that the solubility and crystal formation of salts vary widely over the temperature range relevant to MD. Moreover, the solubility of individual salts may be either positively or negatively correlated with temperature. For instance, the solubility of NaCl increases while those of $CaCO_3$, $Mg(OH)_2$, and $Ca_3(PO_4)_2$ decrease with increasing temperature (Gryta, 2002; Morel and Hering, 1993). This negative correlation might bring the concentrated RO discharged brine even closer to the saturation point during MD, increasing the risk of scaling. Other study shows that a higher temperature also reduces the induction periods of some salts (Stamatakis et al., 2005). To mitigate the inorganic scaling on MD membranes, prevention tools including thermal water softening, use of antiscalants, pH control, and magnetic water treatment have been suggested (Warsinger et al., 2015). In addition, one may tailor MD membrane properties with greater hydrophobicity and smaller surface porosity/pore size (Gryta, 2007). Other fouling contributing factors such as temperature, concentration polarization, residence time, and stagnation areas should be carefully examined when it comes to the membrane module design (Warsinger et al., 2015).

In the context of using the MD discharged brine as a draw solution, robust PRO membranes with high withstanding pressures are crucial to maximize the capacity of harvesting the osmotic energy (Han et al., 2013a, 2013b; Straub et al., 2014; Wan et al., 2018; Zhang and Chung, 2013). Up to date, there are no commercially available membranes or membrane modules specifically designed for the PRO application. Best PRO membranes reported in bench-scale were developed by Straub et al. (2014) for the flat-sheet configuration and Wan et al. (2018) for the hollow fiber configuration. They could achieve the peak power densities of 60 W/m^2 at 48 bar, using 3 M NaCl and deionized water as the feed pair and 38 W/m^2 at 30 bar, using 1.2 M NaCl and deionized water as the feed pair, respectively.

As the discharged brine often carries a certain heat from the MD unit, its temperature effect on PRO membranes needs to be evaluated. A higher temperature benefits water flux and power density mainly owing to the enhanced draw solute diffusivity. Nonetheless, an enhanced diffusivity allows the solute to permeate across the membrane faster than it would at lower temperatures, which might counteract the positives brought by the temperature effect (Anastasio et al., 2015; She et al., 2012). One should also be aware that increasing the temperature will result in physiochemical property changes of both membranes and solutions, directly influencing the ultimate membrane performance. Although fouling on the draw solution side is greatly suppressed, foulants on the feed solution side can be easily brought into the porous membrane substrate by the bulk water flow and accumulate underneath the AL (Han et al., 2016). To effectively control this, strategies of pre-treating the feed solution (Antony et al., 2011; Chen et al., 2015, 2016; Van der Bruggen et al., 2003; Wan and Chung, 2015) and developing antifouling PRO membranes (Han et al., 2015a, 2017, 2018; Li et al., 2016, 2017), as well as membrane processes are normally adopted.

12.2.4 Economic Analysis

The economic analyses by comparing RO, RO–MD, RO–PRO, and RO–MD–PRO were conducted by Choi et al. (2017) using theoretical models. Figure 12.5 shows that the hybrid systems could outperform a RO stand-alone system in terms of reducing water cost and alleviating the disposal and environmental impacts of the RO discharged brine. The electricity cost plays an important role in determining the economic feasibility of hybrid processes (Choi et al., 2017; Wan and Chung, 2016a). If the electricity cost exceeds 0.2 US $/kWh, the water cost of hybrid systems (i.e., RO–MD, RO–PRO, and RO–MD–PRO) becomes lower than that of stand-alone RO. The heat supply cost for MD is another crucial factor affecting the economic feasibility of MD-hybridized systems (i.e., RO–MD and RO–MD–PRO). By further taking the membrane cost and interest rate into consideration, Table 12.2 shows the estimated project costs with different hybrid processes. Overall, RO–MD–PRO is the most cost-effective configuration at the same level of electricity cost owing to its extra freshwater production using waste heat via MD and osmotic energy harvested via PRO.

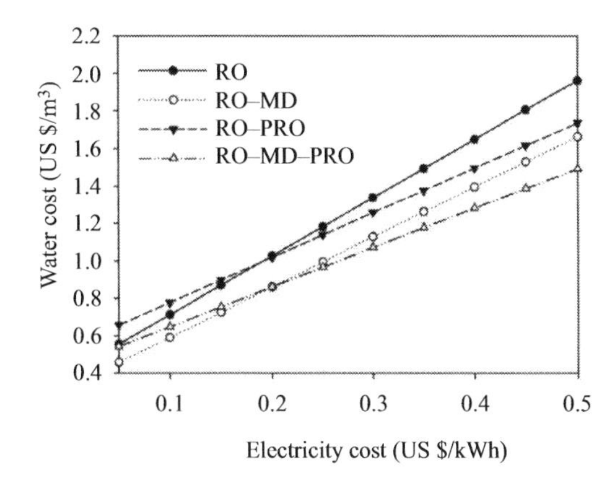

FIGURE 12.5
Effect of electricity cost on water cost for RO, RO–MD, RO–PRO, and RO–MD–PRO (Choi et al., 2017; Lee et al., 2019).

TABLE 12.2

Comparison of estimated project costs based on various hybrid processes (based on 100 m^3/day plant size) (Choi et al., 2017; Lee et al., 2019)

Processes	RO	RO–MD	RO–PRO	RO–MD–PRO
Initial capital cost (US $/$m^3$·day)	115.4	115.7	166.6	144.9
Operation cost (US $/$m^3$)	1.263	1.039	1.233	1.074
Specific energy consumption (kWh/m^3)	3.320	2.809	2.869	2.683

12.3 PRO–FO

12.3.1 Process Configuration

Cheng et al. (2018) conducted the pioneer study on the PRO–FO hybrid and Figure 12.6 shows the corresponding process. By operating under the AL-FS mode, the FO unit extracts water from the feed solution of WWRe to the inter-loop solution. Subsequently, the diluted inter-loop solution would serve as a clean feed to the PRO unit. It should be noted that, in order to meet these goals, a proper concentration of the inter-loop solution has to be selected. Clearly, there is a trade-off relationship existing between the FO and PRO performance. A higher inter-loop concentration will prompt the water flux for FO. This will inevitably reduce the driving force for the PRO process,

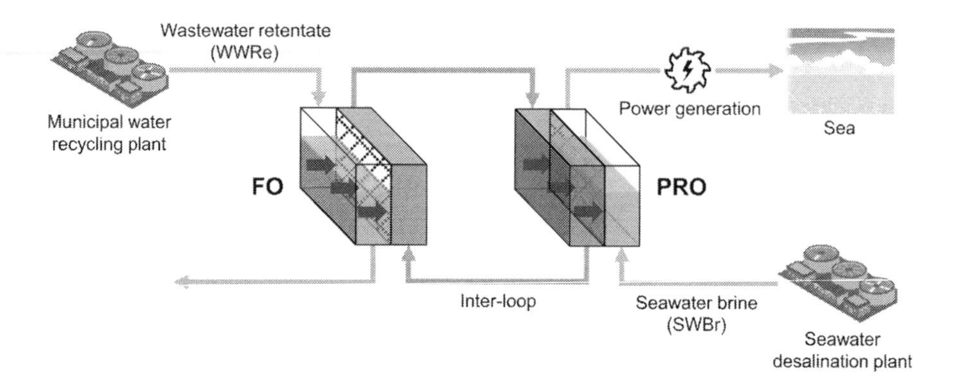

FIGURE 12.6
Illustration of a PRO–FO hybrid system utilizing WWRe as the feed solution and SWBr as the draw solution (Cheng et al., 2018).

resulting in a lower osmotic power production. In contrast, a low inter-loop concentration will bring about a high throughput of PRO, but it may require an extremely large membrane area for the FO process as the water flux would be lower because of the small salinity difference between the feed pair. To examine the effects of the inter-loop concentration toward the whole hybrid system, the corresponding mathematical models (refer to Chapter 12.3.2) have been built to analyze both local (refer to Chapter 12.3.3.1) and full-scale (refer to Chapter 12.3.4) scenarios. In addition, it is assumed that the membrane fouling is negligible in the PRO–FO hybrid process. As a result, its performance can be predicted via the developed models with synthetic solutions as feeds. The synthetic WWRe and SWBr were prepared using 0.011 M NaCl (Wan and Chung, 2015) and 1.2 M NaCl, respectively. The inter-loop solution of 0.2 M NaCl was chosen based on the concentration determined in Chapter 12.3.3.1. For the experimental verification, one can refer to Chapter 12.3.3.2. The required membrane mini-module was fabricated according to the method published elsewhere (Wan and Chung, 2015). The module consists of TFC hollow fibers with a membrane area of ~11 cm^2. The specifications of the TFC hollow fiber membrane are summarized in Table 12.3 and used for subsequent model simulations.

12.3.2 Process Modeling

The local mass transport for PRO membranes has been described in Chapter 4, while that for FO membranes can be found in other studies (Bui et al., 2015; Lin, 2016). To consider the effects of internal concentration polarization (ICP) and reverse salt diffusion, the water flux (J_w) and reverse salt flux (J_s) are expressed by the following equations.

TABLE 12.3

Summary of hollow fiber membrane specifications (Cheng et al., 2018)

Water permeability coefficient, A (LMH/bar)	3.5
Salt permeability coefficient, B (LMH)	0.3
Structural parameter[a], S (μm)	450
Fiber outer diameter (μm)	1025
Fiber inner diameter (μm)	575

[a] Calculated using the equivalent membrane thickness.

For the FO process (i.e., AL-FS mode):

$$J_w = A \frac{\pi_D \exp\left(-\frac{J_w S}{D}\right) - \pi_F}{1 + \frac{B}{J_w}\left[1 - \exp\left(-\frac{J_w S}{D}\right)\right]} \tag{7}$$

$$J_s = \frac{B}{iRT}\frac{J_w}{A} \tag{8}$$

For the PRO process (i.e., AL-DS mode):

$$J_w = A\left\{\frac{\pi_D - \pi_F \exp\left(\frac{J_w S}{D}\right)}{1 + \frac{B}{J_w}\left[\exp\left(\frac{J_w S}{D}\right) - 1\right]} - \Delta P\right\} \tag{9}$$

$$J_s = \frac{B}{iRT}\left(\frac{J_w}{A} + \Delta P\right) \tag{10}$$

where A and B are the membrane water and salt permeability coefficients, respectively. π_D and π_F are the bulk osmotic pressures of the draw and feed solutions, respectively. S is the structural parameter, D is the draw solute diffusivity in water, i is the van't Hoff factor, R is the universal gas constant, T is the absolute temperature of the solution, and ΔP is the applied hydraulic pressure difference across the membrane. We note that although equations (7)–(10) were derived based on the flat-sheet geometry, they are also applicable to hollow fiber membranes if an equivalent thickness of the fiber is used in the determination of S (Cheng and Chung, 2017; Lin, 2016; Sivertsen et al., 2012).

The mass transport phenomena in hollow fiber modules are more complicated than the local scenario. To ensure an efficient FO/PRO process, the counter-current flow configuration is employed to maintain the concentration gradient (i.e., osmotic driving force) throughout the membrane module. As shown in Figure 12.7, the salinities and flow rates of the feed and draw solutions continuously change along their respective flow channels within the module due to the water permeation and reverse salt diffusion. Thus, in the feed channel, the flow rate at the outlet ($V_{F,0}$) is lower than its inlet counterpart ($V_{F,L}$), because water permeates from the feed solution to the draw solution. Meanwhile, the reverse salt diffusion from the draw solution to the feed solution causes the feed outlet ($C_{F,0}$) with a higher salinity than the inlet one ($C_{F,L}$). Similarly, in the draw channel, $V_{D,L} > V_{D,0}$ and $C_{D,L} < C_{D,0}$. For easy reference, we arbitrarily choose the flow direction of the draw solution as the positive direction. The subscripts 0 and L of flow rate and salinity correspond to the module positions 0 (where the draw solution flows in) and L (where the feed solution flows in), respectively. By carrying out material balance in the feed and draw solutions across a differential membrane area (dA_m), a one-dimensional mass transport model can be developed as follows (Sivertsen et al., 2013; Wan and Chung, 2016b; Xiao et al., 2012):

$$d(\rho_F V_F) = \rho_w J_w dA_m \tag{11}$$

$$d(\rho_D V_D) = \rho_w J_w dA_m \tag{12}$$

$$d(C_F V_F) = -J_s dA_m \tag{13}$$

$$d(C_D V_D) = -J_s dA_m \tag{14}$$

where ρ_F, ρ_D, and ρ_w are the densities of the feed solution, draw solution, and pure water, respectively. Since both the feed and draw solutions are rather diluted (i.e., C_F and $C_D << 400\,g/L$), the following equations can be used to correlate ρ_F and ρ_D with ρ_w (Sharqawy et al., 2010).

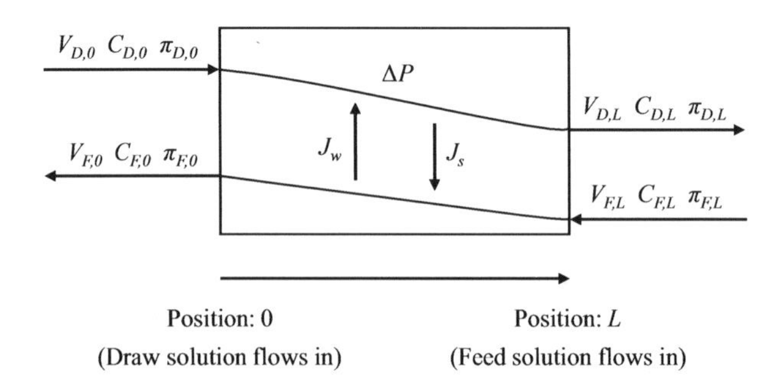

FIGURE 12.7
Schematic drawing of salinity versus module position in a counter-current flow configuration for FO ($\Delta P = 0$) and PRO ($\Delta P \neq 0$) processes (Cheng et al., 2018).

$$\rho_F = \rho_w + C_F \tag{15}$$

$$\rho_D = \rho_w + C_D \tag{16}$$

It should be noted that the units of C_F and C_D are g/L in all the equations.

Before evaluating the module performance, the following dimensionless numbers – recovery (Rec) of the feed solution, initial feed flow rate fraction (ϕ_F), dilutive factor (DF) of the draw solution, and the initial draw flow rate fraction (ϕ_D) – are defined to facilitate the analyses.

For the FO process:

$$\text{Rec} = \frac{V_{D,L} - V_{D,0}}{V_{F,L}} \tag{17}$$

$$\phi_F = \frac{V_{F,L}}{V_{F,L} + V_{D,0}} \tag{18}$$

For the PRO process:

$$\text{DF} = \frac{V_{D,0}}{V_{D,L}} \tag{19}$$

$$\phi_D = \frac{V_{D,0}}{V_{F,L} + V_{D,0}} \tag{20}$$

To analyze the module performance as the draw solution gets diluted, equations (11)–(14) can be rearranged as functions of dV_D (Wan and Chung, 2016b) and integrated with the corresponding boundary conditions as follows:

(1) For the FO process, at the inlet of the draw solution where no dilution occurs (i.e., $V_D = V_{D,0}$), we have $A_m = 0$. At the outlet of the draw solution where dilution occurs (i.e., $V_D = V_{D,L} = V_{D,0}(1 + \text{Rec}\phi_F/(1 - \phi_F))$), we have $C_D = C_{D,L}$, $C_F = C_{F,L}$, and $V_F = V_{D,0}\phi_F/(1 - \phi_F)$.

(2) For the PRO process, similarly, at $V_D = V_{D,0}$, we have $A_m = 0$ and $C_D = C_{D,0}$. At $V_D = V_{D,L} = V_{D,0}/\text{DF}$, we have $C_F = C_{F,L}$ and $V_F = V_{D,0}(1 - \phi_D)/\phi_D$.

The integration is conducted in this special way to evaluate the water flux or the amount of osmotic energy harvested at a certain dilution of draw solution.

The average water flux of the hollow fiber module, $\overline{J_w}$, is expressed as

$$\overline{J_w} = \frac{V_{D,L} - V_{D,0}}{A_m} \tag{21}$$

For the PRO process, the average power density of the module, $\overline{PD}$, at a constant ΔP can be defined in a similar way as the local scenario.

$$\overline{PD} = \overline{J_w}\Delta P \tag{22}$$

We note that the pressure drops of the feed and draw solutions should be minimized during design and operation of the PRO modules. Moreover, one of the important characteristics of the TFC membrane is the low salt permeability and $J_s/J_w << 1$. Thus, for simplification, both the pressure drop and the salinity change of the inter-loop caused by the reverse salt diffusion are neglected when formulating the mathematical models.

12.3.3 Experimental Design and Operation

12.3.3.1 Determination of the Inter-Loop Concentration in Local Scenario

Figure 12.8(a)–(c) summarizes the calculated peak power density and the corresponding water flux and operating pressure of the PRO process as a function of inter-loop (feed) solution concentration using a synthetic SWBr of 1.2 M NaCl as the draw solution. When the inter-loop concentration is low, the PRO membrane enjoys a high water flux. As the inter-loop concentration increases, both peak power density (Figure 12.8(a)) and water flux (Figure 12.8(b)) decline rapidly. This is because of the combined effects of a smaller osmotic driving force and severer ICP when increasing the inter-loop concentration. Moreover, the operating pressure (Figure 12.8(c)) decreases monotonically with an increase in inter-loop concentration. To ensure a reasonable amount of osmotic energy can be harvested in PRO, the commercially viable benchmark of power density 5 W/m^2 set by Statkraft (Gerstandt et al., 2008; Skilhagen, 2010; Thorsen and Holt, 2009) is used as the guidance to draw the upper bound of the inter-loop concentration, which is 0.348 M NaCl, as indicated in Figure 12.8(a). At this salinity, the corresponding water flux (Figure 12.8 (b)) and the operating pressure (Figure 12.8(c)) are 7.7 LMH and 23.3 bar, respectively.

For the FO process operating under the AL-FS mode, Figure 12.8(d) presents the calculated water flux as a function of inter-loop (draw) solution concentration using a synthetic WWRe of 0.011 M NaCl as the feed solution. Similar to Figure 12.8(a) and (b), the water flux shows a prompt rise followed by a mild increase as the inter-loop concentration increases. When the inter-loop salinity reaches 0.348 M (i.e., corresponding to 5 W/m^2 in PRO), the water flux can reach as high as 15 LMH. Under the

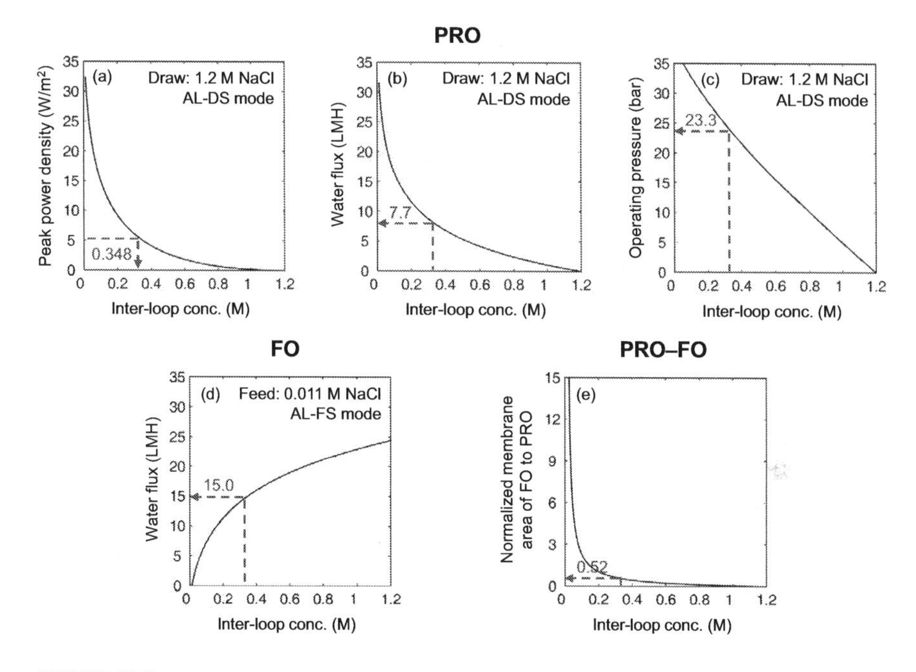

FIGURE 12.8
For the PRO process, (a) peak power density and the corresponding (b) water flux and (c) operating pressure under the AL-DS mode as a function of inter-loop (feed) solution concentration using 1.2 M NaCl as the draw solution. For the FO process, (d) water flux under the AL-FS mode as a function of inter-loop (draw) solution concentration using 0.011 M NaCl as the feed solution; For the PRO–FO hybrid process, (e) normalized membrane area of FO to PRO processes as a function of inter-loop concentration (Cheng et al., 2018).

steady-state operation, the amount of water extracted from WWRe in FO should be equal to that transporting to dilute SWBr in PRO. Based on this material balance, one can estimate the normalized membrane area of FO to PRO processes (Figure 12.8(e)) by dividing the PRO water flux (Figure 12.8(b)) by FO water flux (Figure 12.8(d)). It is 0.52 at the inter-loop salinity of 0.348 M. In other words, to operate the current hybrid system with a power density of 5 W/m^2, the required membrane area ratio between FO and PRO processes is 0.52. As shown in Figure 12.8(e), slightly lowering the inter-loop concentration below 0.1 M leads to a sharp increase in the normalized membrane area of FO to PRO. Therefore, one must choose a reasonable value of the normalized membrane area of FO to PRO in advance, in order to set the lower bound of the inter-loop concentration for practical operations. Overall, an intermediate value of 0.2 M with the normalized area of 1.02 has been selected as the inter-loop salinity to experimentally verify the concept of the PRO–FO hybrid system in Chapter 12.3.3.2.

12.3.3.2 Fouling Resistance of the PRO–FO Hybrid System at ΔP = 0 bar

By using WWRe as the feed solution and an inter-loop solution of 0.2 M NaCl as the draw solution, fouling phenomena and propensity of TFC hollow fiber membranes were first investigated under the AL-FS mode for the FO process. A synthetic WWRe of 0.011 M NaCl was also employed as a feed for comparison. Figure 12.9 compares the normalized water fluxes (with an initial flux of ~7 LMH) as a function of cumulative volume of the feed permeate (or feed recovery). The water flux of the baseline experiment displays a slow and mild decay with a total flux reduction of ~20% at the feed recovery of 50% (i.e., the cumulative permeate volume of 500 mL). This flux decline is mainly caused by the combined effects of reverse salt diffusion, concentration of the feed solution, and ICP. When WWRe is used as the feed solution, the fouling curve only shifts slightly downward against the baseline, suggesting that the polyamide-active layer on the inner surface of hollow fibers can effectively prevent the salts and foulants from entering the porous support to foul the membrane. As a result, it shows a superior antifouling feature than other TFC membranes reported in the literature for osmosis-driven processes (Chun et al., 2017; Chung et al., 2015; Han et al., 2015a; Kim et al., 2015; Klaysom et al., 2013; She et al., 2016; Straub et al., 2016; Touati and Tadeo, 2017).

Figure 12.10 shows a comparison of surface and cross-section morphologies between the fouled (after FO) and pristine membranes, confirming

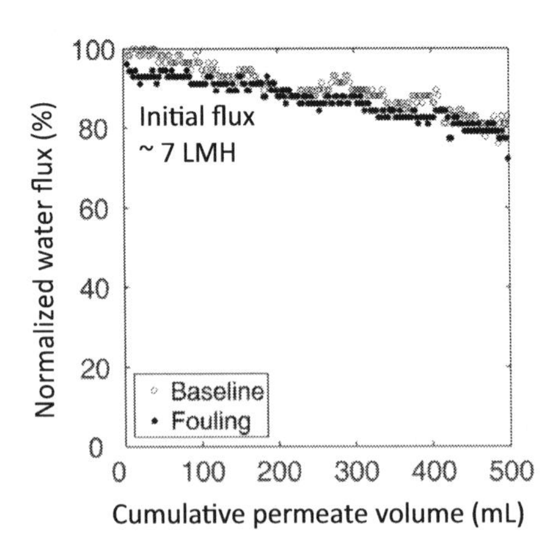

FIGURE 12.9
Normalized water flux of the TFC hollow fiber membrane as a function of cumulative permeate volume in the FO process under the AL-FS mode. 0.011 M NaCl and WWRe were used as feeds for the baseline and fouling tests, respectively, with an initial volume of 1 L. 0.2 M NaCl was used as the draw solution for both tests with an initial volume of 4 L. No concentration adjustment was given to either feed or draw solutions throughout the entire tests (Cheng et al., 2018).

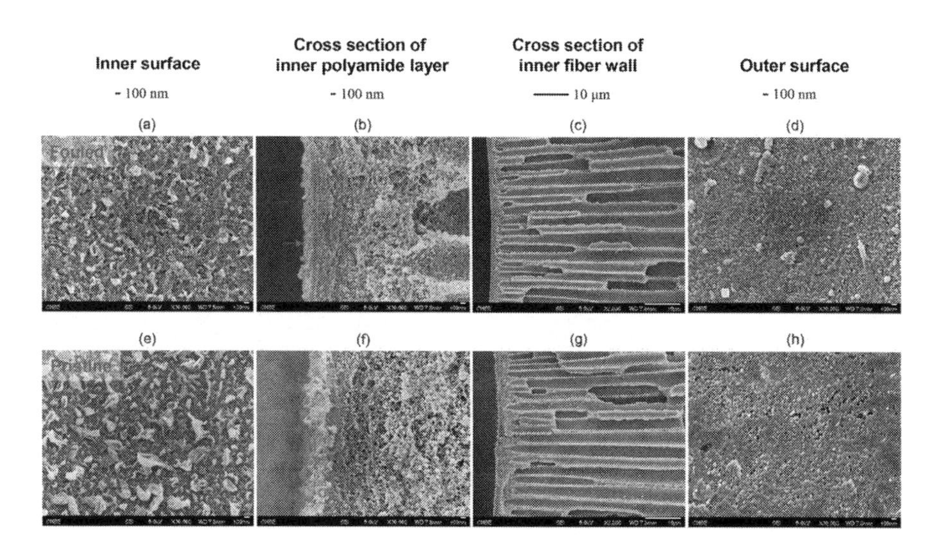

FIGURE 12.10
Inner surface, cross section, and outer surface morphologies of (a–d) the fouled (after the FO process) and (e and f) pristine TFC hollow fiber membranes (Cheng et al., 2018).

very mild fouling associated with the FO process. The pristine membrane has an inner polyamide-active layer with a typical ridge-and-valley structure (Figure 12.10(e)), while the inner surface of the fouled membrane still possesses a similar structure but comprises some fine salt crystals on it (Figure 12.10(a)). Moreover, the newly formed fouling layer does not seem to increase the apparent thickness of the active layer as observed from the cross-section images of Figure 12.10(b) and (f). One possible reason is due to the hydrodynamic shear forces brought by the relatively high velocity (0.64 m/s) in the lumen side of the hollow fiber, which may drive foulants away from the membrane (Lotfi et al., 2017; Mi and Elimelech, 2008). Thus, the impact of the fouling layer on water transport is greatly suppressed, as supported by the fouling data in Figure 12.9. In addition, there are negligible changes on the cross section of fiber wall and outer surface morphologies (Figure 12.10(c), (d), (g), and (h)) before and after the fouling tests. These indicate that the polyamide-active layer can well reject the salts and foulants to enter the membrane and make the fouling mainly occur on the surface. Contrary to fouling on other membrane regions, this type of fouling generally shows great reversibility and is easy to be cleaned (Chun et al., 2017; Han et al., 2017; Mi and Elimelech, 2008; She et al., 2016).

XPS analyses were conducted to characterize the surface chemistry of the fouled and pristine polyamide-active layers. Figure 12.11 shows the signals of Ca, S, P, and Si elements on the fouled membrane surface, which

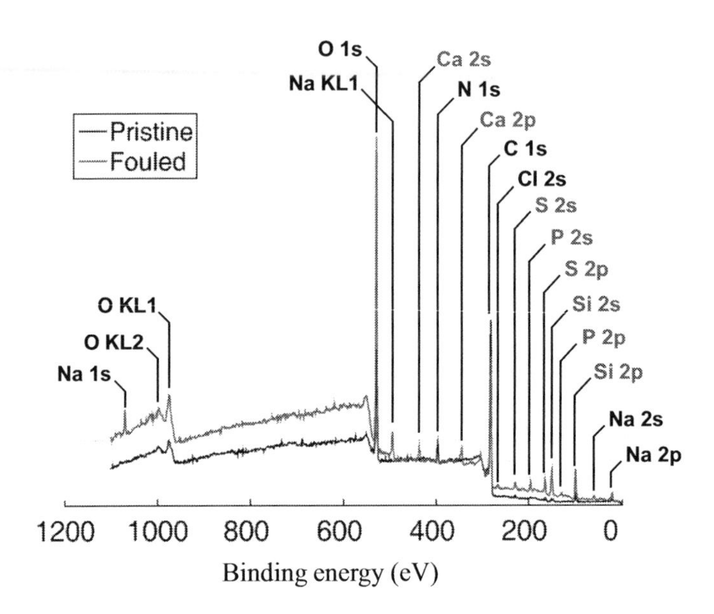

FIGURE 12.11
Inner surface XPS spectra of the fouled (after the FO process) and pristine TFC hollow fiber membranes (Cheng et al., 2018).

are significant as compared to the pristine one, suggesting the co-existence of inorganic fouling and silica scaling. Similar implications have been reported in other studies (Han et al., 2017; Li et al., 2017).

After the FO process, water samples of the concentrated WWRe (i.e., the left 500 mL of the feed solution) and diluted inter-loop solution (i.e., ~4.5 L of the draw solution) as well as the original WWRe were taken for analyses. Table 12.4 summarizes the results. Compared to the original WWRe, the inter-loop solution diluted by FO contains trace amounts of silica and inorganic fouling/scaling ions (i.e., SO_4^{2-} and Ca^{2+}). Clearly, FO is an effective means to extract the clean water from WWRe and the diluted inter-loop solution can serve as a suitable feed to the subsequent PRO process at negligible fouling tendency. Since the concentrated WWRe is obtained at a feed recovery of 50%, it has twice the amount of total organic carbon (TOC), silica, and major ions compared to the original WWRe. Figure 12.12 displays the appearance of those water samples. One can see that the concentrated WWRe looks darker than the original WWRe, while the diluted inter-loop solution seems clean.

Figure 12.13 compares the normalized water fluxes (with an initial flux of ~12 LMH) as a function of cumulative volume of the feed permeate if the diluted inter-loop solution and its equivalent concentrated NaCl solution (0.18 M) were used as the feed solution to PRO, respectively. As

TABLE 12.4

Summary of various water-quality parameters of WWRe, concentrated WWRe, and inter-loop solution (Cheng et al., 2018)

	pH	Conductivity (μS/cm)	Turbidity (NTU)	TOC (ppm)	TDS (ppm)	SO_4^{2-} (ppm)	Ca^{2+} (ppm)	Silica (ppm)
WWRe[a]	7.7	1463	0.39	22.2	966	150	61.1	23.8
Concentrated WWRe[b]	7.9	2404	2.49	42.4	1205	300	122.1	44.8
Diluted inter-loop solution[b]	7.3	17,500	0.15	<1[c]	11,273	<0.10[c]	<1.0[c]	0.06

[a] Measured before the FO process.
[b] Measured after the FO process.
[c] Lower than the detection limits.

FIGURE 12.12
Photos of (a) WWRe before the FO process; (b) concentrated WWRe, and (c) diluted inter-loop solution after the FO process (Cheng et al., 2018).

expected, two curves reach a good agreement. The total flux reduction is about 40% at the end of tests, which is larger than that of the FO experiments presented in Figure 12.9. This is caused by a higher salinity of the feed solution. Overall, the PRO–FO hybrid system shows much better fouling resistance than the stand-alone PRO, which directly uses WWRe as the feed solution (Chen et al., 2016; Wan and Chung, 2015).

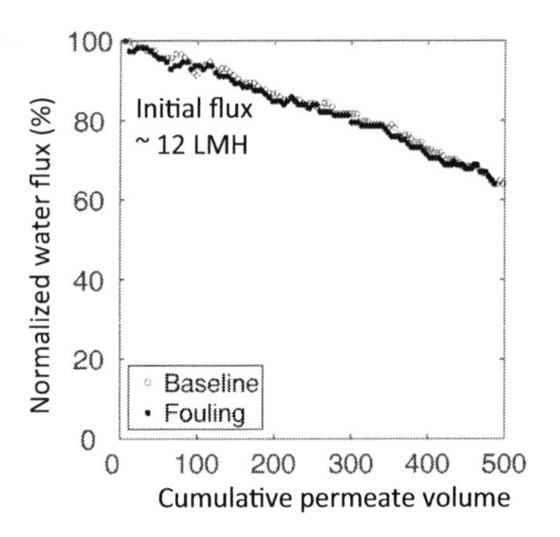

FIGURE 12.13
Normalized water flux of the TFC hollow fiber membrane as a function of cumulative permeate volume in the PRO process under the AL-DS mode at ΔP = 0. 0.18 M NaCl and diluted inter-loop solution (i.e., the draw solution of the FO process when WWRe was used as the feed solution) were used as feeds for the baseline and fouling tests, respectively, with an initial volume of 1 L. 1.2 M NaCl was used as the draw solution for both tests. No concentration adjustment was given to the feed solution while the draw solution concentration was maintained throughout the entire tests (Cheng et al., 2018).

12.3.4 Full-Scale Simulation Results and Discussion

The calculations in Chapter 12.3.3.1 suggest that a considerable amount of osmotic energy can be harvested by the PRO–FO hybrid process at the local scenario where no dilution is concerned, while the designed experiments in Chapter 12.3.3.2 validate our hypothesis that fouling can be significantly suppressed. To further understand its mass transport behavior in large-size modules, full-scale analyses are carried out to consider the dilution effect. These analyses are conducted by numerically solving the material balance of equations (11)–(14) and the governing equations of water flux and reverse salt diffusion in FO and PRO processes (equations (7) and (8) as well as equations (9) and (10), respectively) along the module length from position 0 to *L*. Besides, several dimensionless numbers (i.e., *Rec*, ϕ_F, DF, and ϕ_D) are used for easy quantification of module performance under different operating conditions. To fully explore the potential of the PRO–FO hybrid system, three salinities of inter-loop solution; namely, 0.1, 0.2, and 0.3 M are chosen to simulate the outlet concentration of the draw solution ($C_{D,L}$) for FO and the inlet concentration of the feed solution ($C_{F,}$) for PRO, respectively (Figure 12.7). All these salinities are below the maximum inter-loop concentration (0.348 M) as determined in Chapter 12.3.3.1.

12.3.4.1 FO Process

Figure 12.14 shows the average water flux of the FO process as a function of Rec, ϕ_F, and draw solution concentration at module position L ($C_{D,L}$). The feed solution concentration at module position L ($C_{F,L}$) is fixed to be 0.011 M (i.e., the salinity of WWRe). In general, an increase in $C_{D,L}$ raises the average water flux because of a larger osmotic driving force. The horizontal axis, Rec, describes the percentage of the feed solution extracted by FO. A higher Rec means a greater efficiency of the FO process. The dark triangular regions on the top and right sides of each figure correspond to the cases where $C_{D,0} > 1.2$ M and $C_{F,0} > C_{D,0}$, respectively. The FO operation within these two regions is impractical, thus the average water flux is assigned to 0 by default. Nonetheless, as shown in Figure 12.14, the variations of Rec and ϕ_F do not significantly affect the average water flux for all three cases, probably because of the limited salinity-gradient between the feed pair and the ICP effect owing to the applied AL-FS mode. Overall, the average water flux obtained at the module level is about 10–15 LMH. This value is comparable to others obtained using low-pressure nanofiltration (NF)/RO to pretreat the feed wastewater (Van der Bruggen et al., 2003; Wan and Chung, 2015). However, since FO barely requires any external energy, it may be superior as an alternative to pretreat WWRe for the PRO process.

12.3.4.2 PRO Process

Figure 12.15 elucidates the average peak power density and the corresponding average water flux and operating pressure as a function of DF, ϕ_D, and feed solution concentration at module position L ($C_{F,L}$) for the PRO process. The draw solution concentration at module position 0 ($C_{D,0}$) is

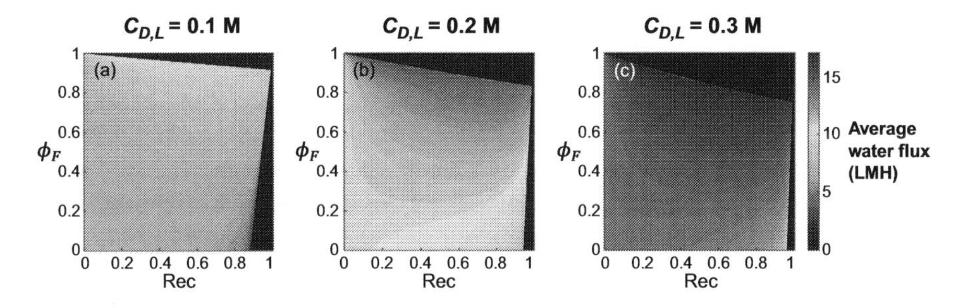

FIGURE 12.14
Average water flux as a function of the recovery of the feed solution (Rec), initial feed flow rate ratio (ϕ_F), and draw solution concentration at module position L ($C_{D,L}$) in the FO process. The feed solution concentration at module position L ($C_{F,L}$) is fixed at 0.011 M NaCl (Cheng et al., 2018).

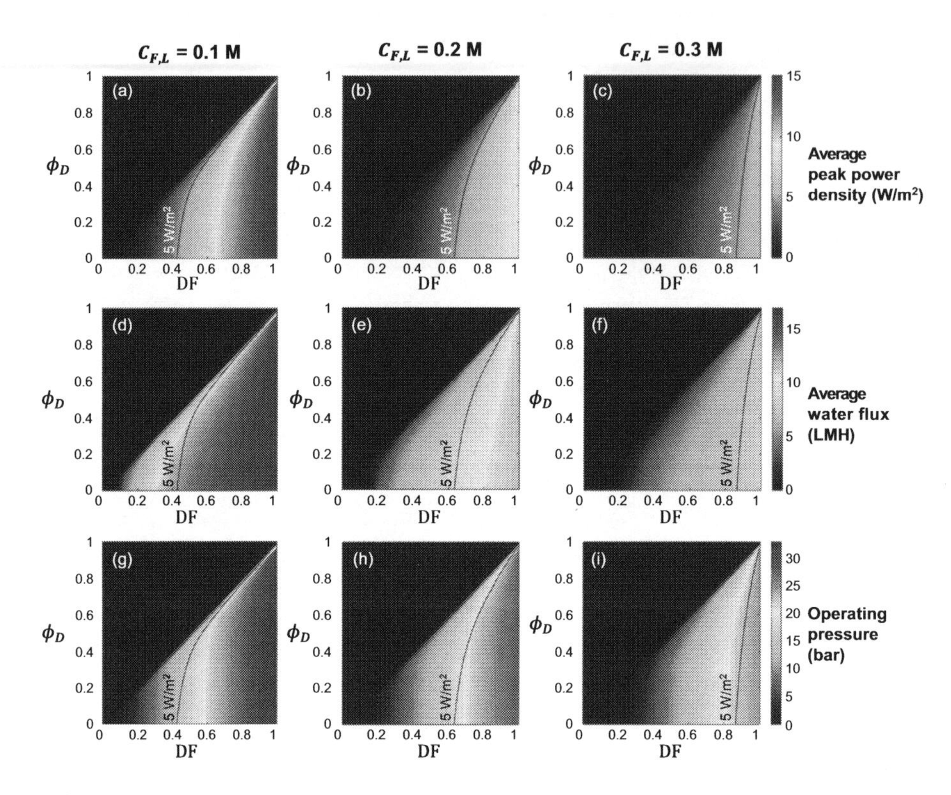

FIGURE 12.15
(a–c) Average peak power density and the corresponding (d–f) average water flux and (g–i) operating pressure as a function of dilutive factor of the draw solution (DF), initial draw flow rate ratio (ϕ_D), and feed solution concentration at module position L ($C_{F,L}$) in the PRO process. The draw solution concentration at module position 0 ($C_{D,0}$) is fixed at 1.2 M NaCl. The contour line of average power density of 5 W/m^2 is plotted for benchmarking purpose (Cheng et al., 2018).

1.2 M (i.e., the salinity of SWBr). The diagonal line in each figure presents the extreme condition where the entire feed solution would permeate across the membrane to dilute the draw solution, above which the water flux and power density turn out to be negative. Therefore, similar to Figure 12.14, those regions are automatically set to 0 and are not recommended for practical operations. The horizontal axis, DF, indicates the fraction of the original draw solution in the diluted draw solution after PRO. A DF approaching 0 means the draw solution is infinitely diluted, while a DF approaching 1 means the draw solution is not diluted at all. Hence, at DF $\rightarrow$ 1, the average peak power densities of PRO reach 15 W/m^2 (Figure 12.15(a)), 9.1 W/m^2 (Figure 12.15(b)), and 6 W/m^2 (Figure 12.15(c)) when the feed solution concentrations ($C_{F,L}$) are 0.1, 0.2, and 0.3 M, respectively.

These power densities are consistent with the readings from Figure 12.8 (a), where only local mass transport is concerned. As DF becomes smaller, a lower average power density could be achieved for all three cases due to the reduced osmotic driving force caused by gradual dilution of the draw solution. Meanwhile, both the average water flux (Figure 12.15 (d)–(f)) and optimal operating pressure (Figure 12.15(g)–(i)) decrease accordingly.

For the benchmark of a PRO process, the contour line of power density of 5 W/m^2 is plotted in each figure of Figure 12.15. To mitigate the environmental hazards of SWBr (1.2 M) toward marine life, it is suggested to operate PRO at a DF close to 0.5. At this value, the diluted SWBr will possess a salinity similar to that of seawater (i.e., ~0.6 M) and can be directly and safely discharged to the sea or recycled back to the upstream seawater RO plant as a potential feed. In order to have a PRO process with a high power density and adequate dilution of SWBr simultaneously, which are two metrics assessing the performance of osmotic power generation, a low feed solution salinity ($C_{F,L}$) is essential. From Figure 12.15, one can see that it is possible to achieve an average peak power density of greater than 5 W/m^2 at DF = 0.5 when $C_{F,L}$ = 0.1 M. If $C_{F,L}$ becomes 0.2 and 0.3 M, DF must be elevated to 0.63 (Figure 12.15(b)) and 0.86 (Figure 12.15 (c)), respectively, at $\phi_D = 0$. In other words, to maintain the average peak power density of 5 W/m^2 with a larger ϕ_D, DF will be further increased to compensate the effect of a less feed solution intake during the dilution of the draw solution.

12.3.4.3 PRO–FO Hybrid Process

So far, we have separately examined the performance of FO and PRO processes under different inter-loop concentrations. To consider the integration of a PRO–FO hybrid process, Figure 12.16 gives its schematic drawing with some newly defined (i.e., r) or redefined (i.e., ϕ_F and ϕ_D) dimensionless numbers to analyze the overall material balance. Consistent with the definitions in Chapter 12.3.2 and Figure 12.7, subscripts 0 and L correspond to the module positions 0 (where the draw solution flows in) and L (where the feed solution flows in), respectively.

$$r = \frac{V_{\mathrm{Int}L,0}}{V_{\mathrm{Int}L,L}} \tag{23}$$

$$\phi_F = \frac{V_{\mathrm{WWRe},L}}{V_{\mathrm{WWRe},L} + V_{\mathrm{Int}L,0}} \tag{24}$$

$$\phi_D = \frac{V_{\mathrm{SWBr},0}}{V_{\mathrm{SWBr},0} + V_{\mathrm{Int}L,L}} \tag{25}$$

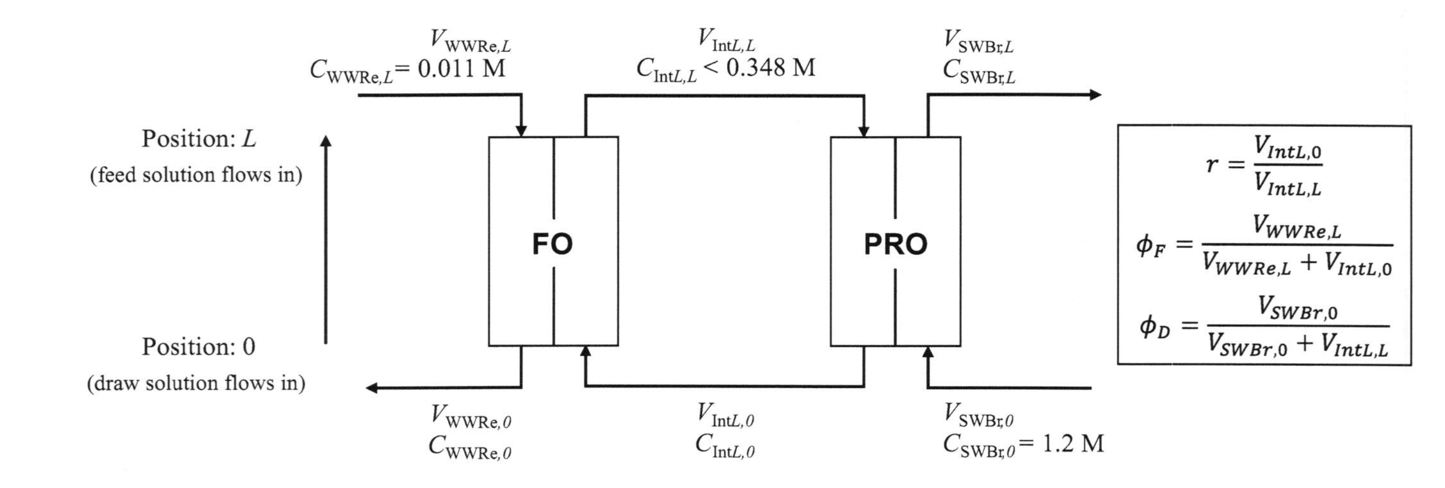

FIGURE 12.16
Schematic drawing of a PRO–FO hybrid system (Cheng et al., 2018).

Where r is the inter-loop flow rate ratio between the stream from PRO to FO ($V_{IntL,0}$) and that from FO to PRO ($V_{IntL,L}$). ϕ_F and ϕ_D remain the same as in equations (18) and (20), respectively, but are expressed differently to match with the notations in Figure 12.16. The salinities of WWRe entering FO ($C_{WWRe,L}$) and SWBr entering PRO ($C_{SWBr,0}$) are 0.011 and 1.2 M, respectively. While the inter-loop concentration at module position L ($C_{IntL,L}$) is predetermined and smaller than 0.348 M.

Figure 12.17 illustrates the respective performance of PRO and FO units when a hybrid system is concerned. An intermediate value of 0.2 M is arbitrarily chosen for $C_{IntL,L}$. Different from Figures 12.14 and 12.15, the horizontal axis in Figure 12.17 utilizes r instead of Rec or DF for easier expression of the integrated process, while the vertical axes are kept as ϕ_D and ϕ_F for PRO and FO, respectively. For the PRO unit (Figure 12.17(a)–(c)), analogous to DF, r indicates the concentrated extent of the feed solution. When $r \to 1$, no concentration occurs on the feed side and the average peak power density of PRO reaches the maximum value of 9.1 W/m^2 since the osmotic driving force could well maintain along the module. When $r \to 0$, a severely concentered feed solution leaving the PRO module would remarkably drag down the osmotic driving force across the membrane, resulting in a very low or even negative water flux and power density. For the practical reason, those negative regions are given as 0 by convention. For the FO unit (Figure 12.17(d)), the diagonal line draws the operating limit analogous to Figure 12.15 and the dark region on the left bottom corner of Figure 12.17(d) is caused by the limit of $C_{IntL,0} \leq C_{SWBr,0} = 1.2\ M$.

Figure 12.17 also displays the quick design steps for a PRO–FO hybrid system. Briefly, one may (1) locate an operating point (e.g., ☆) of PRO (Figure 12.17(a)–(c)) and read off the corresponding ϕ_D and r of the vertical and horizontal axes, respectively; (2) calculate $V_{IntL,L}$ from the given $V_{SWBr,0}$ using the read off ϕ_D; (3) calculate $V_{IntL,0}$ from $V_{IntL,L}$ using the read off r; (4) select an operating point (e.g., ✧) in FO (Figure 12.17(d)) at the same r and read off the corresponding ϕ_F of the vertical axis; and (5) calculate $V_{WWRe,L}$ from $V_{IntL,0}$ using the read off ϕ_F. To estimate the membrane area ratio of PRO to FO, it is inversely proportional to the ratio of average water flux read from the elevation of Figure 12.17(b) for PRO and Figure 12.17(d) for FO, respectively. The same approach has been used to construct Figure 12.8(e) previously.

12.4 Challenges and Future Prospects

12.4.1 PRO–MD

In order to fully maximize the potential of PRO–MD hybrid processes for osmotic power and clean water production, more studies must be dedicated

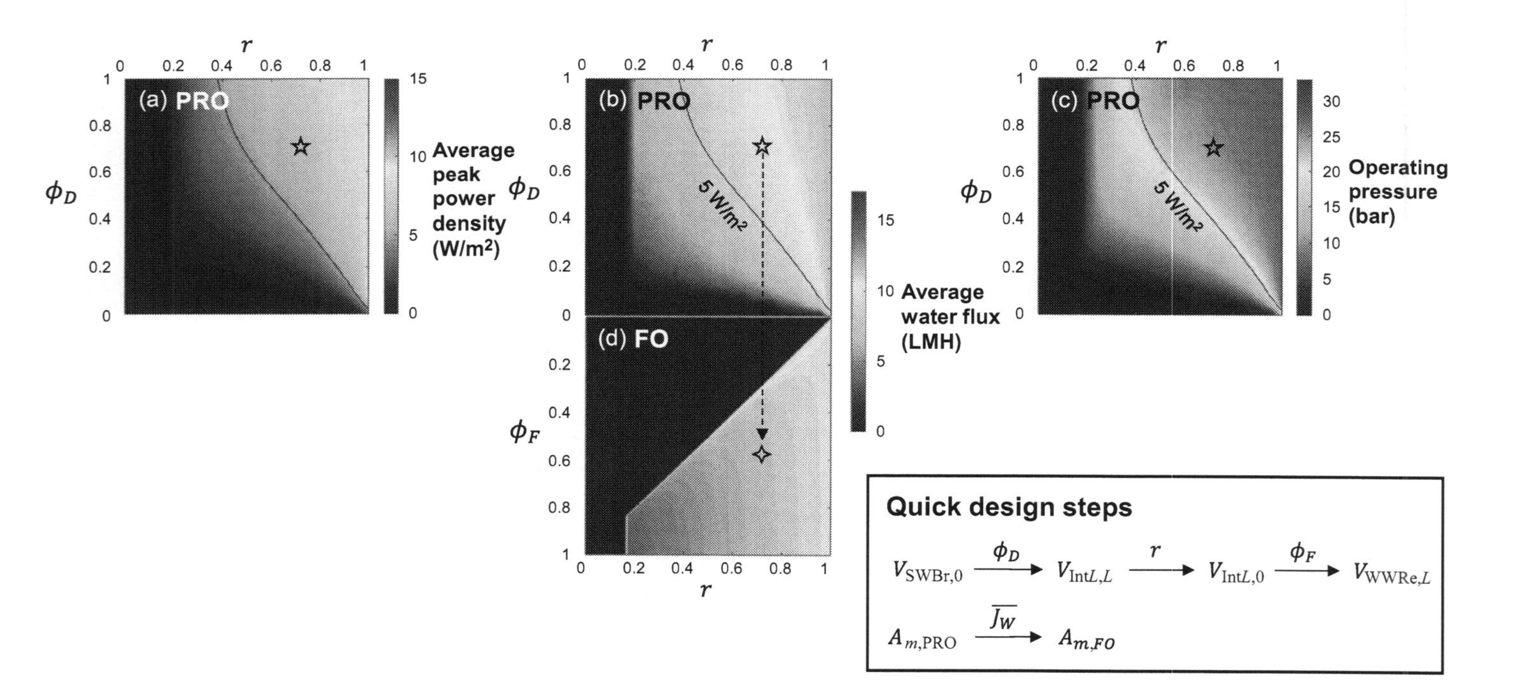

FIGURE 12.17
Design map of a PRO–FO hybrid system. For the PRO unit, (a) average peak power density and the corresponding (b) average water flux and (c) operating pressure as a function of inter-loop flow rate ratio (r) and initial draw flow rate ratio (ϕ_D). For the FO unit, (d) average water flux as a function of inter-loop flow rate ratio (r) and initial feed flow rate ratio (ϕ_F). $C_{\text{Int}L,L}$ is chosen as 0.2 M in the calculations. The contour line of average power density of 5 W/m^2 is plotted for benchmarking purpose (Cheng et al., 2018).

to (1) advance the long-term performance and stability of commercial-scale PRO and MD membrane modules and (2) further improve the efficiency of converting low-cost thermal energy to useful works in MD. As restricted by practical losses in actual processes and thermodynamic limitations, only a fraction of the abundant waste heat can be feasibly converted. The resultant intrinsically low efficiency suggests that it would be more prudent to utilize the heat resources directly, for instance, in heat pumps and space heating (Lin et al., 2014b). Yet such direct uses are geographically confined to the locality and availability of the heat sources. In the absence of a matching need for direct use at the source, converting waste heat to electricity is the only available option to exploit the low-cost thermal energy. Another method to particularly raise the thermal separation efficiency of MD is to use a solvent (e.g., methanol) with higher volatility and lower heat capacity and enthalpy of vaporization than water (Shaulsky et al., 2015). For the PRO aspect, apart from designing highly robust and permselective membranes, employing novel draw solutes of large molecule sizes (Han et al., 2014; Xia et al., 2018; Zhong et al., 2017) in the close-loop process could minimize the leakage while maintaining the osmotic driving force. By successfully establishing an integrated process model to account for the system nonidealities and economic analysis, further works are required to demonstrate the pilot-scale studies based on the optimal conditions to advance the PRO–MD hybrids toward actual implementation.

12.4.2 PRO–FO

Although the proposed PRO–FO hybrid system shows promising results, there are several improvements that can be done to further raise its potential and competency. In the present work of Chapter 12.3, our PES TFC hollow fiber membranes (Wan and Chung, 2015; Zhang et al., 2014b) have been employed in both FO and PRO processes. This type of membranes possesses a thick fiber wall to sustain the high-pressure PRO operations. Therefore, it has a large S value of 450 μm. As ΔP is not present in FO operations, the hollow fiber membranes for FO can be tailored with a smaller S. For example, Sukitpaneenit and Chung have developed high-performance TFC FO hollow fiber membranes with a S value of 220–260 μm (Sukitpaneenit and Chung, 2012). By having a smaller S, a higher water flux is expected for the FO process under the feed pair of the same salinities. As a result, a less concentrated inter-loop (draw) solution can be used to achieve a desirable water flux when WWRe is served as the feed solution. This would boost the osmotic driving force and peak power density of the PRO unit in the hybrid system.

In addition, NaCl is currently chosen as the draw solute of the inter-loop solution because of its reasonably high osmolality, low cost, and easy accessibility. However, when such an inter-loop solution is used as the feed solution to PRO under the AL-DS mode, significant ICP may be triggered and further exacerbated because of the concentration of the feed solution along the module. To eliminate the adverse effect of ICP, one may replace NaCl by smart draw solutes such as hydroacid complex (Han et al., 2014) and magnetic nanoparticles (Ge et al., 2016). These draw solutes can be easily recovered from the inter-loop stream from FO to PRO, making the inter-loop solution as a feed solution of almost zero concentration for the PRO process. Afterward, they are re-dissolved in the inter-loop stream from PRO back to FO, giving sufficient osmosis to the inter-loop solution as a draw solution for the FO process. In this way, simultaneous achievement of high-performance PRO and FO can be realized, breaking the trade-off relationship underlined in Chapter 12.3.3.1.

References

Achilli, A., Prante, J.L., Hancock, N.T., Maxwell, E.B., Childress, A.E., 2014. Experimental results from RO-PRO: A next generation system for low-energy desalination. *Environ. Sci. Technol.* 48, 6437–6443.

Anastasio, D.D., Arena, J.T., Cole, E.A., McCutcheon, J.R., 2015. Impact of temperature on power density in closed-loop pressure retarded osmosis for grid storage. *J. Membr. Sci.* 479, 240–245.

Antony, A., Low, J.H., Gray, S., Childress, A.E., Le-Clech, P., Leslie, G., 2011. Scale formation and control in high pressure membrane water treatment systems: A review. *J. Membr. Sci.* 383, 1–16.

Bajraktari, N., Hélix-Nielsen, C., Madsen, H.T., 2017. Pressure retarded osmosis from hypersaline sources – A review. *Desalination* 413, 65–85.

Bui, N.N., Arena, J.T., McCutcheon, J.R., 2015. Proper accounting of mass transfer resistances in forward osmosis: Improving the accuracy of model predictions of structural parameter. *J. Membr. Sci.* 492, 289–302.

Chae, S.H., Seo, J., Kim, J., Kim, Y.M., Kim, J.H., 2018. A simulation study with a new performance index for pressure-retarded osmosis processes hybridized with seawater reverse osmosis and membrane distillation. *Desalination* 444, 118–128.

Chen, S.C., Amy, G.L., Chung., T.S., 2016. Membrane fouling and anti-fouling strategies using RO retentate from a municipal water recycling plant as the feed for osmotic power generation. *Water Res.* 88, 144–155.

Chen, S.C., Wan, C.F., Chung, T.S., 2015. Enhanced fouling by inorganic and organic foulants on pressure retarded osmosis (PRO) hollow fiber membranes under high pressures. *J. Membr. Sci.* 479, 190–203.

Cheng, Z.L., Chung, T.S., 2017. Mass transport of various membrane configurations in pressure retarded osmosis (PRO). *J. Membr. Sci.* 537, 160–176.

Cheng, Z.L., Li, X., Chung, T.S., 2018. The forward osmosis-pressure retarded osmosis (FO-PRO) hybrid system: A new process to mitigate membrane fouling for sustainable osmotic power generation. *J. Membr. Sci.* 559, 63–74.

Choi, Y., Kim, S.H., Lee, S., 2017. Comparison of performance and economics of reverse osmosis, membrane distillation, and pressure retarded osmosis hybrid systems. *Desalination Water Treat.* 77, 19–29.

Chowdhury, M.R., McCutcheon, J.R., 2018. Elucidating the impact of temperature gradients across membranes during forward osmosis: Coupling heat and mass transfer models for better prediction of real osmotic systems. *J. Membr. Sci.* 553, 189–199.

Chun, Y., Mulcahy, D., Zou, L., Kim, I.S., 2017. A short review of membrane fouling in forward osmosis processes. *Membranes* 7, 30. doi:10.3390/membranes7020030.

Chung, T.S., Li, X., Ong, R.C., Ge, Q., Wang, H., Han, G., 2012. Emerging forward osmosis (FO) technologies and challenges ahead for clean water and clean energy applications. *Curr. Opin. Chem. Eng.* 1, 246–257.

Chung, T.S., Luo, L., Wan, C.F., Cui, Y., Amy, G., 2015. What is next for forward osmosis (FO) and pressure retarded osmosis (PRO). *Sep. Purif. Technol.* 156, 856–860.

Curcio, E., Drioli, E., 2005. Membrane distillation and related operations – A review. *Sep. Purif. Rev.* 34, 35–86.

Edwie, F., Chung, T.S., 2013. Development of simultaneous membrane distillation–crystallization (SMDC) technology for treatment of saturated brine. *Chem. Eng. Sci.* 98, 160–172.

Efraty, A., 2013. Pressure retarded osmosis in closed circuit: A new technology for clean power generation without need of energy recovery. *Desalination Water Treat.* 51, 7420–7430.

Francis, L., Ghaffour, N., Alsaadi, A.S., Nunes, S.P., Amy, G.L., 2014. Performance evaluation of the DCMD desalination process under bench scale and large scale module operating conditions. *J. Membr. Sci.* 455, 103–112.

Ge, Q., Yang, L., Cai, J., Xu, W., Chen, Q., Liu, M., 2016. Hydroacid magnetic nanoparticles in forward osmosis for seawater desalination and efficient regeneration via integrated magnetic and membrane separations. *J. Membr. Sci.* 520, 550–559.

Gerstandt, K., Peinemann, K.V., Skilhagen, S.E., Thorsen, T., Holt, T., 2008. Membrane processes in energy supply for an osmotic power plant. *Desalination* 224, 64–70.

Gryta, M., 2002. Concentration of NaCl solution by membrane distillation integrated with crystallization. *Sep. Sci. Technol.* 37, 3535–3558.

Gryta, M., 2007. Influence of polypropylene membrane surface porosity on the performance of membrane distillation process. *J. Membr. Sci.* 287, 67–78.

Gryta, M., Tomaszewska, M., Grzechulska, J., Morawski, A.W., 2001. Membrane distillation of NaCl solution containing natural organic matter. *J. Membr. Sci.* 181, 279–287.

Han, G., Cheng, Z.L., Chung, T.S., 2017. Thin-film composite (TFC) hollow fiber membrane with double-polyamide active layers for internal concentration polarization and fouling mitigation in osmotic processes. *J. Membr. Sci.* 523, 497–504.

Han, G., Ge, Q., Chung, T.S., 2014. Conceptual demonstration of novel closed-loop pressure retarded osmosis process for sustainable osmotic energy generation. *Appl. Energy* 132, 383–393.

Han, G., Liu, J.T., Lu, K.J., Chung, T.S., 2018. Advanced anti-fouling membranes for osmotic power generation from wastewater via pressure retarded osmosis (PRO). *Environ. Sci. Technol.* 52, 6686–6694.

Han, G., Wang, P., Chung, T.S., 2013a. Highly robust thin-film composite pressure retarded osmosis (PRO) hollow fiber membranes with high power densities for renewable salinity-gradient energy generation. *Environ. Sci. Technol.* 47, 8070–8077.

Han, G., Zhang, S., Li, X., Chung, T.S., 2013b. High performance thin film composite pressure retarded osmosis (PRO) membranes for renewable salinity-gradient energy generation. *J. Membr. Sci.* 440, 108–121.

Han, G., Zhang, S., Li, X., Chung, T.S., 2015a. Progress in pressure retarded osmosis (PRO) membranes for osmotic power generation. *Prog. Polym. Sci.* 51, 1–27.

Han, G., Zhou, J.L., Wan, C.F., Yang, T.S., Chung, T.S., 2016. Investigations of inorganic and organic fouling behaviors, antifouling and cleaning strategies for pressure retarded osmosis (PRO) membrane using seawater desalination brine and wastewater. *Water Res.* 103, 264–275.

Han, G., Zuo, J., Wan, C.F., Chung, T.S., 2015b. Hybrid pressure retarded osmosis-membrane distillation (PRO-MD) process for osmotic power and clean water generation. *Environ. Sci. Water Res. Technol.* 1, 507–515.

Hu, M., Zheng, S., Mi, B., 2016. Organic fouling of graphene oxide membranes and its implications for membrane fouling control in engineered osmosis. *Environ. Sci. Technol.* 50, 685–693.

Kim, J., Jeong, K., Park, M.J., Shon, H.K., Kim, J.H., 2015. Recent advances in osmotic energy generation via pressure-retarded osmosis (PRO): A review. *Energies* 8, 11821–11845.

Kim, J., Lee, J., Kim, J.H., 2012. Overview of pressure-retarded osmosis (PRO) process and hybrid application to sea water reverse osmosis process. *Desalination Water Treat.* 43, 193–200.

Kim, J., Park, M., Shon, H.K., Kim, J.H., 2016. Performance analysis of reverse osmosis, membrane distillation, and pressure-retarded osmosis hybrid processes. *Desalination* 380, 85–92.

Klaysom, C., Cath, T.Y., Depuydt, T., Vankelecom, I.F.J., 2013. Forward and pressure retarded osmosis: Potential solutions for global challenges in energy and water supply. *Chem. Soc. Rev.* 42, 6959–6989.

Le, N.L., Quilitzsch, M., Cheng, H., Hong, P.Y., Ulbricht, M., Nunes, S.P., Chung, T. S., 2017. Hollow fiber membrane lumen modified by polyzwitterionic grafting. *J. Membr. Sci.* 522, 1–11.

Lee, J., Kim, S., 2016. Predicting power density of pressure retarded osmosis (PRO) membranes using a new characterization method based on a single PRO test. *Desalination* 389, 224–234.

Lee, J.G., Kim, Y.D., Shim, S.M., Im, B.G., Kim, W.S., 2015. Numerical study of a hybrid multi-stage vacuum membrane distillation and pressure-retarded osmosis system. *Desalination* 363, 82–91.

Lee, K.L., Baker, R.W., Lonsdale, H.K., 1981. Membranes for power generation by pressure-retarded osmosis. *J. Membr. Sci.* 8, 141–171.

Lee, S., Choi, J., Park, Y.G., Shon, H., Ahn, C.H., Kim, S.H., 2019. Hybrid desalination processes for beneficial use of reverse osmosis brine: Current status and future prospects. *Desalination* 454, 104–111.

Li, X., Cai, T., Amy, G.L., Chung, T.S., 2017. Cleaning strategies and membrane flux recovery on anti-fouling membranes for pressure retarded osmosis. *J. Membr. Sci.* 522, 116–123.

Li, X., Cai, T., Chen, C., Chung, T.S., 2016. Negatively charged hyperbranched polyglycerol grafted membranes for osmotic power generation from municipal wastewater. *Water Res.* 89, 50–58.

Lin, S., 2016. Mass transfer in forward osmosis with hollow fiber membranes. *J. Membr. Sci.* 514, 176–185.

Lin, S., Straub, A.P., Elimelech, M., 2014a. Thermodynamic limits of extractable energy by pressure retarded osmosis. *Energy Environ. Sci.* 7, 2706–2714.

Lin, S., Yip, N.Y., Cath, T.Y., Osuji, C.O., Elimelech, M., 2014b. Hybrid pressure retarded osmosis-membrane distillation system for power generation from low-grade heat: Thermodynamic analysis and energy efficiency. *Environ. Sci. Technol.* 48, 5306–5313.

Loeb, S., Norman, R.S., 1975. Osmotic power plants. *Science* 189, 654–655.

Logan, B.E., Elimelech, M., 2012. Membrane-based processes for sustainable power generation using water. *Nature* 488, 313–319.

Lotfi, F., Chekli, L., Phuntsho, S., Hong, S., Choi, J.Y., Shon, H.K., 2017. Understanding the possible underlying mechanisms for low fouling tendency of the forward osmosis and pressure assisted osmosis processes. *Desalination* 421, 89–98.

Lu, K.J., Cheng, Z.L., Chang, J., Luo, L., Chung, T.S., 2019. Design of zero liquid discharge desalination (ZLDD) systems consisting of freeze desalination, membrane distillation, and crystallization powered by green energies. *Desalination* 458, 66–75.

Madsen, H.T., Nissen, S.S., Muff, J., Søgaard, E.G., 2017. Pressure retarded osmosis from hypersaline solutions: Investigating commercial FO membranes at high pressures. *Desalination* 420, 183–190.

McCutcheon, J.R., Elimelech, M., 2007. Modeling water flux in forward osmosis: Implications for improved membrane design. *AIChE J.* 53, 1736–1744.

McGinnis, R.L., McCutcheon, J.R., Elimelech, M., 2007. A novel ammonia–carbon dioxide osmotic heat engine for power generation. *J. Membr. Sci.* 305, 13–19.

Mi, B., Elimelech, M., 2008. Chemical and physical aspects of organic fouling of forward osmosis membranes. *J. Membr. Sci.* 320, 292–302.

Morel, F., Hering, J.G., 1993. *Principles and applications of aquatic chemistry*, Wiley-Interscience, New York.

Nguyen, A., Azari, S., Zou, L., 2013. Coating zwitterionic amino acid L-DOPA to increase fouling resistance of forward osmosis membrane. *Desalination* 312, 82–87.

Phuntsho, S., Shon, H.K., Hong, S., Lee, S., Vigneswaran, S., 2011. A novel low energy fertilizer driven forward osmosis desalination for direct fertigation: Evaluating the performance of fertilizer draw solutions. *J. Membr. Sci.* 375, 172–181.

Pitzer, K.S., Peiper, J.C., Busey, R.H., 1984. Thermodynamic properties of aqueous sodium chloride solutions. *J. Phys. Chem. Ref. Data* 13(1984), 1.

Ramon, G.Z., Feinberg, B.J., Hoek, E.M.V., 2011. Membrane-based production of salinity-gradient power. *Energy Environ. Sci.* 4, 4423–4434.

Ruiz-Saavedra, E., Ruiz-García, A., Ramos-Martín, A., 2015. A design method of the RO system in reverse osmosis brackish water desalination plants (calculations and simulations). *Desalin. Water Treat.* 55, 2562–2572.

Sharqawy, M.H., Lienhard V, J.H., Zubair, S.M., 2010. Thermophysical properties of seawater: A review of existing correlations and data. *Desalin. Water Treat.* 16, 354–380.

Shaulsky, E., Boo, C., Lin, S., Elimelech, M., 2015. Membrane-based osmotic heat engine with organic solvent for enhanced power generation from low-grade heat. *Environ. Sci. Technol.* 49, 5820–5827.

She, Q., Jin, X., Tang, C.Y., 2012. Osmotic power production from salinity gradient resource by pressure retarded osmosis: Effects of operating conditions and reverse solute diffusion. *J. Membr. Sci.* 401–402, 262–273.

She, Q., Wang, R., Fane, A.G., Tang, C.Y., 2016. Membrane fouling in osmotically driven membrane processes: A review. *J. Membr. Sci.* 499, 201–233.

She, Q., Wong, Y.K.W., Zhao, S., Tang, C.Y., 2013. Organic fouling in pressure retarded osmosis: Experiments, mechanisms and implications. *J. Membr. Sci.* 428, 181–189.

Sikdar, S.K., 2014. An e-conversation with Prof. Neal Chung. *Clean Technol. Environ. Policy* 16, 1481–1485.

Sivertsen, E., Holt, T., Thelin, W., Brekke, G., 2012. Modelling mass transport in hollow fibre membranes used for pressure retarded osmosis. *J. Membr. Sci.* 417–418, 69–79.

Sivertsen, E., Holt, T., Thelin, W., Brekke, G., 2013. Pressure retarded osmosis efficiency for different hollow fibre membrane module flow configurations. *Desalination* 312, 107–123.

Skilhagen, S.E., 2010. Osmotic power – A new, renewable energy source. *Desalin. Water Treat.* 15, 271–278.

Stamatakis, E., Stubos, A., Palyvos, J., Chatzichristos, C., Muller, J., 2005. An improved predictive correlation for the induction time of $CaCO_3$ scale formation during flow in porous media. *J. Colloid Interface Sci.* 286, 7–13.

Straub, A.P., Deshmukh, A., Elimelech, M., 2016. Pressure-retarded osmosis for power generation from salinity gradients: Is it viable? *Energy Environ. Sci.* 9, 31–48.

Straub, A.P., Yip, N.Y., Elimelech, M., 2014. Raising the bar: Increased hydraulic pressure allows unprecedented high power densities in pressure-retarded osmosis. *Environ. Sci. Technol. Lett.* 1, 55–59.

Sukitpaneenit, P., Chung, T.S., 2012. High performance thin-film composite forward osmosis hollow fiber membranes with macrovoid-free and highly porous structure for sustainable water production. *Environ. Sci. Technol.* 46, 7358–7365.

Thelin, W.R., Sivertsen, E., Holt, T., Brekke, G., 2013. Natural organic matter fouling in pressure retarded osmosis. *J. Membr. Sci.* 438, 46–56.

Thorsen, T., Holt, T., 2009. The potential for power production from salinity gradients by pressure retarded osmosis. *J. Membr. Sci.* 335, 103–110.

Touati, K., Tadeo, F., 2017. Green energy generation by pressure retarded osmosis: State of the art and technical advancement – Review. *Int. J. Green Energy* 14, 337–360.

Van der Bruggen, B., Vandecasteele, C., Van Gestel, T., Doyen, W., Leysen, R., 2003. A review of pressure-driven membrane processes in wastewater treatment and drinking water production. *Environ. Prog. Sustain. Energy* 22, 46–56.

Wan, C.F., Chung, T.S., 2015. Osmotic power generation by pressure retarded osmosis using seawater brine as the draw solution and wastewater retentate as the feed. *J. Membr. Sci.* 479, 148–158.

Wan, C.F., Chung, T.S., 2016a. Maximize the operating profit of a SWRO-PRO integrated process for optimal water production and energy recovery. *Renew. Energy* 94, 304–313.

Wan, C.F., Chung, T.S., 2016b. Energy recovery by pressure retarded osmosis (PRO) in SWRO–PRO integrated processes. *Appl. Energy* 162, 687–698.

Wan, C.F., Yang, T., Gai, W., Lee, Y.D., Chung, T.S., 2018. Thin-film composite hollow fiber membrane with inorganic salt additives for high mechanical strength and high power density for pressure-retarded osmosis. *J. Membr. Sci.* 555, 388–397.

Wang, P., Chung, T.S., 2012. A conceptual demonstration of freeze desalination-membrane distillation (FD-MD) hybrid desalination process utilizing liquefied natural gas (LNG) cold energy. *Water Res.* 46, 4037–4052.

Warsinger, D.M., Swaminathan, J., Guillen-Burrieza, E., Arafat, H.A., Lienhard V, J. H., 2015. Scaling and fouling in membrane distillation for desalination applications: A review. *Desalination* 356, 294–313.

Xia, L., Arena, J.T., Ren, J., Reimund, K.K., Holland, A., Wilson, A.D., McCutcheon, J. R., 2018. A trimethylamine-carbon dioxide draw solution for osmotic engines. *AIChE J.* 64, 3369–3375.

Xiao, D., Li, W., Chou, S., Wang, R., Tang, C.Y., 2012. A modeling investigation on optimizing the design of forward osmosis hollow fiber modules. *J. Membr. Sci.* 392–393, 76–87.

Yip, N.Y., Elimelech, M., 2013. Influence of natural organic matter fouling and osmotic backwash on pressure retarded osmosis energy production from natural salinity gradients. *Environ. Sci. Technol.* 47, 12607–12616.

Yip, N.Y., Tiraferri, A., Phillip, W.A., Schiffman, J.D., Hoover, L.A., Kim, Y.C., Elimelech, M., 2011. Thin-film composite pressure retarded osmosis membranes for sustainable power generation from salinity gradients. *Environ. Sci. Technol.* 45, 4360–4369.

Zhang, M., Hou, D., She, Q., Tang, C.Y., 2014a. Gypsum scaling in pressure retarded osmosis: Experiments, mechanisms and implications. *Water Res.* 48, 387–395.

Zhang, S., Chung, T.S., 2013. Minimizing the instant and accumulative effects of salt permeability to sustain ultrahigh osmotic power density. *Environ. Sci. Technol.* 47, 10085–10092.

Zhang, S., Sukitpaneenit, P., Chung, T.S., 2014b. Design of robust hollow fiber membranes with high power density for osmotic energy production. *Chem. Eng. J.* 241, 457–465.

Zhong, Y., Wang, X., Feng, X., Telalovic, S., Gnanou, Y., Huang, K.W., Hu, X., Lai, Z., 2017. Osmotic heat engine using thermally responsive ionic liquids. *Environ. Sci. Technol.* 51, 9403–9409.

Index

T